Springer Complexity

Springer Complexity is a publication program, cutting across all traditional disciplines of sciences as well as engineering, economics, medicine, psychology and computer sciences, which is aimed at researchers, students and practitioners working in the field of complex systems. Complex Systems are systems that comprise many interacting parts with the ability to generate a new quality of macroscopic collective behavior through self-organization, e.g., the spontaneous formation of temporal, spatial or functional structures. This recognition, that the collective behavior of the whole system cannot be simply inferred from the understanding of the behavior of the individual components, has led to various new concepts and sophisticated tools of complexity. The main concepts and tools – with sometimes overlapping contents and methodologies – are the theories of self-organization, complex systems, synergetics, dynamical systems, turbulence, catastrophes, instabilities, nonlinearity, stochastic processes, chaos, neural networks, cellular automata, adaptive systems, and genetic algorithms.

The topics treated within Springer Complexity are as diverse as lasers or fluids in physics, machine cutting phenomena of workpieces or electric circuits with feedback in engineering, growth of crystals or pattern formation in chemistry, morphogenesis in biology, brain function in neurology, behavior of stock exchange rates in economics, or the formation of public opinion in sociology. All these seemingly quite different kinds of structure formation have a number of important features and underlying structures in common. These deep structural similarities can be exploited to transfer analytical methods and understanding from one field to another. The Springer Complexity program therefore seeks to foster cross-fertilization between the disciplines and a dialogue between theoreticians and experimentalists for a deeper understanding of the general structure and behavior of complex systems.

The program consists of individual books, books series such as "Springer Series in Synergetics", "Institute of Nonlinear Science", "Physics of Neural Networks", and "Understanding Complex Systems", as well as various journals.

Understanding Complex Systems

Understanding Complex Systems

Future scientific and technological developments in many fields will necessarily depend upon coming to grips with complex systems. Such systems are complex in both their composition (typically many different kinds of components interacting with each other and their environments on multiple levels) and in the rich diversity of behavior of which they are capable. The Springer Series in Understanding Complex Systems series (UCS) promotes new strategies and paradigms for understanding and realizing applications of complex systems research in a wide variety of fields and endeavors. UCS is explicitly transdisciplinary. It has three main goals: First, to elaborate the concepts, methods and tools of self-organizing dynamical systems at all levels of description and in all scientific fields, especially newly emerging areas within the Life, Social, Behavioral, Economic, Neuro- and Cognitive Sciences (and derivatives thereof); second, to encourage novel applications of these ideas in various fields of Engineering and Computation such as robotics, nano-technology and informatics; third, to provide a single forum within which commonalities and differences in the workings of complex systems may be discerned, hence leading to deeper insight and understanding. UCS will publish monographs and selected edited contributions from specialized conferences and workshops aimed at communicating new findings to a large multidisciplinary audience.

M. Ausloos M. Dirickx (Eds.)

The Logistic Map and the Route to Chaos

From The Beginnings to Modern Applications

With 178 Figures

Springer

Professor Dr. Marcel Ausloos
Institut de Physique
Université de Liège
B5, Sart Tilman
4000 Liège
Belgium
E-mail: ausloos@gw.unipe.ulg.ac.be

Dr. Michel Dirickx
Royal Military Academy/ELEC
Avenue de la
Renaissancelaan 30
1000 Brussels
Belgium
E-mail: michel.dirickx@rma.ac.be

Library of Congress Control Number: 2005932565

ISSN 1860-0840
ISBN-10 3-540-28366-8 Springer Berlin Heidelberg New York
ISBN-13 978-3-540-28366-9 Springer Berlin Heidelberg New York

Springer is a part of Springer Science+Business Media
springeronline.com

Printed in The Netherlands

Typesetting: by the authors, the editors and TechBooks using a Springer LaTeX macro package
Cover design: Erich Kirchner Heidelberg

Printed on acid-free paper SPIN: 11517627 54/TechBooks 5 4 3 2 1 0

To our families

Preface

Pierre-François Verhulst was born in 1804 in Brussels. He became famous through the quadratic equation he introduced in order to describe population evolution subject to spontaneous creation but under constraints due to its own accumulation or size. This led to the famous sigmoid curve used in many fields of science. The discrete version of this equation has been found to be exceptionally interesting following the superb numerical and analytical work of M. J. Feigenbaum. The discrete Verhulst equation has become the paradigm of nonlinear equations implying various states of the system including chaos, depending on the value of the control parameter.

During the last decades, mainstream nonlinear physics was almost exclusively concerned with the properties of equilibria under strong rational assumptions. Statistical physics was a little bit more ambitious, since through the development of models and the search for the so-called universality, it could touch upon fields which were not standard physics ones, but have sometimes been called exotic. A few thought that nonlinear physics could do nothing more or different than insure extension of rational equilibrium and the steady state, and is even barely of interest for first-year undergraduate students. The idea that the very structure of a physical system might evolve in a complex manner as a result of nonlinearity and adaptation/feedback processes remained largely confidential, and sometimes frightening to traditional teachers. This is rapidly changing. In the past few years, the concepts of nonlinearity and complexity increasingly gained legitimacy in the eyes of the profession. Extensive efforts have been made both on the experimental and theoretical viewpoints.

The chapters of this book provide state-of-the-art solutions to some of the most important issues and questions at stake in the current research on nonlinear physics, we would say "science", and on the evolution of quantities in complicated fields or environments. They include important advances on specific topics, while giving a fairly wide and balanced introduction to the general field of investigation.

The idea behind this book was to make a path for nonlinear and statistical physics to possibly impact the research and teaching of physical concepts to many.

There are 22 chapters, grouped into five parts. Part I is a general and historical introduction, the second one deals with life-relevant physics, part III is about econophysics, the fourth one deals with condensed matter and the last one with miscellaneous topics.

Several of these chapters stem from a coordinated aim which started from the organization of the "Verhulst 200 on Chaos" meeting, at the Royal Military Academy (RMA) of Belgium, in Brussels, in September 2004. The meeting was held under the patronage of His Majesty King Albert II. The outgoing and the entering European commissioners in charge of scientific research as well as six Belgian ministers were members of the Honorary Committee.

At the same time a "Risk of Chaos" workshop was held in the framework of the COST P10 European program.

There were 12 keynote speakers, and about 40 invited papers or contributions, presented orally or as posters, about one third of them having been selected for contributing to this book, which goes much beyond proceedings style and goals.

We would like to thank the sponsors of the above events:

- the Belgium Minister of Defence, through the Renaissance Association and the RMA:
- the European Union through the 6th framework Program, in particular the COST P10 "Physics of Risk" action:
- the Fonds National de la Recherche Scientifique:
- the Fonds voor Wetenschappelijk Onderzoek-Vlaanderen:
- the University of Liège Research Council:
- the City of Brussels.

Sometimes sponsorship cannot be quantified; therefore lastly, but not the least, we would like to thank again the RMA staff (especially Prof. Hugo Pastijn, Dr. Dave Faulconer, and Lieutenant Colonel Jean-Paul Salmon) and hierarchy for providing facilities for rooms, transportation, and so on. Their enthusiasm and understanding made it possible to have very lively sessions and fruitful meetings, thereby enhancing friendship (or friendly relations) for the good of humanity and the world fraternity.

Liège, Brussels
June 2005

Marcel Ausloos
Michel Dirickx

Contents

Part I General and Historical Introduction

Part IV Condensed Matter

Part V Miscellaneous

List of Contributors

Christof M. Aegerter
Universität Konstanz
Universitätstrasse 10
P.O. Box 5560
D-78457 Konstanz, Germany
christof.aegerter@uni-konstanz.de

Lucas M. Alves
Universidade Estadual de Ponta Grossa
Av. Gal. Carlos Calvalcanti 4748
Ponta Grossa, CEP.84030.000
Brazil lmalves@uepg.br

Marcel Ausloos
University of Liège, B5
B-4000 Liège, Belgium
Marcel.Ausloos@ulg.ac.be

Katy Austin
Brunel University
UB8 3PH Uxbridge
United Kingdom
Katy.Austin@brunel.ac.uk

Mark Ya. Azbel'
Tel Aviv University
Ramat Aviv - Tel Aviv, 69978
azbel@post.tau.ac.il

Denis Constales
Ghent University
Galglaan 2
B-9000 Gent, Belgium
Denis.Constales@UGent.be

Dana Constantinescu
University of Craiova
A.I. Cuza Street 13
Craiova 200585, Romania
dconsta@central.ucv.ro

Sergey A. Demin
Kazan State Pedagogical University
Mezhlauk Street 1
Kazan 420021, Russia

Michel Droz
University of Geneva
Quai E. Ansermet 24
CH-1211 Geneva 2, Switzerland
Michel.Droz@physics.unige.ch

Daniel Dubois
University of Liège
Grande Traverse 12
B-4000 Liège, Belgium
Daniel.Dubois@ulg.ac.be

Adam Gadomski
U.T.A. Bydgoszcz
al. Kaliskiego 7/222
PL-85796 Bydgoszcz, Poland
agad@atr.bydgoszcz.pl

Mohammad Hasan Ghaffari Saadat
Amirkabir University of Technology
Department of Mechanical Engineering
Hafez, Tehran, Iran
Ghaffari@aut.ac.ir

Jason A.C. Gallas
Universität Stuttgart
Pfaffenwaldring 27 – Stuttgart
D-70569
jgallas@icp.uni-stuttgart.de

Peter Hänggi
University of Augsburg
Universitätsstrasse 1 - Augsburg
D-86135
Hanggi@Physik.Uni-Augsburg.de

Emilio Hernández-García
Instituto Mediterráneo de Estudios Avanzados
Campus Universitat Illes Balears
E-07122 Palma de Mallorca Spain
emilio@imedea.uib.es

Hans J. Herrmann
Universität Stuttgart
Pfaffenwaldring 27
D-70569 Stuttgart, Germany
hans@ica1.uni-stuttgart.de

Stefan Hutzler
Trinity College
Dublin 2, Ireland shutzler@tcd.ie

Jos Kint
Ghent University
Eeuwfeestlaan 38
B-9840 De Pinte, Belgium
jos.kint@belgacom.net

Helmut Kröger
Université Laval
Cité Universitaire
Québec
Québec G1K 7P4, Canada
hkroger@phy.ulaval.ca

Pierre Kunsch
Vrije Universiteit Brussel
Pleinlaan 2
B-1050 Brussels, Belgium
pkunsch@vub.ac.be

Jean-François Laprise
Université Laval
Cité Universitaire
Québec
Québec G1K 7P4, Canada
jlaprise@phy.ulaval.ca

Pedro G. Lind
Universität Stuttgart
Pfaffenwaldring 27
D-70569 Stuttgart, Germany
lind@ica1.uni-stuttgart.de

Adam Lipowski
Adam Mickiewicz University
Ul. Umultowska 85
PL-61614 Poznań, Poland
lipowski@amu.edu.pl

Rui F.M. Lobo
Universidade Nova de Lisboa
2829-516 Caparica, Portugal
rfl@fct.unl.pt

Cristóbal López
Instituto Mediterráneo de Estudios Avanzados
Campus Universitat Illes Balears
E-07122 Palma de Mallorca,
Spain clopez@imedea.uib.es

Kinga A. Lőrincz
Vrije Universiteit
De Boelelaan 1081
1081HV Amsterdam, The Netherlands
ka.lorincz@few.vu.nl

Jakob Mann
Risø National Laboratory
DK-4000 Roskilde, Denmark
jakob.mann@risoe.dk

Estela Mayoral
Universidad Nacional Autónoma de México
Apartado Postal 20-364
México 01000, D.F.
México
estela@eros.pquim.unam.mx

Gurgen G. Melkonyan
Université Laval
Cité Universitaire
Québec
Québec G1K 7P4, Canada
gmelkony@phy.ulaval.ca

Janusz Miśkiewicz
University of Wrocław
pl M. Borna 9
PL-50204 Wrocław, Poland
jamis@ift.uni.wroc.pl

Suzana Moss de Oliveira
Universidade Federal Fluminense
Av. Litorânea S/N, Boa Viagem
Niterói 24210-340
Rio de Janeiro, Brazil
suzana@if.uff.br

Eimear Ni Dhuinn
Trinity College
Dublin 2, Ireland
nidhuine@tcd.ie

Soeren Ott
Risø National Laboratory
DK-4000 Roskilde, Denmark
soeren.ott@risoe.dk

Hugo Pastijn
Royal Military Academy
Renaissancelaan 30
B-1000 Brussels, Belgium
Hugo.Pastijn@rma.ac.be

Hans L. Pécseli
University of Oslo, Institute of Physics
Box 1048 Blindern
N-0316 Oslo Norway
hans.pecseli@fys.uio.no

Przemysław Repetowicz
Trinity College
Dublin 2, Ireland
repetowp@tcd.ie

Peter Richmond
Trinity College
Dublin 2, Ireland richmond@tcd.ie

Alberto Robledo
Universidad Nacional Autónoma de México
Apartado Postal 20-364
México 01000, D.F.
México
robledo@fisica.unam.mx

Geoff J. Rodgers
Brunel University
UB8 3PH Uxbridge
United Kingdom
G.J.Rodgers@brunel.ac.uk

Giulia Rotundo
University of Tuscia
via del Paradiso 47
I-01100 Viterbo, Italy
Giulia.Rotundo@uniroma1.it

Houman Safaai
Amirkabir University of Technology
Department of Physics
Azadi, Tehran, Iran
Houman1359@yahoo.com

Dietrich Stauffer
Cologne University
Zülpicherstr. 77
D-50923 Köln, Germany
stauffer@thp.uni-koeln.de

Jan Trulsen
University of Oslo, Institute of Theoretical Astrophysics
Box 1029 Blindern, N-0315 Oslo, Norway
jan.trulsen@astro.uio.no

André Vanderbauwhede
Ghent University
Krijgslaan 281
B-9000 Gent, Belgium
Andre.Vanderbauwhede@UGent.be

Boris Weyssow
Université Libre de Bruxelles
Bvd du triomphe 231
B-1050 Bruxelles, Belgium
bweyssow@ulb.ac.be

Rinke J. Wijngaarden
Vrije Universiteit
De Boelelaan 1081
1081HV Amsterdam, The Netherlands
rw@nat.vu.nl

Renat M. Yulmetyev
Kazan State Pedagogical University
Mezhlauk Street 1
Kazan 420021, Russia
rmy@dtp.ksu.ras.ru

Reza Zomorrodi
Université Laval
Cité Universitaire
Québec
Québec G1K 7P4
rzomor@phy.ulaval.ca

Part I

General and Historical Introduction

Chaotic Growth with the Logistic Model of P.-F. Verhulst

Hugo Pastijn

Department of Mathematics, Royal Military Academy B-1000 Brussels, Belgium
Hugo.Pastijn@rma.ac.be

Summary. Pierre-François Verhulst was born 200 years ago. After a short biography of P.-F. Verhulst in which the link with the Royal Military Academy in Brussels is emphasized, the early history of the so-called "Logistic Model" is described. The relationship with older growth models is discussed, and the motivation of Verhulst to introduce different kinds of limited growth models is presented. The (re-)discovery of the chaotic behaviour of the discrete version of this logistic model in the late previous century is reminded. We conclude by referring to some generalizations of the logistic model, which were used to describe growth and diffusion processes in the context of technological innovation, and for which the author studied the chaotic behaviour by means of a series of computer experiments, performed in the eighties of last century by means of the then emerging "micro-computer" technology.

1 P.-F. Verhulst and the Royal Military Academy in Brussels

In the year 1844, at the age of 40, when Pierre-François Verhulst on November 30 presented his contribution to the "Mémoires de l'Académie" of the young Belgian nation, a paper which was published the next year in "tome XVIII" with the title: "Recherches mathématiques sur la loi d'accroissement de la population" (mathematical investigations of the law of population growth), he did certainly not know that his work would be the starting point for further research by Raymond Pearl and Lowell J. Reed [10, 11], by the famous A.-J. Lotka [8] and independently by Volterra [16] and later by V.A. Kostitzin [5], in the fields of mathematical biology, biometry, and demography. It was in these Mémoires that he introduced a growth model for a closed population (no immigration, no emigration) facing a living environment with limited resources for the subsistence of its members. The purpose was to predict the demographic evolution of the young Belgian society, and to answer the question about the maximum population size sustainable with these limited resources. He certainly did not expect that more than one century later, the study of a discrete version of this model would give rise to a new field in science: chaos theory. By the time he presented his paper, he was already a member of the "Académie" (Academy of Sciences) and "Professeur d'Analyse à l'Ecole Militaire de Belgique" – Professor of Calculus at the Military Academy – which had been founded in 1834, and which became later our present Royal Military Academy (Ecole Royale Militaire - Koninklijke Militaire School) in Brussels.

In 1844 he had already a remarkable career behind him. Pierre-François Verhulst was born in Brussels, October 28, 1804. This year we commemorate the 200th anniversary of his birth. He was a member of a family that neglected nothing to facilitate study opportunities for the young boy. In the Athénée (secondary school – high school) of Brussels he received in August 1822 the first prize of mathematics, shared with Plateau and Nerenburger, becoming later his colleagues in the class of sciences of the Académie. Quetelet will remind us later the excellent reputation of this Athénée of Brussels, which previously as "Lycée" during the "Empire" period, had been the school of several future polytechnicians. In September 1822, the young Verhulst wanted to register for the "exact sciences" at the university of Ghent, although he had not yet finished the complete curriculum at the Athénée in Brussels. In Ghent too he completed his studies successfully after three years, with a doctoral dissertation about the reduction of binomial equations. He obtained his doctoral degree on August 3, 1825 (see also "Pierre-François Verhulst's final triumph" by J. Kint et al. in the present book).

After some teaching duties at the "Musée des Sciences et des Lettres" in Brussels, he went to Italy to recover from fatigue and exhaustion, just before the Belgian Revolution of 1830 broke out. In the short period 1830-1831 he hardly thought about mathematics. He came back to Belgium and in spite of his illness, he enrolled in the army to participate in the battle against Holland. In 1832 he agreed to help Quetelet to establish the mortality tables for the young Belgian state. This collaboration with Quetelet, who was one of the first professors of the newly founded Military Academy in Brussels, led in 1834 to join him to the Military Academy first as a "Répétiteur" of "Analyse" (calculus) without financial remuneration. Very quickly he became professor at the Military Academy. Quetelet, in his "Notice sur Pierre-François Verhulst" published in the "Annuaire de l'Académie royale des Sciences, des Lettres et des Beaux-Arts de Belgique" [12], is mentioning the care with which Verhulst prepared and permanently updated the lecture notes. Unfortunately to our knowledge no copies of these notes are still available in the present archives of the Royal Military Academy.

In 1841, after he bought at a public sale, an old issue of a book by Legendre about elliptic functions, he published a compilation of what was currently known about that subject in his book "Traité élémentaire des fonctions elliptiques". This book is still available in the present collection of the Royal Military Academy. After the publication of this book he is appointed as "correspondant de la section des sciences" of the Académie royale on May 7, 1841. In December of the same year he is appointed member of the Académie royale, to replace Garnier, his former professor at the University of Ghent. From that year on he developed an ever increasing interest for the application of mathematics in a political context. It was probably after the publication of

Quetelet's "Essai de physique sociale" that he got convinced about the idea that the sum of obstacles against an unlimited growth of a closed population is increasing proportionally to the square of the population level currently reached. It was the famous Fourier ("Théorie de la Chaleur") who made an analogous proposal in the introduction of the Tome I of his "Recherches statistiques sur Paris" (1835), and Quetelet urged Verhulst to submit this hypothesis to the investigation of empirical data available for Belgium. The results of the early research of Verhulst on this subject have been published already in 1838, in Tome X of the journal "Correspondance Mathématique et Physique", with Quetelet as chief editor. The idea of what Verhulst called "the logistic growth model" ("la courbe logistique") was born.

In 1845 his communication to the Académie with the title "Recherches mathématiques sur la loi d'accroissement de la population" (Mathematical investigations about the law of population growth) was published in Tome XVIII of the report of the Académie. A second version of a growth model is presented by him on May 15, 1846, to the Académie, in which he is actually criticizing the logistic model presented by himself about one year earlier. The text was published in Tome XX of the Mémoires de l'Académie in 1847. The self-criticism about the logistic model in this publication, and the emphasis Quetelet later puts in his "Notice sur Pierre-François Verhulst" [12] on Verhulst's hesitation and his own reluctance to accept the applicability of the logistic model, are probably the main reasons why after Verhulst's death the logistic model was entirely forgotten for a long time.

In 1848 the King of the Belgians appointed him as President of the Académie royale des Sciences, des Lettres et des Beaux-Arts de Belgique. Although his health condition became ever worse, he continued to deliver his lectures at the Military Academy and to take office as the President of the Académie royale de Belgique. On February 15, 1849, P.-F. Verhulst died at the age of 44. One of his last publications of which there is still a copy available in the library of the Royal Military Academy is his very modest booklet of 1847 "Leçon d'Arithmétique dédiée aux candidats aux écoles spéciales" (Lesson in Arithmetics to the candidates of the "special schools") on 72 pages.

2 The Exponential Growth Process

Until the end of the 18th century human and raw material resources were seemingly so unlimited from a Western point of view, that really no obstacles were supposed to exist for the development of human activities and for the growth of human population. This mental state was at the basis of the industrial revolution. Engineers and other scientists considered almost everything to be known and almost everything could be achieved by man without the need for ecological considerations, related to constraints on natural resources. With these ideas in mind it is quite natural to make very optimistic forecasting about the growth of an economic system and about human population.

So, when dealing with human population growth, Malthus suggested that the growth speed of a population is proportional to the current population level. When no other external constraints on the growth speed are considered, then the continuous time model corresponding to this hypothesis is: $\mathrm{d}x/\mathrm{d}t = rx$, with x denoting the population level at time t. The solution of this differential equation is an exponential function $x(t) = x(0)\exp(rt)$. The exponential growth model is clearly not appropriate to describe the evolution of a population over a long period of time, even if it approximates sufficiently the growth phenomenon during a certain period (for instance during the start-up episode). This is essentially the consequence of the hypothesis of a constant growth rate r during the whole lifetime of the process, and of independence of this growth rate with regard to the current population level at time t.

3 Limited Growth Models

If we consider the coefficient r as roughly the difference between the birth rate and the death rate $(B - D)$, then this means that natality and mortality in a population are independent both on the age and on the level (density) of this population. This seems not to be true in the real world. The natality rate is mostly decreasing for higher population densities, whereas the mortality rate is generally increasing for higher densities. The most simple assumption about these relationships is (with positive coefficients): $B = b' - bx$ and $D = d' + dx$.

Substitution into the Malthusian equation yields $\mathrm{d}x/\mathrm{d}t = ex - fx^2$ with $e = b' - d'$ and $f = b + d$.

This means that the obstacle against an unlimited growth rate is proportional to the square of the current population level at time t. Another way to obtain the same form for the differential equation is to consider the coefficient r in the Malthusian equation as dependent (and decreasing) on x, instead of being constant. The most simple dependence would then be: $r = g - hx$. In this case we obtain:

$$\frac{\mathrm{d}x}{\mathrm{d}t} = g\left(1 - \frac{h}{g}x\right)x \ ,$$

with the same observation about the obstacle against unlimited growth. Of course it is possible to imagine more sophisticated slowdown functions relating the growth rate to the current population level at time t. For example such a slowdown function could be:

$$r = sg\left[1 - \left(\frac{x}{k}\right)^{m-1}\right] ,$$

with

$$s = -1 \text{ for } m < 1 \ ,$$

$$s = 1 \text{ for } m > 1 \ ,$$

and g, k, m real valued parameters.

Notice that $m = 2$ is yielding the same quadratic right hand side of the equation as we mentioned previously. The particular case for m tending to 1 is known as the well-known Gompertz model. For all values of m different from 1 and 2, the model is now known as the generalized logistic growth model. For other old variants and generalizations we refer for instance to Lebreton and Millier [6], to De Palma et al. [1] and to Kinnunen and Pastijn [4]. For $m = 2$ we actually obtain the model introduced by P.-F. Verhulst in 1844 and which he called the logistic growth model ("la courbe logistique").

4 The Logistic Growth Process

For continuous time, this process is described by a differential equation, which is a special case of the Riccati type. The solution is straightforward:

$$\text{for } \frac{\mathrm{d}x}{\mathrm{d}t} = gx\left(1 - \frac{x}{k}\right)$$

the closed form solution is

$$x(t) = \frac{k}{1 + C\exp(-gt)} \text{ with } C = \frac{k - x(0)}{x(0)} .$$

The parameter k is the maximum size of the population, or the asymptotic value of $x(t)$. This is the closed form of the continuous time logistic growth curve. Although there are two parameters, certain morphological aspects of this curve are rather rigid. So, for instance the only existing inflexion point, when $x(0)$ is less than a certain value related to the equation parameters, has always the same ordinate. This was one of the main reasons for introducing variants and generalizations of this simple model in the late previous century, in order to have more flexibility for fitting the model to experimental data.

The reason why Verhulst called this curve "la courbe logistique" in his communication of November 30, 1844, is not clear. He does not give any explanation. One might guess that he refers to the term logistics, related to transportation and distribution in the supply chain of an army, analogous to the supply of subsistence means of a population which he considered to be limited. The term logistic was then already to a certain extent in use in the military environment. He could have been familiar with it, through his military contacts in the Military Academy in Brussels. Another possible root of the term logistic could have been the French word "logis" (place to live) which was of course related to the limited resources for subsistence of a population, Verhulst was dealing with in his model. It is however pure speculation, although the term was still in use in the Belgian army until the mid 20th century as a rank of a non-commissioned officer called "Maréchal des Logis". Another explanation – probably the most likely one – is related to the Greek word $\lambda o\gamma\sigma\pi\kappa o\zeta$, which means "the art of computation" (see also the

"Dictionnaire Quillet de la langue française" of 1961 for this meaning of the French word "logistique"). With his high school education, where Greek and Latin were key subjects at that time, he certainly must have known this term. When we adopt this explanation, Verhulst simply called his curve "logistique" because it enabled him to predict the future population of Belgium – during the era without computers – by simple computations.

In the second degree right hand side of the equation, the slowdown term $-(g/k)x^2$ can be interpreted as the result of the interacting competition between the individuals of the population. This competition is proportional to the number of potential encounters per time unit, and is therefore proportional to x. This interpretation is of course a bit simplistic because it doesn't take into account that the major reason for slowing down the growth speed is exogenously imposed by the limited capacity of the closed "adiabatic" world we are focusing on. A more chemico-physical interpretation of the right hand side is that the relative growth rate for this logistic model is proportional to the currently available non transformed resources $(k - x)$. This idea stems from the dynamics of autocatalytic chemical reactions. Therefore in chemistry, the logistic curve is often referred to as the autocatalytic function. This last interpretation is perhaps a more fundamental one.

This theoretical justification and the marvellous fit of this model to real world data of some first applications in economics and demography, let Kostitzin in 1937 [5] write: "Une population fermée tend vers une limite qui ne dépend que des coefficients vitaux ; elle est indépendante de la valeur initiale $x(0)$". This optimistic view on the self-regulating mechanism of human population growth is inspired by the conviction that the logistic model is of a universal validity and also by the bare mathematical fact that the asymptotic attractor of this model is always locally stable, when these "vital" coefficients are positive – which is no restricting condition for real world growth processes. However, the simple outlook of this logistic equation makes us forget the complexity of the mechanisms in evolutionary processes.

With the model Verhulst introduced in 1844 he predicted that the maximum size of the Belgian population would be six million and six hundred thousand individuals. Presently Belgium has a population of roughly eleven million. In his communication of 1846, he adapts his logistic growth model. The solution of the new differential equation is no logistic function any more. Now his prediction of the maximum size of the Belgian population is about nine million and four hundred thousand people, which is remarkably closer already to the present population level of Belgium. The main difference between both models is the following in Verhulst's own wordings. In the logistic model the sum of obstacles against unlimited growth is proportional to the excess population. This excess population is the difference between the current population at time t and a minimum level of the population which is sustainable by means of a given number of available resources, which are considered as constant. In the model of 1846 he considers the obstacles against unlimited growth to be proportional to the ratio of the excess population and the total population at time t. It is now obvious that the logistic model was not the most effective to predict the long-term evolution of the Belgian population. This continuous time model is finally not as universally valid as it was sometimes considered. In addition, it is now recognized that the continuous time model does not always reflect reality. When there are for instance jump-wise simultaneous behavioural changes of all the individuals of the population, the structural dynamics of the population may fundamentally change. This has been "re-discovered" in 1974 and published in 1976 by R. May [9]. When we construct a discrete version of the logistic differential equation, for instance by applying the common Euler method for numerical integration, then we obtain a discrete form of the logistic growth process:

$$x(t+1) = x(t)\left(1 + g\frac{1 - x(t)}{k}\right) .$$

This equation is describing the evolution of a population which is progressing jump-wise with equally time spaced intervals. The memory of this system is only one time unit. This means that the future of this system only depends on the "now and here", and that the role of the grandparents is instantaneously neglected. All individuals of the population have the same and synchronized reproductive behaviour. The importance of this model is due to its peculiar behaviour for some values of its parameters. This model has been extensively studied by May [9], and was one of the first simple models to illustrate the phenomenon of chaos. With the advent of the so-called "micro-computers" of the eighties of last century, it became very easy to generate illustrations of this "chaotic" behaviour.

5 Attractors for the Discrete Logistic Model

If a dynamic system defined by difference equations is allowed to evolve over a long time, starting from different initial conditions, the information about these initial conditions may disappear as time is going on. From a set of different initial conditions the system may tend to the same restricted region. This restricted region is called an attractor, whereas the set of initial points that is "attracted" by this attractor is called the basin of attraction. We know that there are three types of attractors: static or fixed point attractors, limit cycles or periodic attractors, and "chaotic" attractors. These three types of attractors have been illustrated for the equation

$$x(t+1) = rx(t)[1 - x(t)]$$

which is a simplified form of the discrete version of the logistic model. These illustrations are widely present in the literature of the eighties of last century, and showed the existence of what we now call chaos, for values of the parameter r beyond 3. This logistic model is a member of a quadratic family [the right hand side is of the second degree in $x(t)$]. It is also a member of a larger family of single peaked functions in the right hand side of the equation, for which general properties with respect to the chaotic behaviour have been studied.

Last century many variants and generalizations of the logistic model have been introduced to describe the diffusion of new products and of technological innovations. These models have been summarized by De Palma et al. [1]. The most well-known are those of Gompertz (see supra), Blumberg with the differential equation $\mathrm{d}x/\mathrm{d}t = rx^a(1 - x^b/k)$, Bertalanffy with the differential equation $\mathrm{d}x/\mathrm{d}t = r_1x^m - r_2x^n$, Bass with the differential equation $\mathrm{d}x/\mathrm{d}t = (a + bk/x)(K - x)$, all with positive parameters.

The Bass model was describing the evolution of several consumer goods markets in the USA (refrigerators, TV-sets, air-conditioners,...) during the second half of last century. Several discrete versions of these models have been studied by the author, and their chaotic behaviour illustrated [4]. It was then also announced that variants and generalizations of these models, used for the description of the substitution process of old by new technologies (Blackman–Fisher–Pry), and for the evolution of commercial naval transportation and railways in the USA (Sharif–Kabir), have chaotic attractors.

In the meantime, the study of chaos has achieved a certain degree of maturity, conditions for its generation having been discovered in a wide category of discrete models [7] and chaos itself having been considered in the general context of fractal geometry.

6 Conclusion

The maturity of the field of chaos theory, and the fact that chaotic behaviour now pervades almost all the sciences, is an argument to include this topic in the future curricula of our engineering and science students. This inclusion is possible in a very early stage of the student's curriculum. The minimal prerequisites are related to basic calculus. The logistic model of Verhulst still nowadays plays an important role in the first introduction of chaos theory to undergraduate students ("Encounters with chaos", Denny Gulick, 1992). We can be confident that through these undergraduate courses of chaos theory, the ideas of Verhulst will survive in another format however than for the purpose they were originally introduced. But that happens quite often in the history of science.

References

1. A. De Palma, F. Droesbeke, Cl. Lefevre, C. Rosinski: *Modèles mathématiques de base pour la diffusion des innovations, Jorbel*, Vol 26, No 2, 1986, pp 37–69

2. D. Gulick: *Encounters with chaos* (McGraw–Hill, New York 1992)
3. T. Kinnunen, H. Pastijn: Chaotic Growth – attractors for the logistic model of P.-F. Verhulst. In: *Revue X* (RMA Brussels 1986), 4, pp 1–17, 1986
4. T. Kinnunen, H. Pastijn: The chaotic behaviour of growth processes, *ICOTA proceedings*, Singapore, 1987
5. V.A. Kostitzin: *Biologie mathématique* (Armand Colin, Paris 1937)
6. J.D. Lebreton, C. Millier: *Modèles dynamiques déterministes en biologie* (Masson, Paris 1982)
7. T.-Y. Li, J. Yorke: *Period three implies chaos, American mathematical monthly (82)*, pp 985–992, 1975
8. A.J. Lotka: *Elements of physical Biology* (Williams & Wilkins, Baltimore 1925)
9. R. May: Nature **261**, 459 (1976)
10. R. Pearl: *Introduction to Medical Biometry and Statistics* (W.B. Saunders, Philadelphia London 1923)
11. R. Pearl, L.J. Reed: Metron **5**, 6 (1923)
12. A. Quetelet: Notice sur Pierre-François Verhulst. In: *Annuaire de l'Académie royale des Sciences, des Lettres et des Beaux-Arts de Belgique* (Impr. Hayez, Brussels 1850) pp 97–124
13. P.-F. Verhulst: *Traité élémentaire des fonctions elliptiques* (Impr. Hayez, Brussels 1841)
14. P.-F. Verhulst: Recherches mathématiques sur la loi d'accroissement de la population. In: *Mem. Acad. Royale Belg.*, vol 18 (1845) pp 1–38
15. P.-F. Verhulst: Deuxième mémoire sur la loi d'accroissement de la population. In *Mem. Acad. Royale Belg.*, vol 20 (1847) pp 1–32
16. V. Volterra: *Leçons sur la théorie mathématique de la lutte pour la vie* (Gauthier-Villars, Paris 1931)

Pierre-François Verhulst's Final Triumph

Jos Kint[1], Denis Constales[2], and André Vanderbauwhede[3]

[1] Ghent University, Faculty of Medicine and Health Sciences, Department of Pediatrics and Medical Genetics GE02
jos.kint@belgacom.net
[2] Ghent University, Faculty of Engineering, Department of Mathematical Analysis TW16
Denis.Constales@UGent.be
[3] Ghent University, Faculty of Sciences, Department of Pure Mathematics and Computer Algebra WE01
Andre.Vanderbauwhede@UGent.be

The so-called *Logistic function* of Verhulst led a turbulent life: it was first proposed in 1838, it was dismissed initially for being not scientifically sound, it became the foundation of social politics, it fell into oblivion twice and was rediscovered twice, it became the object of contempt, was subsequently applied to many fields for which it was not really intended and it sank to the bottom of scientific philosophy. Today it is cited many times a year. And last but not least, during the past three decades it has been claimed as the prototype of a chaotic oscillation and as a model of a fractal figure.

It is only now, 155 years after Verhulst's death, that it becomes clear that his logistic function transcends the importance of pure mathematics and that it plays a fundamental role in many other disciplines. The logistic curve has lived through a long and difficult history before it was finally and generally recognised as a universal milestone marking the road to unexpected fields of research. Only at the end of the 20th century did Verhulst's idea enjoy its definitive triumph. But let us start at the beginning.

On August 3, 1825 the magnificent auditorium of Ghent University was still under construction. It would only be completed early 1826. However, at 11 a.m. of that particular August 3, a small function was held in the provisional hall of the university. In the presence of the then rector of the university, Professor Louis Raoul, a mathematician of scarcely 21 years old defended his doctorate's thesis. Even in those days, twenty-one was very young to take one's PhD. It was clear that, from that moment on, Pierre-François Verhulst would not go through life unnoticed.

1 His Life

He was born in Brussels on October 28, 1804 as the child of wealthy parents. As a pupil at the Brussels Atheneum, where Adolphe Quetelet was his mathematics teacher, he already excelled, and not only because of his knowledge of mathematics. He also had linguistic talents. Twice he won a prize for Latin poetry. However, he had a distinct preference for mathematics. His desire to study exact sciences was so strong that in September 1822, without even having completed his grammar high school, Verhulst enrolled as a student at the University of Ghent. Evidently, his lack of formalism caused some problems when he tried to enrol, although, in those days such matters could easily be resolved with some negotiating and argumentation. It was here that he met Quetelet again, this time as his algebra professor. Just like his studies at the Brussels Atheneum, his academic performance at the University of Ghent was a success. In less than a year, between February 1824 and October 1824, he was honoured with two prizes, one at the University of Leiden for his comments on the theory of maxima, and a second time he won the gold medal of the University of Ghent for a study of variation analysis [1].

In 1825, after only three years of study, Verhulst took his PhD in mathematics with a thesis entitled *De resolutione tum algebraica, tum lineari aequationum binominalium*, in other words, with a thesis in Latin on reducing binomial equations (Fig. 1).

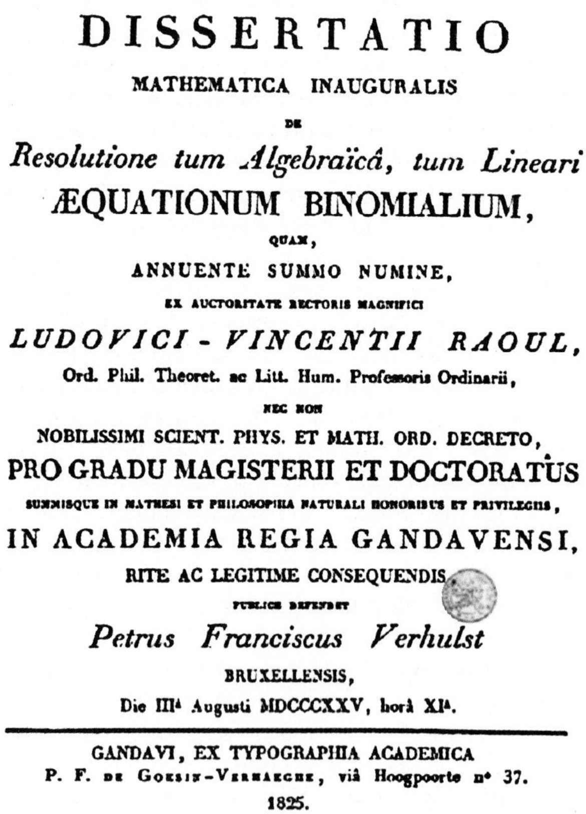

DISSERTATIO

MATHEMATICA INAUGURALIS

DE

Resolutione tum Algebraïcâ, tum Lineari

ÆQUATIONUM BINOMIALIUM,

QUAM,

ANNUENTE SUMMO NUMINE,

EX AUCTORITATE RECTORIS MAGNIFICI

LUDOVICI - VINCENTII RAOUL,

Ord. Phil. Theoret. ac Litt. Hum. Professoris Ordinarii,

NEC NON

NOBILISSIMI SCIENT. PHYS. ET MATH. ORD. DECRETO,

PRO GRADU MAGISTERII ET DOCTORATÛS

SUMMISQUE IN MATHESI ET PHILOSOPHIA NATURALI HONORIBUS ET PRIVILEGIIS,

IN ACADEMIA REGIA GANDAVENSI,

RITE AC LEGITIME CONSEQUENDIS,

PUBLICE DEFENDET

Petrus Franciscus Verhulst

BRUXELLENSIS,

Die IIIª Augusti MDCCCXXV, horâ XIª.

GANDAVI, EX TYPOGRAPHIA ACADEMICA

P. F. DE GOESIN-VERHAEGHE, viâ Hoogpoorte nº 37.

1825.

Fig. 1. Doctorate's thesis of Pierre-François Verhulst from 1825

After his studies Verhulst returned to Brussels. He took a keen interest in the calculus of probability and in political economy, an interest which he shared with Quetelet. From then on Quetelet's influence on Verhulst is marked. Indeed, on several occasions Verhulst did some computations to support research carried out by Quetelet.

Moreover, Quetelet's influence was not limited to passing on ideas and stimulating research. It was through his agency that Verhulst was entrusted with a teaching assignment at the "Musée des Sciences et des Lettres" in Brussels in April 1827. A job which he soon had to give up on account of his poor health. Verhulst would be in bad health all of his life as a result of a chronic illness, the nature of which could not be retrieved from the documents that are left from that period. A brief stay in Italy, shortly after his promotion, did not help much to improve his state of health. During his stay in Rome in September 1830, the Belgian Revolution broke out in Brussels. In the mind of Verhulst, who was 26 at that time, a rather peculiar idea began to take shape. An idea only conceivable by young people who in their youthful exuberance and audacity let their imaginations run free. Verhulst always consistently acted upon the consequences of his principles with the self-confidence of a profound conviction. He conceives the rather original idea that the papal state could use a constitution, just like Belgium, his own country which had just become independent. And of course he is not satisfied with the idea alone, but immediately prepares a draft constitution. It seems incredible, yet it is true: the draft constitution was given some consideration by a few cardinals of the papal Curia and was sent to various foreign ministries. However, the matter came to the attention of the Roman bourgeoisie who was not at all pleased with someone from Brussels lecturing the Italians on how to deal with their political matters. The Roman police ordered him to leave the country at once. Verhulst retired to his residence for a couple of days and tried to barricade himself, expecting a siege by the police. But in the end, after having discussed the matter with some friends, he decided to obey the expulsion order and left Italy. Queen Hortense of Holland – at that time living in Rome – made in her memoirs a lively account of the affair. Translated from French: "... A young Belgian savant, Mr Verhulst, had come to Rome for his health. He came very often to my house in the evening; we had frequent discussions together. He asked to speak to me one morning, and brought a plan for a constitution for the Papal States, which he wished to submit to my criticism before giving it to the cardinal-vicar to submit to the pope. I could not help laughing at the singularity of my position. I [the exiled Queen of Holland] to revise a constitution, and for the pope! That seemed to me like a real joke. But my young Belgian friend did not laugh. 'I was talking yesterday evening,' he said to me, 'with several cardinals; their terror is great. I told them of the only way to save the church and the state. They agreed with all my observations. And one of them wishes to submit them to the pope himself. Here is the constitution of which I have sketched the basis ... '" [2]

Back in Brussels, in 1831, he writes a document on behalf of the recently established Congress – the present Belgian parliament – in which he deplores the situation at the university and formulates a way to resolve it [3].

He complained about the political favouritism in the appointment of university professors and the poor standard of the lectures. In spite of his rebellious attitude he is appointed professor at the Royal Military Academy in 1835, and in the same year he is also appointed professor of mathematics at the Université Libre of Brussels, both newly established teaching institutes. However, Verhulst had to give up his professorship at the Université Libre of Brussels in 1840, following a decision of the then Minister of War, which stipulated that professors at the Military School were not allowed to teach in other education institutes. It is not unlikely that Quetelet had a part in the appointments of Verhulst. In 1837 he married a miss Debiefve, who would bear him a daughter about a year later.

Verhulst and Quetelet were closely associated in their life and work [4]. They were both professors at the Military School, they were both members of the Académie royale des Sciences et des Belles Lettres de Bruxelles and they were both interested in mathematical statistics which could be the key to revealing the "natural laws" of human society. Although Verhulst hardly made any general statements regarding the purpose and methodology of these statistics, his practical routine was in line with the theories of Quetelet. The application of mathematics was an essential feature. In both Quetelet's and Verhulst's opinion scientific statistics should be based on a precise mathematical formula to make the accurate incorporation of statistical data possible. However, gradually a significant difference arose in the approach of Verhulst and Quetelet. Verhulst was not in the least interested in what Quetelet called "applied statistics". Verhulst was of the opinion that the calculations were only applicable if there was a direct relation between cause and effect. Quetelet himself did not feel so strongly about such reservations. In contrast he always preferred to find some analogy between physical laws and social phenomena. The debate on this problem, which must have been going on between Verhulst and Quetelet for several years, came to a sudden end with Verhulst's untimely death [4]. It is difficult to determine the precise nature of their relationship from the available documents of that period. Adolphe Quetelet (1796–1876) was eight years older than Verhulst. It is true that Quetelet called Verhulst "successively my pupil, my fellow-worker, my colleague at the Military School, my confrere at the university and the Academy and my friend". However, according to several authors, the relationship between both men was not always as serene as it appeared at first sight. There is one thing we know for sure: they were both interested in mathematical statistics capable of explaining the so-called natural laws of society. Quetelet spoke highly of Verhulst's work, but he had more regard for his compilations than for his original ideas. On one particular occasion, at a public sale, Verhulst managed to get hold of a valuable edition of the complete works of the French mathematician Legendre (1752–1833). The satisfaction of having acquired these works inspired Verhulst to study the "Traité des fonctions elliptiques" and to read the works of the German Abel (1802–1829) and the Norwegian Jacobi (1804–1851), with the intention of making a compilation of all aspects related to elliptic functions. He read and summarized the works of these three famous mathematicians as well as every other document on this subject. Quetelet was full of praise about the result of this study entitled "Traité élémentaire des fonctions elliptiques", which, in fact, was nothing more than a critical résumé of the works of others. However, Quetelet did not approve of what was in fact Verhulst's most original achievement, i.e., the logistic function. After the publication of his "Traité élémentaire des fonctions elliptiques" Verhulst was admitted as a member of the "Académie royale" in 1841. In 1848 Verhulst is appointed director of the scientific department and later, in spite of his deteriorating health, the king appointed him chairman of the Academy. He died a couple of months later on February 15, 1849, at the age of 44.

According to Quetelet, Verhulst was somewhat of an "enfant terrible" [1]. He was self-willed, a man with a social conscience and a man of principle, controversial and often an advocate of extreme ideas, but he also had a strong sense of justice and acted from a deep feeling for his duty. He was straightforward and consistent in his thinking, but on the other hand also conciliatory. As chairman of the Academy he shrank from anything that might have caused dissension. He was never offensive, and the higher his position the more unassuming he became. Although he himself did not have the slightest inclination for losing his temper, he respected the short-temperedness of others. Although he loved taking part in debates, it was more out of a craving for knowledge than in a spirit of contradiction or with the intention of imposing his own views. He was noted for his unperturbed equanimity. It would have been difficult to find a man more conscientious. According to Quetelet's testimony, this sense of duty was marked during the last years of his life, when he still went to work every day. It took him more than an hour to walk the short distance from his house to his office. People saw him trudge along the streets, resting with every step he took, to arrive finally at the academy, panting heavily and completely exhausted.

2 His Work in the Field of Population Growth

Verhulst's first research in the field of population growth dates from shortly after the independence of Belgium. In order to grasp the full import of the research on population growth in the nineteenth century, one must recall the social climate of those days. During the first half of the nineteenth century Flanders went through the worst economic depression in its entire history. Although under the "Ancien régime" in the 18th century it had been one of the most prosperous regions of Europe, it became a backward and shattered region with an impoverished and destitute population in only a few decades' time. In addition to sheer destitution, the pauperization of the population also resulted in demoralization, moral degeneration and social unrest. The same confusion was also seen in other European countries. The correlation between poverty and population was first demonstrated by Thomas Robert Malthus, in his famous Essay on the Principle of Population, which was published in 1798. Malthus stated that poverty is only the inevitable result of overpopulation. In turn, overpopulation was the natural result of the fundamental laws of human society. The ideas of Malthus were the subject of heated debates in the nineteenth century. The necessity of conducting a social policy to curb the pauperization of the population turned the study of the laws of population growth into a scientifically respectable subject. A new discipline, political economics, found enthusiastic adherents everywhere. A demographic study of the population was initially impeded by a lack of statistical material or, even worse, by the unreliability of the available material. It was only in 1820 that progress was made in the methods of compiling and processing statistical data on which demographic conclusions could be based. In Belgium it was again Adolphe Quetelet who organized the collection of data with regard to population figures. He was the initiator of the first census carried out in 1829, the results of which were published in 1832. As chairman of the "Commission centrale de statistique" Quetelet was in charge of the general censuses of 1846, 1856, and 1866. Quetelet also laid the foundations of the international conferences of statistics, the first of which took place in Brussels in 1853.

It was against this background that Verhulst started his research on population growth. His research was based on the ideas of Malthus. In his opinion it could not be denied that the population grew according to a geometric sequence. On the other hand it was incontestable that a number of inhibiting factors also increase in strength as the population grows. Verhulst argued that, as a consequence, the growth of the population was bound by an absolute limit, if only because of the limited availability of habitable land and food supplies. This was an original interpretation, but also a deviation from the original concept of Malthus. Malthus' hypothesis can be formulated by means of a differential equation (with p for the population figure)

$$\frac{\mathrm{d}p}{\mathrm{d}t} = mp \, .$$

Integration of this equation produces the well known exponential growth curve, on which economic Malthusianism is founded. Verhulst did not accept this and considered an alternative. In order to implement the check on population growth, Verhulst had to subtract a still unknown factor from the right-hand side of the equation; a factor which, according to Verhulst, is dependent on the population figure itself. He started from the most obvious hypothesis, namely that the growth coefficient m is not constant but in proportion to the distance of the population size from its saturation point.

In other words Verhulst introduced an inhibitory term, proportional to the square of the population size. Consequently, Verhulst stated that

$$\frac{\mathrm{d}p}{\mathrm{d}t} = mp - np^2 .$$

The solution of this differential equation gave rise to a function which was to project the population growth

$$p = \frac{mp_0 \mathrm{e}^{mt}}{np_0 \mathrm{e}^{mt} + m - np_0} ,$$

where p_0 represents the population figure at a given time $t = 0$. Verhulst verified this formula by comparing the real population figures of France, Belgium, Essex and Russia with the result of his calculations. The correspondence was striking, although the available figures related to a period of only twenty years. Verhulst created a new term for his equation and called it the logistic function.

Verhulst never explained why he chose the term "logistique". Yet, in the nineteenth century this French term was used to designate the art of computation, as opposed to a branch of theoretical mathematics such as the theory of proportions and relations. The term was also frequently used in connection with logarithms in astronomic calculations.

As a matter of fact the military meaning of the word "logistic" also found its origin around that period. The third supplement to the sixth edition of the etymological dictionary of the Académie Française first mentions the term in 1835. The military meaning of the word also comprises the calculation of the provisionment of an army or of a population. The "logistic problem" par excellence is the provisioning of the population. Through his contacts at the Military School, Verhulst must have been familiar with military terminology. Verhulst probably used this term to launch the idea of an arithmetical strategy that could be used to calculate the saturation point of a population as well as the time at which that point would be reached within a given percentage.

Verhulst's results were published in 1838 [5] as a modest "Notice sur la loi que la population suit dans son accroissement" in the "Correspondance Mathématique et Physique", a journal of which Quetelet was editor-in-chief. Verhulst regarded his work as a first step towards a much more elaborate study which would be published in 1845 and 1847 in the form of a "Mémoire de l'Académie royale des Sciences et Belles-Lettres de Bruxelles" [6, 7]. For more details on the life of Verhulst, see [8] and its references.

3 The Logistic Function After 1849

From then on this logistic principle of Verhulst led a most peculiar life. It may be said that after Verhulst's death his principle was completely forgotten. One can only guess why this was the case. But Quetelet's rather ambiguous eulogy [1] on Verhulst at the Academy a few months after his death had something to do with it. In a condescending, almost contemptuous tone Quetelet expresses his reservations with regard to Verhulst's principle and even with regard to his former "friend" himself. Quetelet had previously considered another principle regarding population growth, founded on the analogy with a falling stone in a viscous medium which encounters more resistance as its speed of fall increases. Verhulst considered this concept too dogmatic and had always rejected it strongly. For in Verhulst's mind there was only one thing that mattered: to find a correspondence between his calculations and the real population figures, whereas Quetelet attached greater importance to a formal analogy between the laws of physics and the behavioural pattern of a population: much more than Verhulst, Quetelet was obsessed with the notion – which was popular in the nineteenth century – to presuppose exact causal mechanisms without which the world would not be able to function. The title of his magnum opus "La Physique sociale" already outlines Quetelet's tendency to compare human social behaviour to the laws of physics. However, to state that Quetelet's attitude was the decisive factor in the scarce dissemination of Verhulst's ideas in the nineteenth century, would be a limited representation of the facts. At least as important was the fact that Verhulst's work never developed into a practicable theory that could be tested by demographers. John Miner of Johns Hopkins University translated Quetelet's French eulogy on Verhulst into English and published it in 1933 [9].

Whatever the reason may be, it is a fact that Verhulst's work was completely ignored during the whole nineteenth century. The logistic curve was rediscovered only in 1920. In that year two renowned American demographers, Raymond Pearl and Lowell Reed [10], who were not acquainted with Verhulst's publications, formulated the sigmoid growth curve a second time. It was only when their manuscript was already at the printer's that they were informed of Verhulst's work which had been published 75 years earlier. In later publications they recognise their omission and they adopt the term "logistic" from Verhulst [11].

The data of the United States census available to Pearl and Reed only made up half of a logistic curve, and the population level was far from reaching its saturation point. Nevertheless, they endeavoured to make an extrapolation and stated that the American population – at that time only 80 million people – would grow to a saturation point of 198 million people and that this saturation point would only be reached by the end of the twentieth century. Unlike Verhulst, Pearl and Reed did not deduce the curve's equation from any preliminary thinking. On the contrary, reflexions on the inhibitive effect of diminishing ambient factors as a result of the population growth only appear towards the end of the article, and only to support the application of the sigmoid curve. In other words, Pearl and Reed start from the idea that population growth follows a sigmoid curve. In addition they regard the sigmoid curve of population growth as a genuine principle of population growth. On the one hand this was based on the fact that the logistic curve supported the data fairly well, and on the other hand on the fact that, based on reasonable assumptions, it provided a fairly accurate picture of the future evolution of the population. In 1924, Pearl [12] compared his curve "in a modest way" with Kepler's law of planetary motion and with Boyle's law of gases... For many years, the emphasis which Pearl and Reed put on the systematic nature of the logistic curve led to many heated and bitter discussions which would only come to an end with Pearl's death in 1940. In spite of, or maybe thanks to, these fierce discussions, the logistic curve is sometimes also called the Verhulst–Pearl curve.

A first sign of real recognition of Verhulst's merits came in 1925 [13], when the English statistician Udny Yule recognised that Verhulst was far ahead of his time: "...Probably owing to the fact that Verhulst was greatly in advance of his time, and that the then existing data were quite inadequate to form any effective test to his views, his memoirs fell into oblivion; but they are classics on their subject..." But even that was not sufficient to make Verhulst's reputation and his name was lost again. Verhulst's formula got its final victory only after 1965. From then on scientists from various countries and domains start to refer to Verhulst's publications (Fig. 2). There are at least five reasons for this.

First of all there is the major breakthrough of ecology as a new scientific discipline: on account of the scope of their research ecologists are particularly interested in the growth and the evolution of populations. Verhulst's formula appeared to be an excellent basis for calculating ecological growth problems. A second aspect of Verhulst's formula was that it required a considerable

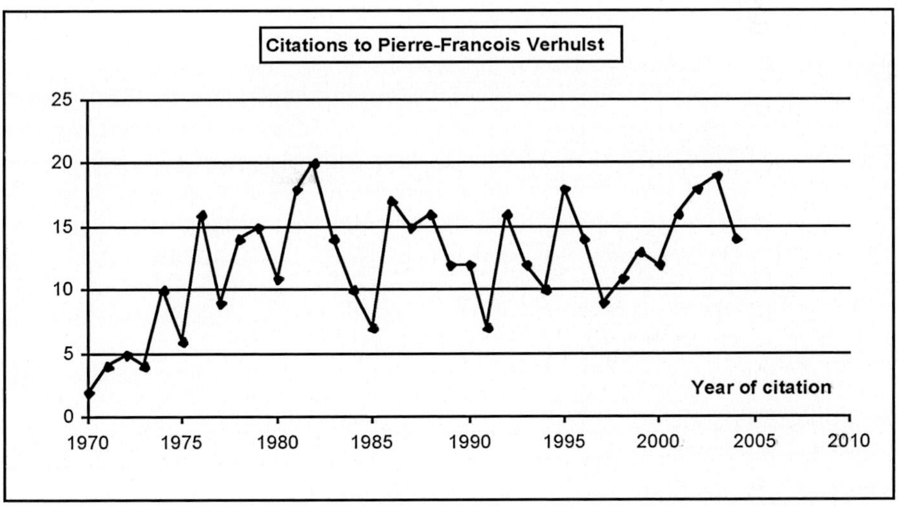

Fig. 2. Citations to the publications of Pierre-François Verhulst

degree of computation. It was only with the advent of the electronic calculator and later the computer that the laborious job of making endless calculations could be carried out with a minimum of effort.

A third factor was the discovery that the S-shaped logistic function could also be applied to a wide variety of other fields, such as chemical autocatalysis, Michaelis–Menten kinetics, cancer chemotherapy, the Hill equation, the Langmuir isotherm, velocity equations of the first and second order of magnitude, oxidation-reduction potentials, erythrocyte haemolysis, the flow of streaming gases, etc. Verhulst's principle was even applied to economics and sociology. It seemed as if everything could be defined using the same sigmoidal curve. Many scientists carried it beyond the limit and applied Verhulst's formula, whether it was relevant or not. This led to a situation in which over the past thirty years Verhulst's work was cited in just about every country of the world, from Brazil to the People's Republic of China, from the Soviet Union to the United States of America. His publications are now cited about 15 times a year, which is quite remarkable considering that his work goes back more than one hundred and sixty years. It is quite amusing in this context to see that each year several authors mention 1938 and 1945 as the year of publication of his works, thinking that 1838 or 1845 must have been a printing error. The journal "Correspondance Mathématique et Physique" ended its publications in 1841. It was in fact published and edited by Quetelet himself on behalf of the Belgian mathematicians. It would reappear only at the end of Quetelet's life from 1874 to 1880 under the name of "Nouvelle Correspondance Mathématique et Physique" and from 1881 to 1961 as "Mathesis".

4 Verhulst's Principle and Chaos Theory

But there is a fourth reason why the work by Verhulst received so much attention all of a sudden: its implication in chaos theory. Already in 1963 Edward Lorenz used a one-dimensional mapping equivalent to the Verhulst mapping to explain certain aspects of his by now famous simplified weather forecast model. In 1976 the biologist Robert May [14] stated explicitly that the logistic model should be studied as early as possible in one's scientific education in order to start understanding nonlinear phenomena. Since the work of May, Feigenbaum [15], and others the Verhulst model has become the paradigm for the period-doubling route to chaos, as is for example nicely illustrated in "The Beauty of Fractals" by H.O. Peitgen and P.H. Richter [16] (one of the first mathematical "coffee table books").

Meanwhile several authors have adopted this idea and it seems to be generally acknowledged now that Verhulst's logistic function is the basis of modern chaos theory, although Verhulst himself had absolutely no idea that something like that lay hidden in his formula.

To obtain deterministic chaos from Verhulst's formula one has to replace the continuous logistic differential equation by its discrete form

$$p_{n+1} - p_n = rp_n(1 - p_n)$$

or equivalently

$$p_{n+1} = p_n + rp_n(1 - p_n).$$

In this difference equation p_n denotes the population size at time n, and $r > 0$ is still the growth coefficient; the carrying capacity has been normalized to 1. Using this prototype of a nonlinear iterative process one calculates the evolution of a population by starting with some initial population p_0 (between 0 and 1) and by applying the formula again and again, thus obtaining successively p_1, p_2, p_3, and so on.

When carrying out this iteration scheme one finds that the resulting evolution of the population depends strongly on the value of the growth parameter r (Fig. 3):

1. For $r < 2$ the population sequence tends to the limit value 1. For $r < 1$ this happens in a monotone way, similar to the behaviour in the differential equation (Fig. 3(a)), but for $1 < r < 2$ in an oscillatory way (Fig. 3(b)). As r increases to 2 these oscillations also increase, both in amplitude and length: for $r = 1.95$ the limit is reached only after more than 2000 steps!
2. For values of r between 2 and 2.5699... the sequence displays, after some initial steps, a periodic behaviour with a period which depends on r. When r increases one first observes an oscillation between a maximum and a minimum (period 2, Fig. 3(c)), then an oscillation between 4 different local extremes (period 4, Fig. 3(d)), and subsequently oscillations with period 8 (Fig. 3(e)), period 16, and so on. Such a period-doubling cascade has been identified as one of the typical ways in which a system can go from orderly to chaotic behaviour.
3. For most values of r larger than 2.5699... (and less than 3) the sequence shows no regularity (periodicity) any more (Fig. 3(f)). For such values of r the system is "chaotic", a regime which is mainly characterized by a few hallmarks as described in the next paragraph.

The main characteristic of a chaotic system is its extreme susceptibility to a change in the initial condition (illustrated for the Verhulst model in Fig. 4). Two sequences with almost identical values for p_0 will at first behave in a virtually identical manner, but then suddenly diverge so that from then on there is no correlation between the two oscillations. A similar sensitivity is also observed with respect to a change in the growth parameter r. Another phenomenon is that a chaotic system sometimes seems to behave regularly for a number of steps in the iteration. For example, for $r = 2.7$ and $p_0 = 0.05$ there is an apparent regularity (a fixed point) between step 590 and step 670 (Fig. 5(a)); with $r = 2.7001$ and $p_0 = 0.05$ there is an apparent period-two behaviour between step 298 and step 316 (Fig. 5(b)). Under further iteration these apparent regularities disappear again. Predictability and chaos alternate with each other, but in a basically unpredictable manner.

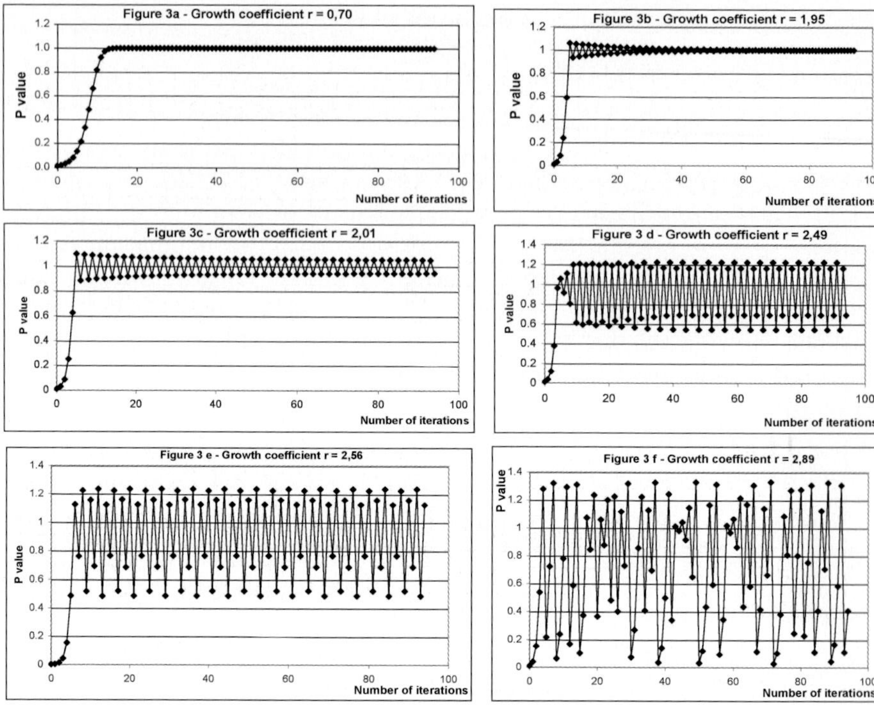

Fig. 3. Deterministic chaos obtained from Verhulst's formula

At the moment when the system becomes chaotic, the size of the population at each step in the iteration will be different from its value at any of the previous steps. There is no stability or regularity any more. Moreover, the long-term evolution of the population will strongly depend on the chosen initial value p_0. Even the smallest deviation – say in the hundredth or thousandth decimal – from the initial value will have a significant effect and in the end, result in a totally different evolution. It is important to notice that also our computers which work with a fixed number of decimals, are subject to this type of unpredictability, however powerful they may be.

5 Logistic Fractal of Verhulst

And finally, a fifth factor can be identified which contributes to the late triumph of Verhulst's logistic function. Indeed, using the logistic formula, one can produce fractal figures comparable to the well-known Mandelbrot fractal. For that purpose we consider again the discrete Verhulst iteration,

$$p_{n+1} = p_n + rp_n(1 - p_n),$$

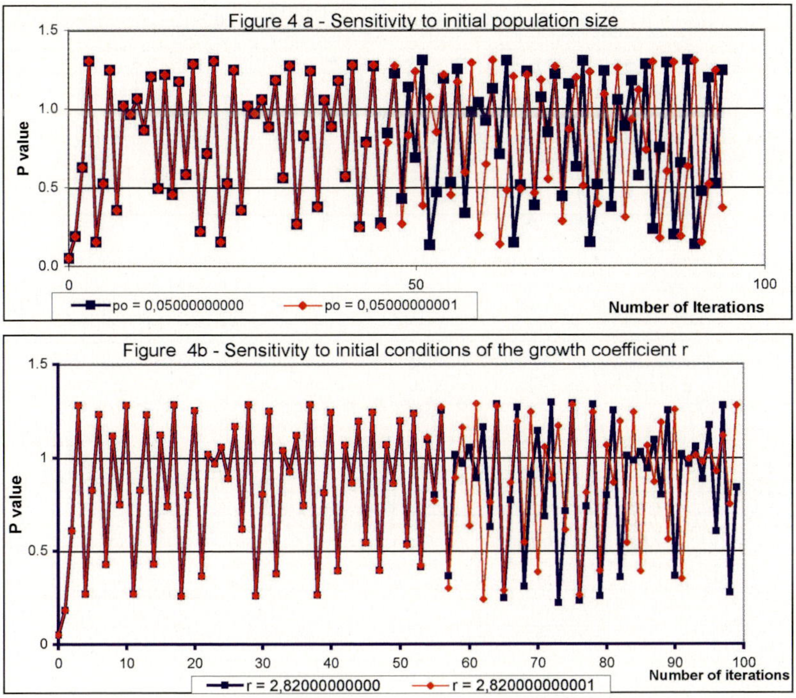

Fig. 4. Example of extreme susceptibility to the initial condition in Verhulst's formula

but this time we allow p and r to be complex, and therefore related to points in the plane. More precisely, p and r take values of the form $a + bi$, and are then identified with the point (a, b) in the plane. The iteration is started by fixing a nonzero value for p_0, for instance $0.01 + 0.01i$. For each value of r one can then calculate the resulting iteration sequence. One finds that there are two possible results: either the sequence stays bounded, or it diverges to infinity. The r-values for which the sequence stays bounded form a set which we call a Verhulst fractal; observe that this Verhulst fractal depends on the choice of the initial value p_0. In a similar way as for the Mandelbrot set, such Verhulst fractals are easily generated on a computer: points not belonging to the fractal evolve towards infinity at different speeds, and by assigning different colors to different speeds one obtains patterns such as in Fig. 6. In this figure the black points form the Verhulst fractal; each picture in the sequence is an enlargement of part of the preceding picture. What we learn from these pictures is that the boundary of the Verhulst set has a fractal structure, in the sense that however much we enlarge this boundary, it will never become a simple line or curve. At each scale new details appear, and the figure never reaches a limit.

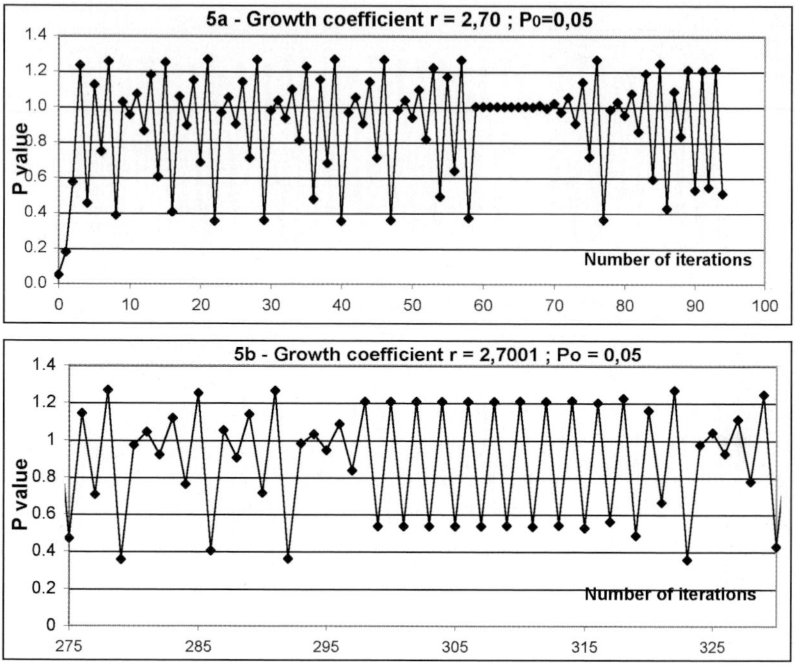

Fig. 5. Predictability and chaos alternating in Verhulst's formula

6 Conclusion

Hence, twice in the past three decades Verhulst's logistic function obtained a new, additional meaning. The first time as a model of a chaotic oscillation and the second time as an example of a fractal figure. The realization that complex phenomena can be represented by means of a simple algebraic equation has radically changed our way of thinking in the past years. Robert May was one of the first people to understand its broader social significance: "Not only in research, but also in the everyday world of politics and economics, we would all be better off if more people realized that simple non-linear systems do not necessarily possess simple dynamical properties."

Verhulst's function is but one of the many examples of a non-linear, chaotic system, although it clearly illustrates the essence of deterministic chaos. It also illustrates how a discovery can go through a real evolution of its own and how the underlying significance of a discovery can change radically as a result of the evolution of its scientific context. Some scientific ideas have to wait for a long period before they come to their final triumph. Verhulst's logistic function is certainly one among them.

Fig. 6. The Logistic fractal of Verhulst for the value $p_0 = -10^{-7}$; each figure to the right and downwards is an enlargement of the preceding figure

References

1. A. Quetelet: Notice sur Pierre-François Verhulst, In: *Annuaire de l'Académie royale des Sciences, des Lettres et des Beaux-Arts de Belgique* (Impr. Hayez, Brussels 1850) pp 97–124
2. *Mémoires de la Reine Hortense, publiés par le Prince Napoléon* (Librairie Plon, Paris 1927) pp 210–212
3. P.-F. Verhulst: *Mémoire sur les abus dans l'enseignement supérieur actuel* (Hayez, Brussels 1831)
4. G. Vanpaemel: Quetelet en Verhulst over de mathematische wetten der bevolkingsgroei, Academiae Analecta, Klasse der wetenschappen **49**, 96 (1987)
5. P.-F. Verhulst: Correspondance Mathématique et Physique **10**, 113 (1838)
6. P.-F. Verhulst: Mem. Acad. Roy. Belg. **18**, 1 (1845)
7. P.-F. Verhulst: Mem. Acad. Roy. Belg. **20**, 1 (1847)
8. J. Mawhin: Les héritiers de Pierre-François Verhulst: une population dynamique, *Bull. Cl. Soc. Acad. R. Belgique*, in press

9. J.R. Miner: Human Biology **5**, 673 (1933)
10. R. Pearl, L.J. Reed: Proc. Natl. Acad. Sci. **6** (1920) pp 275–288
11. R. Pearl, L.J. Reed: Metron **3**, 6 (1920)
12. R. Pearl: *Studies in Human biology* (Williams and Wilkins, Baltimore 1924) p. 585
13. G. Udny Yule: J. Roy. Statist. Society **88**, 1 (1925)
14. R. May: Nature **261**, 459 (1976)
15. M. Feigenbaum: J. Statistical Physics **19**, 25 (1978)
16. H.O. Peitgen, P.H. Richter: *The Beauty of Fractals* (Springer Verlag, Berlin 1986)

Limits to Success. The Iron Law of Verhulst

P.L. Kunsch

Vrije Universiteit Brussel, MOSI department, Pleinlaan 2 BE-1050 BRUSSELS
pkunsch@vub.ac.be

Summary. In this chapter we develop the point of view that Verhulst is a major initiator of systems thinking. His logistic equation is a system archetype, i.e. a simple system built with few feedback loops. In the Fifth Discipline [19] Peter Senge calls this particular archetype "Limits to Success". It can also be called the "Iron law of Verhulst", expressing that trees can never grow to heaven. In a deeper analysis this equation illustrates the shifting loop dominance, one of the basic principles of system dynamics. The basic message is that the combination of some few archetypes, like the logistic growth, can afford valuable insight into many complex systems such as the economy, environment, organisations, etc. This fruitful concept is illustrated by a simple model in behavioural finance describing the equity price evolution, and based on the interplay of three main growth archetypes: "Limits to Success", "Tragedy of the Commons", and "Balancing Loop with Delay".

1 Introduction

Chaos theory is said to have been founded by the 1-D logistic equation. This is certainly true although, as it is well known, the merit of discovering chaos in the discrete formulation of this formula may be given to May [16] in 1976, more than one century later. In its original continuous format the logistic equation is unable to generate chaos. This is a consequence of the Poincaré–Bendixon theorem, which says that there is no chaos on the line, or on the plane, thus at least 3-D is needed. In this chapter we develop the point of view that Verhulst, more directly, started "systems thinking" applicable to complex systems. There is clearly a straight line between Verhulst's germane ideas and the feedback-centred thinking of System Dynamics (SD), developed by J.W. Forrester [6, 7] in the 1960's, and used by the early Club of Rome in its famous book Limits to Growth [17]. What Verhulst's equation simply says, is that there is shifting loop dominance between two feedback loops (FBL): a positive FBL initiates growth; it is brought into balance by a negative FBL with growing importance, incorporating the limits to growth in a finite world. The association of FBL's of different polarities and the shifting dominance between them is indeed the central thought of SD to model complex reality in population dynamics, ecology, economy, organisations, etc. These ideas have been later translated into management recipes by Peter M. Senge in his famous book *The Fifth Discipline* [19]. Simple archetypes are presented there as elementary building blocks, pervasive in all organizational problems. All archetypes result in the association of one to three FBL's with different polarities. Senge argues that most dynamic patterns can be reproduced from the association of some of them.

In Sect. 2 we develop some basic concepts of systems thinking from this perspective. We use as a starting point the logistic equation as an important growth archetype in SD. In Sect. 3 we present two other growth archetypes, "Tragedy of the Commons", and "Balancing Loop with Delay", developed along similar lines to Verhulst's logistic equation. In Sect. 4 we present a simple behavioural model of stock-price evolution by combining the basic mechanisms imbedded in these archetypes. Three families of investors are interacting on the equity market: fundamentalists, opportunists and long-term traders. This model comprises at least three stocks, and, therefore, chaotic dynamics is possible, contrary to the case of the continuous 1-D logistic equation. A conclusion relative to systems thinking and its links to the Iron Law of Verhulst is given in Sect. 5.

2 The Logistic Equation, a Prototype of Systems Thinking

Figure 1 reproduces a possible influence diagram of the logistic equation of Verhulst in the very framework in which it was originally published, i.e., population dynamics. It represents a one-stock, two-flow System-Dynamics (SD) model of the evolution of a deer population; the latter is submitted to a food availability constraint. The only stock is represented by a rectangular reservoir, according to the tradition introduced by J.W. Forrester, the initiator of SD, in the early sixties of the last century.

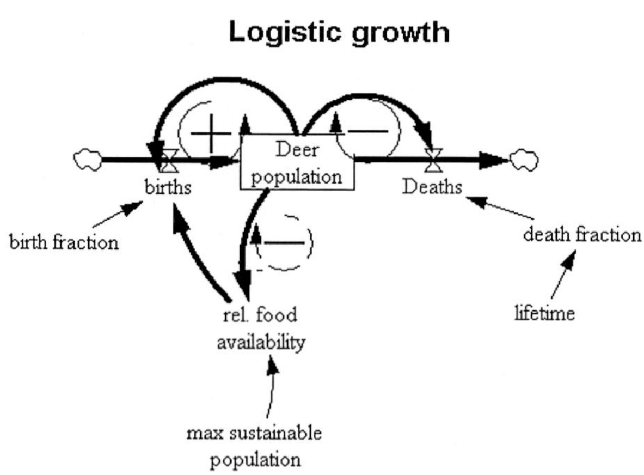

Fig. 1. The influence diagram of the logistic growth of a deer population

Calling P the population, its logistic growth is represented by the Verhulst equation in a modernized form, and slightly modified to explicitly include the deer death rate:

$$\frac{\mathrm{d}P}{\mathrm{d}t} = rP\left(1 - \frac{P}{K}\right) - DP\,. \qquad (1)$$

According to the usage in ecology, r represents the fractional growth rate corresponding to the r-strategy in a biotope, and K the limiting population size at maturity, corresponding to the K-strategy; D is the fractional death rate per unit of time, such that $D = 1/\text{Lifetime of deer}$.

Figure 2 shows the evolution of the population and of the two flows, "Births" and "Deaths". At logistic equilibrium the two flows become equal, so that the net flow vanishes. Figure 3 is the representation in the phase plane (deer population, net growth rate). The equation of the 1-D flow on the r.h.s. of (1) is a parabola. All this is of course well known. The influence diagrams and the computations originate from the SD-code VENSIM ® [23].

Let us spend some more time examining the two feedback loops (FBL) in Fig. 1. The positive FBL in the influence diagram represents the growth process. The induced growth pattern is exponential; it corresponds to the r-strategy.

Except for the natural death rate, the only negative influence is between "deer population" and "relative food availability": both variables move in opposite directions. Assuming that less food means less non-lethal births

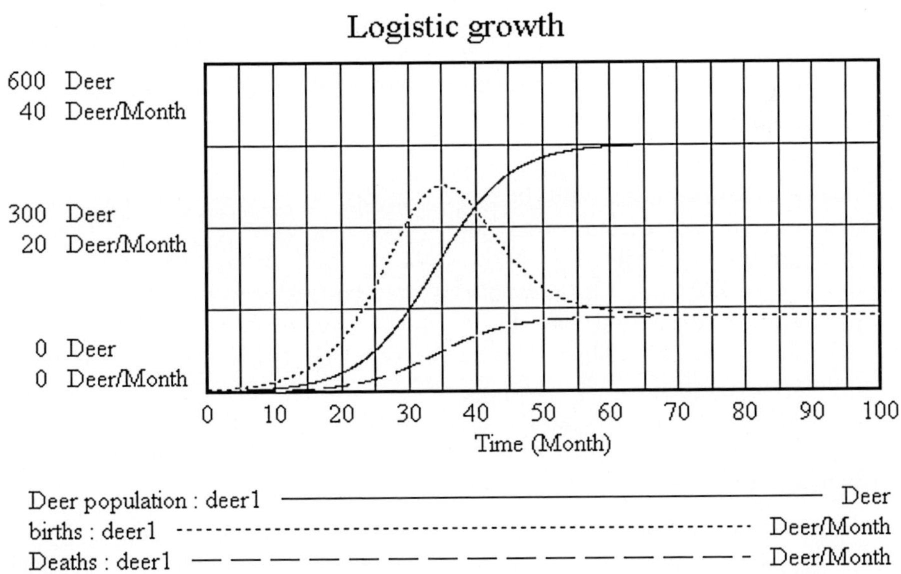

Fig. 2. The evolution of the stock and of the two flows in the logistic-growth model of Fig. 1

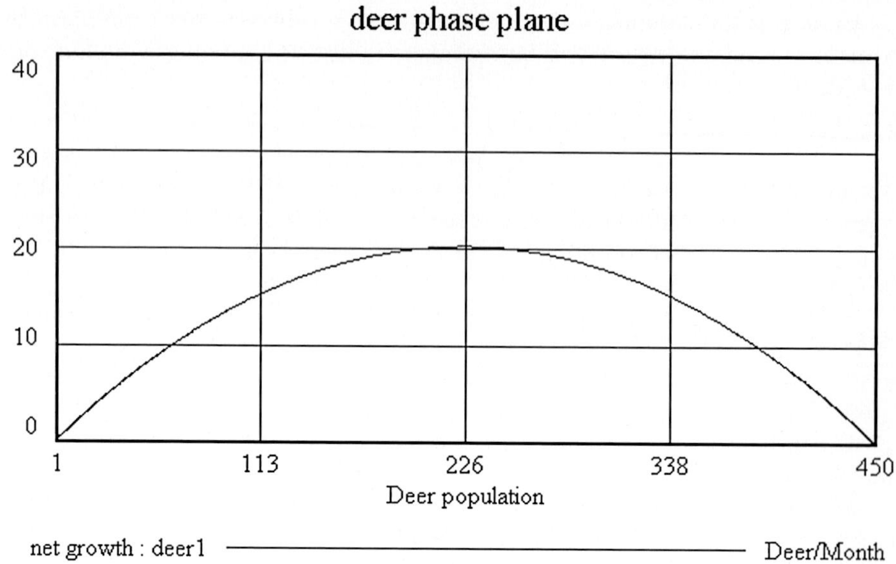

Fig. 3. The phase plane $(P, \mathrm{d}P/\mathrm{d}t)$ of the logistic equation showing the parabolic function on the r.h.s. of (1)

of fawns, a negative FBL is obtained. The induced growth pattern is goal seeking with a resulting equilibrium population size K; it corresponds to the K-strategy.

The dynamic behaviour of this simple dynamic system is dictated by "shifting loop dominance" between the two FBL's in the left part of the diagram:

- First the (+) FBL activates the r-strategy, i.e. nearly exponential growth, the (−) FBL remains weak because it is driven by the term $rP(P/K)$ in (1), which is still second-order, and nearly negligible;
- As P grows this latter term becomes larger, and progressive shifting loop dominance appears. This concept has been introduced by Forrester [6–8]. In this specific case this simply means that the weaker (−) FBL becomes increasingly active with respect to the (+) FBL. In the growth curve, an inflection point is visible when $P = K/2$;
- At equilibrium, both loops are equally active, and thus exactly in balance, and the nonlinear process of shifting loop dominance is then complete to realise the asymptotic equilibrium at $P = K$.

Shifting loop dominance is the central idea of FBL-thinking, and thus of SD [8]. The properties of nonlinear systems are changing in the phase space. Some loops are dominant, or simply active, while some other ones are dormant, or practically inactive. So that there are no universal properties any more, contrary to what happens in linear systems.

Even if all FBL's are present from the beginning in the influence diagram of the model, much different behaviour can be observed by numerical integration as the relative strengths of several FBL's change along the way. This explains why nonlinear systems often show counterintuitive behaviours as already stressed by Forrester in his Urban Dynamics [7]. This complexity can be observed with only few FBL's, but it increases when there are many possible combinations of interacting FBL's present in the model. Given n FBL's there are $n(n+1)/2$ FBL pairs to be compared. A larger system can have hundreds, or thousands FBL's!

This counterintuitive behaviour is a different concept from deterministic chaos. It has to do with the co-existence of many possible attractors of different nature (strange attractors are just one family). Another complication arises because of the possible bifurcations when parameters in the system (like the birth fraction) change value. This further increases the unpredictability and in fact the complexity of the system behaviour.

The 1-D logistic equation is unable to generate chaos, when the integration is done properly. This is because of the Poincaré–Bendixon theorem, which states that there can be no chaos on the line or the plane (see for example [9], Chap. 5.8, on stability properties in nonlinear systems). Chaos is thus only potentially observable in nonlinear systems with three stocks and more.

In the 1-stock case, chaos will only be observed as the result of an improper choice of the integration time step, and in this case it is thus a mere mathematical artefact (see [15]). Equation (1) indeed needs first to be numerically integrated, with introducing of a discrete time step. The Euler integration scheme in time t can be written as follows:

$$P(t+\Delta t) = P(t)\left[1 + r\left(1 - \frac{P(t)}{K}\right)\Delta t\right] . \qquad (2)$$

Assume that the initial condition is such that $0 < P(t=0) < K$. Because for all finite t, the exact solution of (1) is such that $P(t) < K$, if Δt is small enough, $P(t)$ will be increasing from $P(0)$ without ever exceeding K, except when $P(t)$ comes very close to K from underneath. One should then observe that for small enough Δt's:

$$P(t+\Delta t) > P(t) > 0 \text{ when } K - P(t) > \epsilon > 0 , \qquad (3)$$

where $\epsilon < \Delta t$ is a very small number. Numerically, for t sufficiently large $P(t)$ will slightly exceed K, so that the flow of the r.h.s. becomes negative; $P(t)$ will then gently oscillate with hardly observable amplitude around K. It can be intuitively understood that for larger Δt steps, oscillations will become of larger amplitude; once situations arise wherein $P(t)$ becomes significantly larger than K, overshoots of larger amplitude then occur, making $P(t)$ swinging hence and forth passing the K-value; the place where the population size P crosses the horizontal line at the boundary value K then changes at each period. Chaos arises when the set of crossing points becomes infinite. This

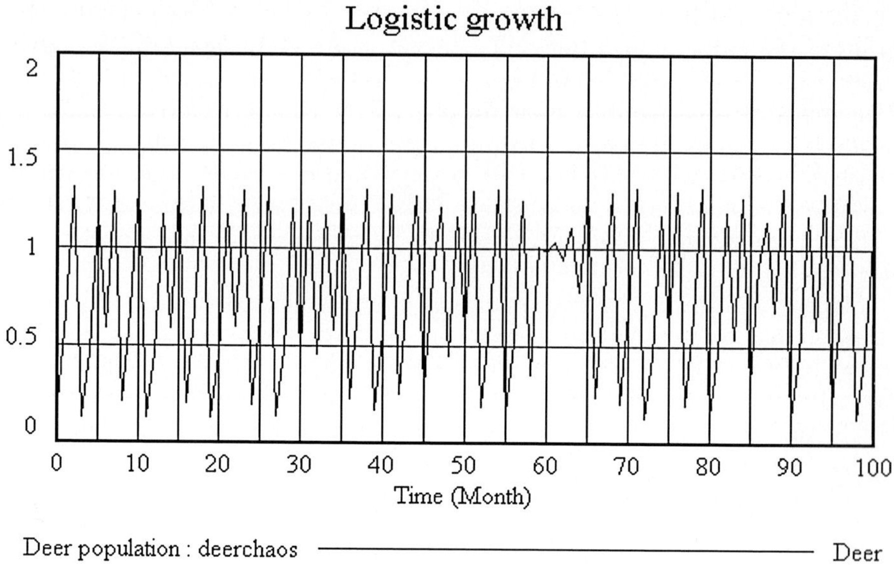

Fig. 4. When the time step is too large, the integration of the logistic equation with the Euler scheme generates similar pattern as in the logistic mapping, including chaos

situation is shown in Fig. 4. It is observed that, contrary to expectations from the continuous (2), P swings widely above the $K = 1$ boundary value of the population. Similar evolutions appear to the logistic map when the growth parameter increases. Several authors have established a correspondence between this latter parameter, and the time step, obtaining herewith the bifurcation diagrams in function of Δt. This discussion does not need to be reproduced here (see for example a review paper in [13]).

3 Archetypes

System Dynamics (SD) is a quantitative simulation technique; many authors in many different fields, such as economics and finance, organisation, environment, macroeconomics, etc. use it. A recent handbook is Sterman [22]. Soft modelling with SD is also a possibility. This approach limits the elaboration of models to the first qualitative step of establishing the influence diagram, and analysing the feedback-loop (FBL) structure to deduce some consequences for the system and to derive possible improvement strategies. This approach has some merit, though it is sometimes of limited predictive value: as mentioned before, systems often behave in a counterintuitive way due to the complex feedback interactions, and numerical simulations are necessary to test the actual behaviour patterns. Peter M. Senge is the author of *The Fifth discipline* [19]. His main message is that Systems Thinking is indispensable for understanding and curing organizational problems. Unfortunately the human mind has difficulty in abandoning linear thinking, which was well adapted to local conditions of human societies in the past, but becomes far less adapted to global societies today. Senge further argues that, in numerous cases, simple systems often consisting of two to three basic feedback loops (FBL) provide a sufficient insight on what is going wrong in the enterprise. These elementary systems, called archetypes, thus can be assembled as building blocks for modelling more complex situations. According to Senge's convictions most situations of crisis are reducible to a small number of archetypes. His book enumerates ten main archetypes. Additional ones have been developed in later books of Senge [20] on the basis of the work of Kim [12]. The most important archetypes are centred on three main growth patterns:

1. Logistic growth of (1) is described as combining exponential growth embodied in a (+) FBL, and goal-seeking growth, embodied in a (−) FBL. It represents the "Limits to Success" archetype in Senge's book. The interplay between the two FBL's leads to the described shifting loop dominance, as has been illustrated in the basic Verhulst model. The archetype is shown in a more general way in Fig. 5: the whole model rests on the assumption that some resource is limited and becomes inadequate at some point. A more business-oriented case is shown in Fig. 6, called "Doctor's Practice" [18]. It illustrates the interplay between on the one hand the growth process of the (+) FBL, around the mouth-to-mouth publicity of satisfied patients, and, on the other hand, the constraints of the time resource. The latter is impeding the further growth because of the (−) FBL related to the diminishing acceptability of time spent in the waiting room.

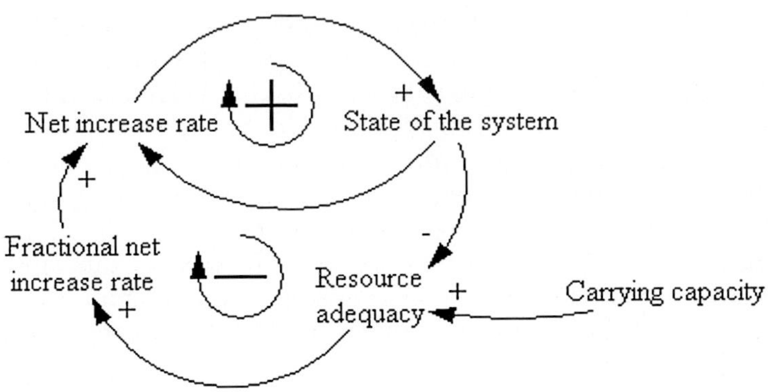

Fig. 5. The archetype "Limits to Success" as a generic model of Verhulst's Iron Law (according to [22])

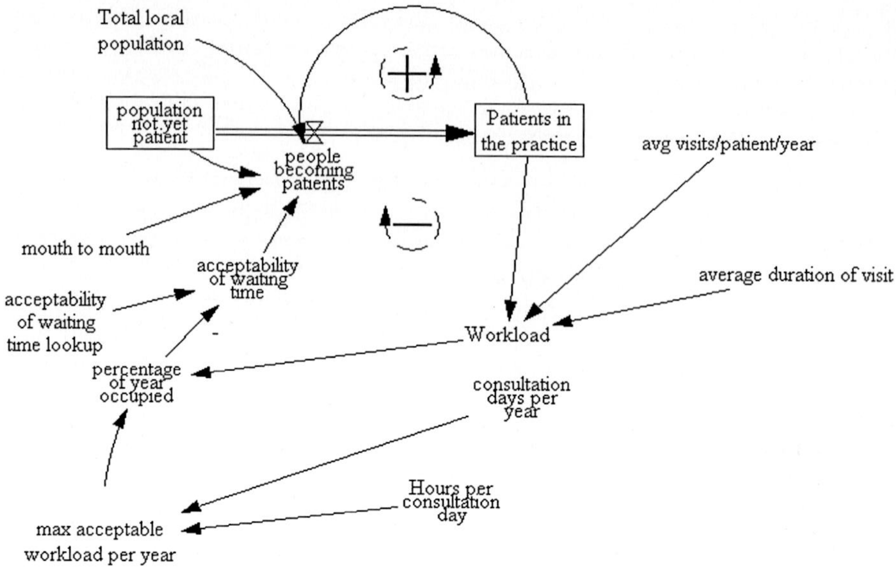

Fig. 6. The logistic growth in the doctor's surgery, as a further illustration of Verhulst's Iron Law. The limited resource is here the time that the doctor can devote to his patients (according to [18])

2. Overshoot and collapse growth appears in a second archetype "Tragedy of the Commons", according to the economist Garrett Hardin [10]. This type of growth is quite pervasive in complex systems (traffic congestion, exhaustion of depletable resources, collapsing of biotopes, etc.). It is obtained from Verhulst's logistic growth by adding just one more (−) FBL, as shown in Fig. 7. In the first archetype of logistic growth in Fig. 5, the resource is in some way an external parameter to the model, embodied in the constant K in (1). The second (−) FBL on the right of the drawing now includes the limiting resource in the model. It corresponds to an erosion mechanism. The growth goal K, instead of being constant, will now be suddenly and often unexpectedly be collapsing through the internal nonlinear forces in the system. In human systems, the erosion is caused by the inadequate use of a common good or resource (highway, oil, etc.) in the egoistic search for individual advantage.
   The model in Fig. 8 has two stocks, i.e. two ordinary differential equations, for representing both state variables, in this example deer population P and vegetation level V. These equations look as follows:

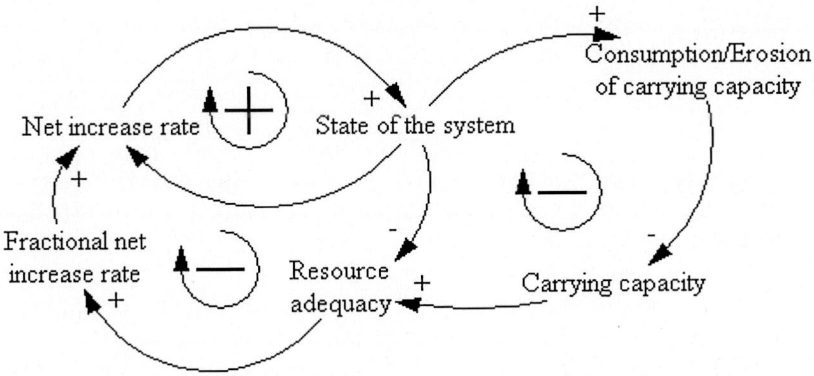

Fig. 7. The extension of the logistic model to a two-stock model representing the erosion of the food resource in the archetype "Tragedy of the Commons"

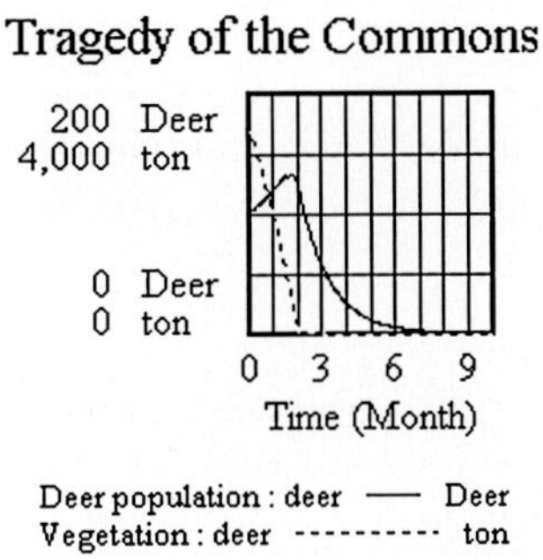

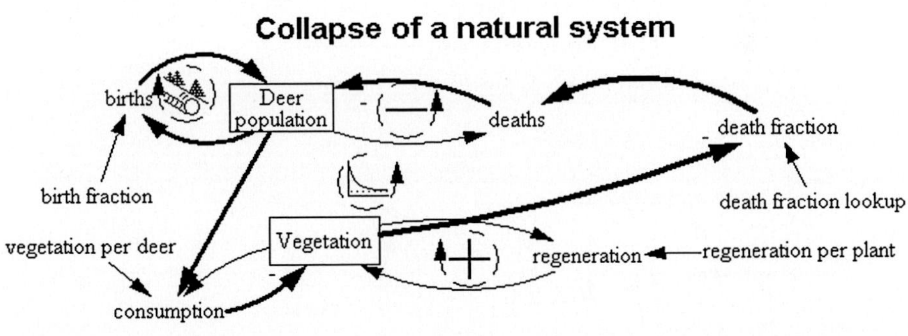

Fig. 8. The extension of the deer model of Fig. 1, including the resource *Vegetation* into the model. The deer population collapses when the food resource is eroded away

$$\frac{\mathrm{d}P}{\mathrm{d}t} = rP\left(1 - \frac{P}{K}\right) - D(V)P \tag{4}$$

$$\frac{\mathrm{d}V}{\mathrm{d}t} = sV\left(1 - \frac{V}{L}\right) - CPV\ . \tag{5}$$

The variables and parameters in (4), representing the deer-population dynamics, have the same meaning as in (1). $D(V)$, the death flow, is a declining nonlinear function of its argument V to be represented by a lookup table. It is of course equal to the natural death rate when the food is abundant, and it grows to 100 % mortality when food is disappearing. In (5), representing the vegetation dynamics, s and L are constants, and they correspond to r and K in (4). The parameter C represents the specific consumption of food per deer and time period.

3. The third archetype is called in Senge's original work "Balancing Loop with Delay". It is basically a goal-seeking loop. Delays may be present at several stages: when information is collected or processed to take action, or before action leads to a change in the state of the system. The generic archetype is shown in Fig. 9. All loops are negative, because each information delay corresponds to one or several one-stock systems with

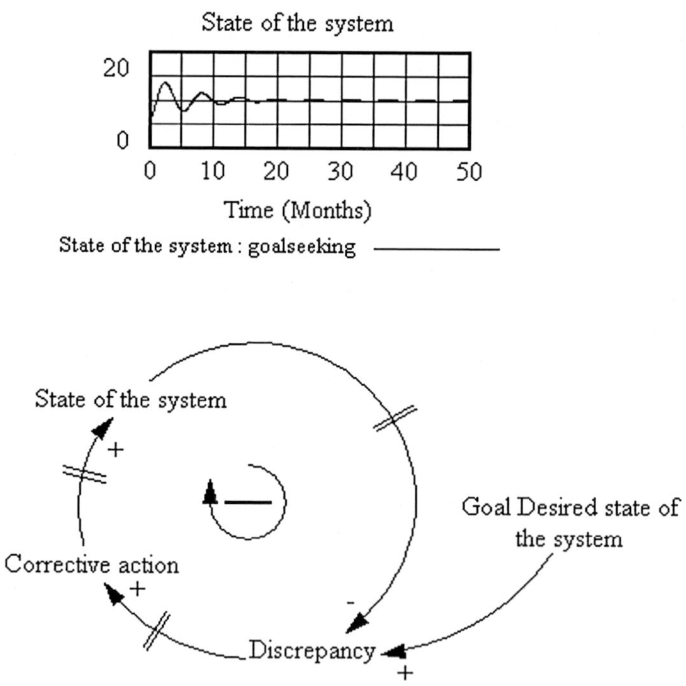

Fig. 9. The archetype "Balancing Loop with Delay" in which an information signal within a negative goal seeking FBL is submitted to delays causing overshooting and oscillations

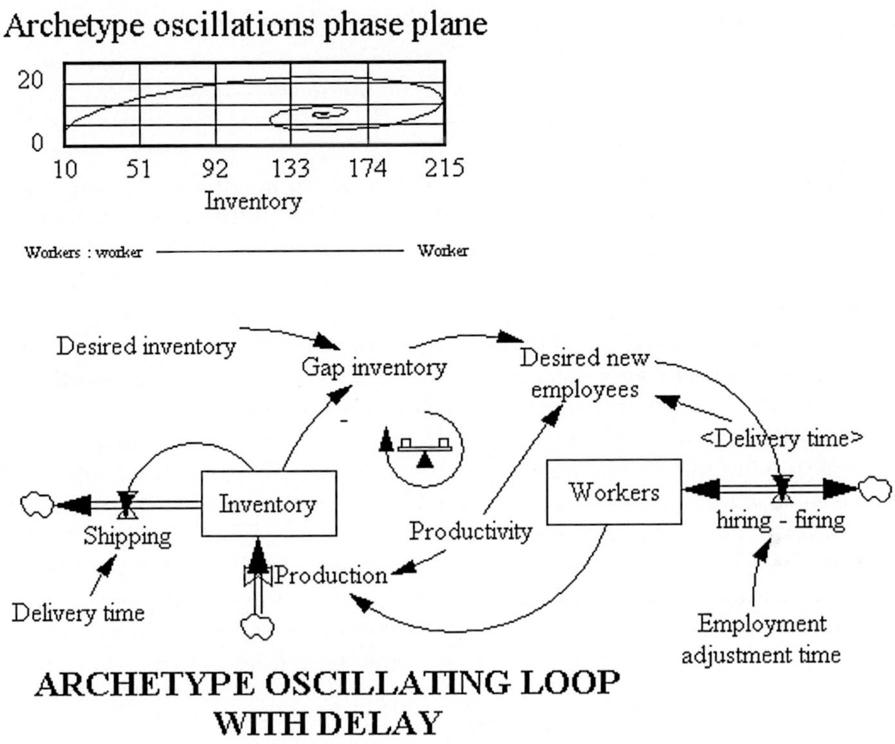

Fig. 10. In this archetype the oscillations are caused by the information delay between the inventory state and the manpower hiring

an outgoing flow (see [22], Chap. 11). This archetype is typical for the existence of business cycles. An example is shown in Fig. 10, representing a manpower-management problem; it consists in a goal seeking loop, which is itself a first-order delay, embedded in a two-stock system: inventory and manpower, as follows:

$$\frac{\mathrm{d}S}{\mathrm{d}t} = \frac{\mathrm{Goal} - S}{T_{\mathrm{adj}}} , \tag{6}$$

where S is the stock due to achieve the goal, and T_{adj} is the time constant necessary for the goal-adjustment process; it also represents the time delay constant. Equation (6) is the equation of a linear proportional controller (e.g., a thermostat) used in engineered devices to bring the state variable (e.g., the room temperature) to a desired goal. More complex nonlinear controllers, used in engineering, can be developed for the same purpose.

4 Modelling a Bubble on the Stock Market

In this section we discuss the modelling of the development and crash of a speculative bubble on the equity market (EM). The recent history of the high-tech bubble mainly in the years 1997 to 2003 provides a good example.

The archetypes presented in the previous section are the starting basis for modelling, because of their characteristic growth patterns. When a financial bubble first builds up, exponential growth is observed; later on temporary plateaux appear reminding of logistic growth equilibrium; there are also pseudo-random oscillations reminding of the "Balancing Loop with Delay"; finally crashes resemble the patterns in the Tragedy of the Commons.

Financial crashes were qualitatively modelled as cusp catastrophes by Zeeman [24]. R.H. Day [4] was one of the first authors who intensively worked in quantitative non-equilibrium models inspired from chaos theory in discrete nonlinear systems. Following ideas of Shiller [21], Day postulates two families of investors, smart and ordinary investors. The formers are called α-investors; the latter are called β-investors. α-investors use quantitative valuations from fundamental analysis, they are basically goal-seekers and they stabilise the market. Their investment profile as a function of the price has a reverse shape, because they are contrarians. β-investors, by contrast, remain in phase with the price trend by using simple investment rules: they overreact to sudden price moves or to fads, creating volatility. Day combines both investors' profiles to define iterative 1-D mappings of the stock price: $p(t+1) = f[p(t)]$. The patterns he observes show phases of high volatility, betraying the existence of deterministic chaos like in the logistic mapping. Unfortunately, with those models it is much more difficult to generate more representative evolutions typical for EM, like bullish or bearish behaviours, bubble formation and crashes, etc.

The idea developed by Kunsch et al. [14] is to use a continuous model with at least three stocks, in order to have the possibility to observe chaos, and a number of FBL's able to generate representative and more realistic EM signatures. The objective of considering at least three stocks is easy to achieve by considering several investors' families and information delays. Each first-order delay requires one stock. Additional budget stocks represent the financial constraints of investors.

Of course an important literature exists dealing with nonlinear dynamic modelling of the EM (see for example [3], [2] from [11]), or with artificial stock markets [1]. The ambition is not to present an up-to-date review of these models generally placed in the field of behavioural finance. Rather it is planned here to show how archetypes, inspired from Verhulst's ancient contribution, are still a source of inspiration for complexity modelling.

The universe in the EM model presented here is very simple. There are only two assets: a risky asset quoted at a variable homogeneous price P (it could represent a common equity index like Eurostoxx 50), and a risk-free asset, e.g., a high-rating bond. This universe is frozen for a given simulation run. This means first that the total number of equity shares n is fixed in all scenarios. Second, all economic parameters of the model are constant, including the growth rate of the fundamental value, and the risk-free rate. Some fixed constraints are imposed on the available budget of the investors and their borrowing capacity.

Three homogeneous groups of investors are considered, instead of two in Day's model. They are called α-, βS-, and βL-investors, where "S" stays for short-term, and "L" stays for long-term. Influence diagrams can be drawn to represent each investor's behaviour. Several important feedback loops are identified. They assist the understanding of basic behavioural rules developed in the investors' minds. Some characteristics are important to understand. Negative loops assist the goal-seeking approach of fundamentalists. α-investors therefore help stabilizing the stock prices. By contrast positive loops, activated by short-term traders (βS-investors) are responsible for amplifying perturbations or rumours. Sometimes such a loop can act as a virtuous circle, in case it triggers a desired growth effect in prices thanks to long-term strategies of βL-investors. Sometimes the loop acts as a vicious circle, because it amplifies the market volatility, or it triggers crashes. In the EM model, the two roles will be played in turn.

The three families of investors are now described in more detail; it is shown in each case in which way they are representative of the previously introduced archetypes. Note beforehand that this model is very simplified because many variables are considered as exogenous parameters, to be held constant: the relative proportions of the different investor types, and the risk-free interest rate among others. These assumptions could be removed at the cost of higher complexity (e.g. including these parameters as model variables into additional FBL's), but that would be beyond the scope of the present work.

The same presentation is adopted as in [14].

Note first that the total equity price P is split up into the three components representing the contributions of investors from different groups:

$$
\begin{aligned}
P(t) &= \frac{1}{n}\left(M_{\alpha} + M_{\beta \mathrm{S}} + M_{\beta \mathrm{L}}\right) \\
&= P_{\alpha} + P_{\beta \mathrm{S}} + P_{\beta \mathrm{L}} \,. \qquad (7)
\end{aligned}
$$

$M_{\alpha}, M_{\beta \mathrm{S}}, M_{\beta \mathrm{L}}$ represent the amounts of money invested by the three investor types in the EM; n is the constant number of shares; $P_{\alpha}, P_{\beta \mathrm{S}}, P_{\beta \mathrm{L}}$ represent the three components of the total price P, attributed to the three investor types.

4.1 Family of α-Investors

α-investors are "smart investors" behaving in a similar way to the rational goal-seekers assumed in Day's model. Their sole aim is to achieve convergence

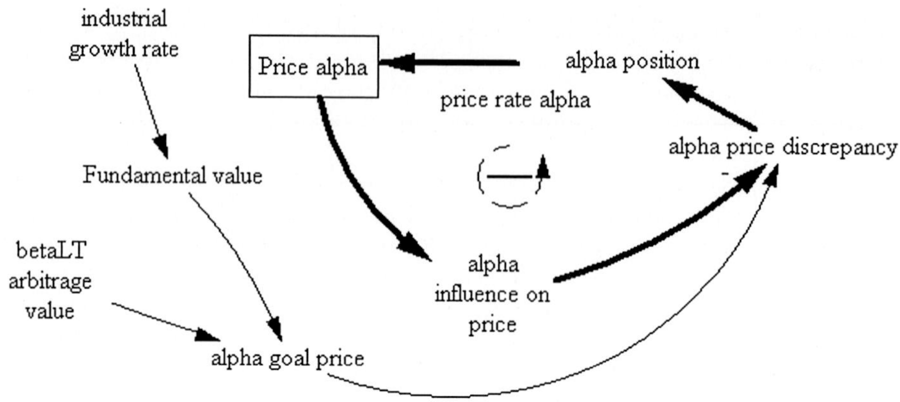

Fig. 11. Negative Feedback loop of α-investors (goal-seeking behaviour)

towards a current goal price G_α for the stock price P. It is why the unique feedback loop visible in Fig. 11 is a negative goal-seeking FBL. The price component P_α obeys to a similar equation to (6):

$$\frac{\mathrm{d}P_\alpha}{\mathrm{d}t} = \frac{G_\alpha - P}{T_\alpha} \tag{8}$$

As said above, more complex goal-seeking formulations can be adopted. Because there are possibly information delays in the price adjustment, damped oscillations caused by slight overshooting above the goal price may be observed.

In our model, the assumed current goal price G_α is the sum of two terms: fundamental value g_α, and risk premium $A_{\alpha,\beta\mathrm{L}}$ resulting from the investing behaviour of βL-investors, to be described later:

$$G_\alpha = g_\alpha + A_{\alpha,\beta\mathrm{L}} \tag{9}$$

- The fundamental value g_α results from fundamental analysis, e.g., Dividend Discount Model (DDM). Deterministic dividends are assumed here, because stochastic changes do not bring more understanding on causal mechanisms. Dividends are growing with the given constant industry growth rate: g_α is growing at the same rate.
- The risk premium depends on the arbitraging behaviour of βL-investors between the stock return and the risk-free rate; this is explained below. In case of a positive gap, they invest more money into the EM, creating herewith a price increase $\Delta P_{\beta L}$, i.e. a risk premium above the fundamental value. In this case α-investors also adjust their long-term expectations, and they follow the observed positive trend over the fundamental value. In practice, the premium is incorporated into the goal price by α-investors only up to a certain point; this occurs with a time delay τ. In the model it

is assumed that, in a bullish market mainly driven by the premium term, α-investors will cap their goal price by a maximum arbitrage value $\Delta g_\alpha^{\max}$. The latter corresponds to an acceptable risk level. Thus the actual risk premium is given by the following equation:

$$A_{\alpha,\beta\mathrm{L}} = \min\left(\Delta g_\alpha^{\max}, \mathrm{delay}_\tau[\max(0, \Delta P_{\beta\mathrm{L}})]\right) . \tag{10}$$

In the model α-investors do not experience any liquidity constraints. This is a reasonable assumption, as they stop anticipating further price growth, as the risk premium above the fundamental value becomes exceedingly large.

In conclusion, α-investors behave according to the goal-seeking part in the logistic equation (Verhulst's Iron Law). Because the goal is changing under the effect of βL-investor strategies, there is a need for information collecting: damped oscillations due to overshoots may be observed, as in the archetype "Balancing Loop with Delay".

4.2 Families of β-Investors

The β-investors are "ordinary investors" in Shiller's sense [21]. They are not entirely rational with respect to the use of information coming from the market. They use different approaches to process the information, from rules of thumb to advanced technical analysis. An important aspect is the time horizon of anticipation, covering a continuum between short-term to long-term. The model only considers two extreme cases in a continuum: βS-investors have a short-time horizon (S); βL-investors have a long-term horizon (L). Also proportions of the two types are kept constant. More sophisticated models may consider intermediate investors' profiles or varying proportions within the model.

Family of βS-Investors

βS-investors are opportunistic traders who are following immediate price movements; they buy in case of a price increase, and they sell when the price is going down. Therefore they destabilise the goal-seeking efforts of α-investors, who are contrarians, and they cause permanent noise. The presence of a positive feedback loop, visible in the right part of Fig. 12, confirms the existence of this destabilizing investment approach. The driver in this loop is the first derivative of the price, initiating a vicious circle of growth or decay. A negative loop is visible in the left part of the diagram. It becomes active as the available budget drops to zero, forcing βS-investors to limit their stock position or even to liquidate part of their portfolio. The dynamic equation of βS-investors has been assumed to be the following:

$$\frac{\mathrm{d}P_{\beta\mathrm{S}}}{\mathrm{d}t} = S_{\beta\mathrm{S}}\, f\left(\frac{\mathrm{d}\bar{P}}{\mathrm{d}t}, P\right) - R_{\beta\mathrm{S}}(B_{\beta\mathrm{S}}) , \tag{11}$$

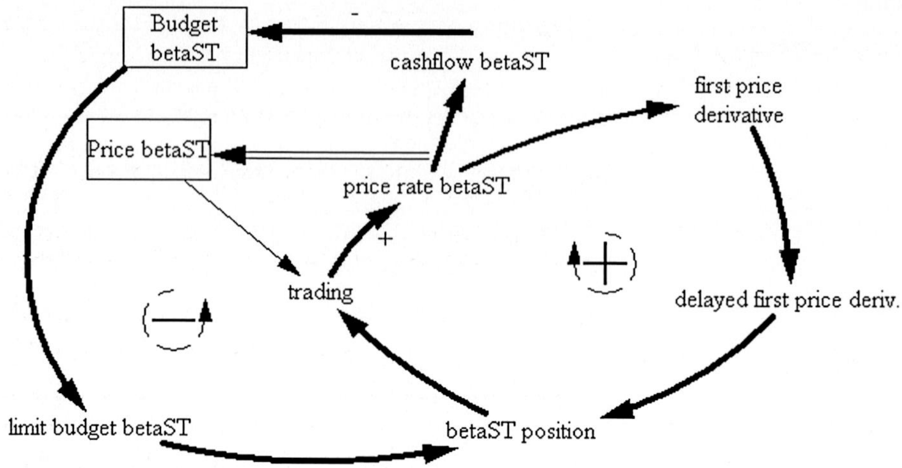

Fig. 12. Feedback structure of short-term traders, i.e. βS-investors. The budget acts as a control mechanism; the presence of information delays causes oscillations. There is also a partial correspondence between this diagram and the archetype "Tragedy of the Commons"

where $S_{\beta S} > 0$ represents the strength of βS-investors on the market (assumed to be constant). The function $f(.,.)$ depends in a nonlinear way on the stock price P and the smoothed value of its first derivative $\mathrm{d}\bar{P}/\mathrm{d}t$. Its sign is the same as the sign of the latter, indicating that βS-investors are trend-followers modifying their positions according to increasing or declining prices. A smoothed signal is calculated as an information delay as in (6), so that this may be the cause of oscillating behaviour, as in the archetype "Balancing Loop with Delay". Overshoots may also be observed, which bring the budget to negative values. In such situations the βS-investors have to liquidate part of their portfolios. This appears in the last term on the r.h.s. of (11): $R_{\beta S}(B_{\beta S})$ represents the reimbursement rate to bring the budget $B_{\beta S}$ back to balance, in case it becomes negative. In conclusion, the delay mechanism and the budget constraints make that βS-investors are a source of instability and create pseudo-random oscillations in the search for price equilibrium on the EM.

Family of βL-Investors

βL-investors rather have a long-term perspective. They permanently compare the long-term stock return and the risk-free interest rate (i_{rate}, assumed to be constant in this simple model). In case of a positive spread, in favour of risky asset positions, they invest additional money, curbing on the growth of the stock price. Therefore a positive feedback loop is visible in the upper part of the diagram in Fig. 13. It is driven by the positive return spread between

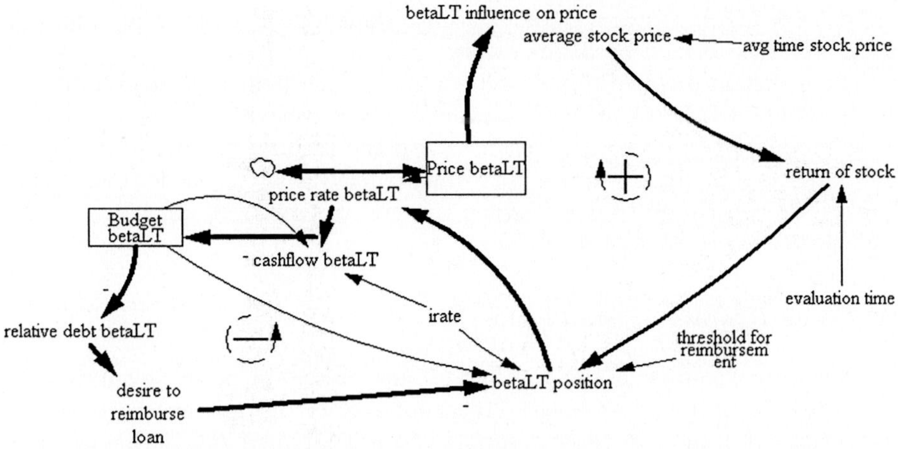

Fig. 13. Feedback structure of long-term traders, i.e. βL-investors. The budget fuels growth up to a certain point just before collapse. There is a clear correspondence between this diagram and the archetype "Tragedy of the Commons"

risky and risk-free assets, creating a risk premium. As indicated in (8) and (9), α-investors will adjust in part their goal price to follow the growing price trend caused by βL-investors. In contrast to βS-investors, βL-investors have a borrowing capacity. They invest the borrowed money reinforcing herewith the growing trend, and transform it progressively into a vicious circle. Of course at some point there is shifting-loop dominance in favour of negative FBL's like in the archetypes "Limits to Success", and the "Tragedy of the Commons". Such a loop is visible in the lower part of the diagram in Fig. 1: it relates to the available money resources. βL-investors have an initial budget and borrowing capacity up to a given permissible debt level. In any case, their willingness to reimburse their loans will grow with the relative level of their debt expressed as a percentage of the value of their stock position. As long as some borrowing capacity remains, βL-investors further strengthen their stock positions. Above some debt threshold, they experience an incentive to liquidate at least part of their positions. Another negative loop is not directly visible in Fig. 13, however. It finds its origin in the cap imposed by α-investors on the permitted price growth above fundamentals, according to (10). The simplified equation representing the dynamics of βL-investors is as follows:

$$\frac{\mathrm{d}P_{\beta\mathrm{L}}}{\mathrm{d}t} = S_{\beta\mathrm{L}}(r - i) - R_{\beta\mathrm{L}}\left(B_{\beta\mathrm{S}}, V_{\beta\mathrm{L}}\right) , \tag{12}$$

where $S_{\beta L} > 0$ represents the strength of βL-investors on the market (assumed to be constant in this simple model). r represents the smoothed stock return, and $i = i_{\text{rate}}$, the risk-free interest rate. The growth pattern of the first term on the r.h.s. of (12) is thus exponential. $R_{\beta\mathrm{L}}$ represents the reimbursement rate of loans in case the current budget $B_{\beta\mathrm{L}}$ is becoming negative and exceeds the borrowing capacity, which depends on the current value $V_{\beta L}$ of the portfolio owned by βL-investors.

In conclusion βL-investors develop a growth mechanism that is similar to the archetype "Tragedy of the Commons". The growth is fuelled on a basis of an artificial money-borrowing resource and the landing can be quite sudden and hard, as some limits in the borrowing capacities are exceeded. Note that α-investors contribute to define when the bubble crash will start by capping the risk premium, as shown in (10).

4.3 Some Results of Simulation

In short some results of the EM model are presented in the form of time diagrams. The latter represent the total stock price (upper curve represented by a thick line), and its three components, indicated as P_{alpha} (heavy line), P_{betaST} (medium-heavy line), and P_{betaLT} (thin line), according to (7). There are many possible choices for the parameters in the model, but we shall limit our discussion to a few typical cases for bubble growth and crash. We again adopt the presentation from [14].

A first scenario represents a boundary situation, helpful to calibrate the model. The risk-free rate is assumed to be very high; it is then expected that the risk premium is vanishing, i.e. no βL-investors will be present on the market, and thus no bubble can appear. In this case the price will gently

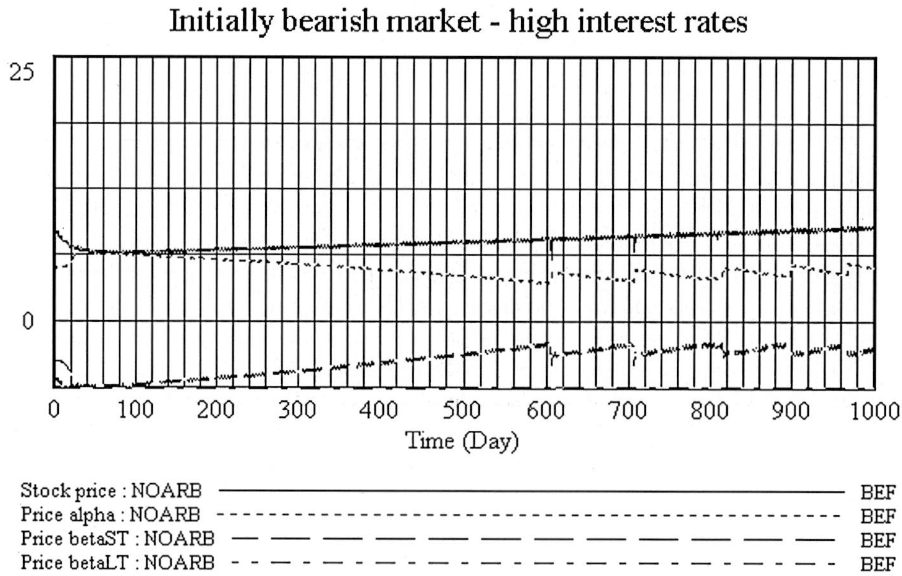

Fig. 14. (From [14]). A calibration calculation in case there is no risk premium, because of high risk-free rates

follow the industry growth rate. The price components and the total price are shown in Fig. 14 confirming our expectations. The total price rapidly comes to the fundamental value. The lower curve shows the contribution of βS-investors following the growing trends. Sometimes budgets become negative, so that, according to (11), a broken line and oscillations back to positive budget values are observed.

The following figures show two situations, in which the risk-free interest rate is low, so that βL-investors can arbitrage. In Fig. 15 the market is initially bearish. The return spread is negative, i.e. in favour of risk-free investment. For quite a long time, the market is in near-equilibrium at the goal price set by α-investors; the growth rate is equal to the industrial growth rate (7% p.a.). However, the steady industrial growth brings about a fresh-born wave of βL-investors. As a result the goal price also shifts up. βL-investors soon find their limits. The borrowing capacity is reached. At this point, βL-investors have to rapidly liquidate the largest part of their stock portfolio in order to bring down their loan debt to an acceptable level. This reimbursement constraint has the same effect on the price as a reflecting barrier. The price bounces back creating chaotic ups and downs of the price in search for a new equilibrium value. βS-investors amplify the appeared volatility. The price volatility becomes so large that at some point the long-term return drops below the risk-free rate. βL-investors disappear from the scene after a crash of limited amplitude. The market moves to a new equilibrium following the natural trend of fundamental values.

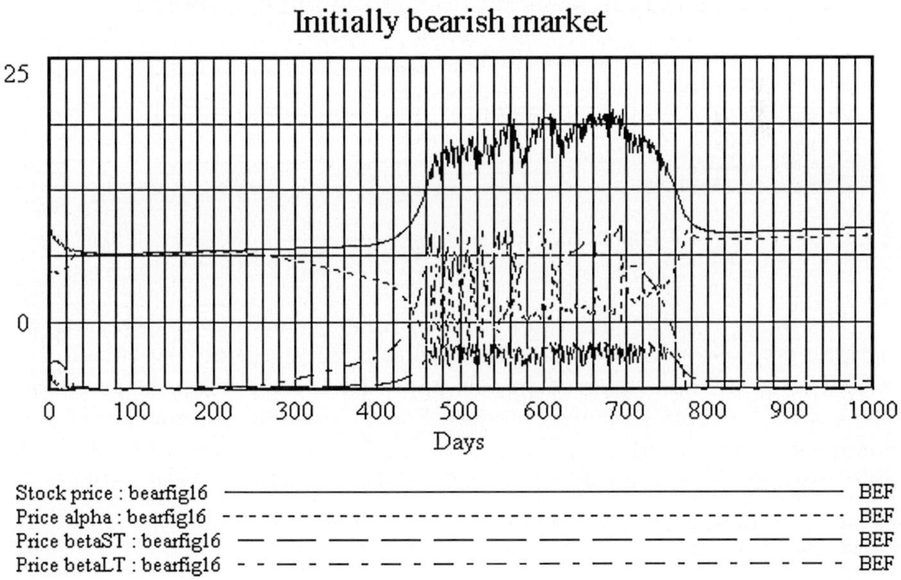

Fig. 15. (from [14]). The case of low interest rates, and an initially bearish market

In Fig. 16 the same assumptions as in the previous figure are used for simulating an initially bullish market. A burst in price sets up immediately; it is accompanied by high volatility. The bouncing back of the price against the debt barrier induces still more volatility than in the previous case. This turbulent behaviour cannot maintain itself very long. The market moves to its fundamental equilibrium as in the previous figure. After a while, a new price upsurge is observed with still more volatility than previously in the growing phase. It lasts for quite some time, exhibiting swings of considerable amplitude. As before, a crash brings back the price to its natural equilibrium. When pursuing the computation, regular replicas with similar shapes are periodically observed.

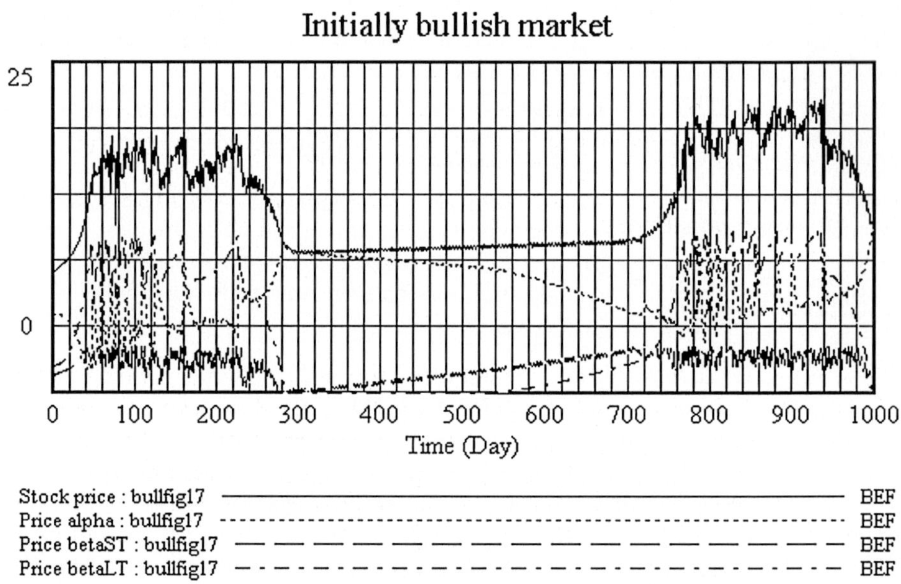

Fig. 16. (from [14]) The case of low interest rates, and an initially bullish market

The more detailed paper contains additional runs with other choices of parameters, and a comparison of simulations with real observations on the EM. The readers are referred to this paper for more details. As a source of inspiration for more advanced models, Fig. 17 presents, without any further comments, the evolution of the Eurostoxx 50 index between October 1994 and September 2004 during the lifetime of the recent high-tech bubble.

Fig. 17. The evolution of the DJ Eurostoxx 50 from October 1994 to September 2004 (from the website wallstreet-online)

5 Conclusions

The author has attempted in this chapter to present Verhulst's contribution in a somewhat different light: discussions generally go about the links between the logistic equation and chaos theory. It is argued here that in some way Verhulst was a pioneer of nonlinear system theory. Systems thinking is today becoming a necessity for survival: the world globalization forces us to think in terms of causality networks rather that in terms of isolated cause-effect links [5]. Verhulst's ancient contribution thus remains modern, and it is still needed.

With the logistic growth, Verhulst introduced for the first time in history an influence diagram, in which two feedback loops are competing for dominance. This simple system teaches us that exponential growth is impossible in the natural world, because constraints on resources must be taken into consideration. The Iron Law of Verhulst remains an important message, which is unfortunately not yet accepted by all. Without this insight it is impossible to start a reflection on how to remove the "Limits to Success", which are today threatening the very long-term existence of mankind in the limited spaceship earth.

References

1. W.B. Arthur, J.H. Holland, B. LeBaron, R. Palmer, P. Tayler: Asset Pricing Under Exogeneous Expectations in an Artificial Stock Market. In: *The Economy as an Evolving Complex System II. Proceedings Vol. XXVII* ed by W.B. Arthur, S.N. Durlauf, D.A. Lane (Santa Fe Institute Studies in the Sciences of Complexity, Addison Wesley, Reading MA 1997)
2. W.A. Brock, C.H. Hommes: Models of Complexity in Economics and Finance. In C. Heij et al. (Eds) op. cit. (John Wiley and Sons, Ltd., Chichester 1997) pp 3–44
3. J. Creedy, V.L. Martin (Eds): *Chaos and Non-Linear Models in Economics – Theory and Applications* (Edward Elgar Publ. Co., Brookfield Vermont 1994)
4. R.H. Day: *Complex Economic Dynamics, Vol. I An Introduction to Dynamical Systems and Market Mechanisms* (MIT Press, Cambridge MA 1994)
5. D. Dörner: *Die Logik des Misslingens. Strategisches Denken in komplexen Situationen* (rororo Science, Rowohlt Verlag, Reinbek bei Hamburg 2004)
6. J.W. Forrester: *Industrial Dynamics* (Wright-Allenn Press, Cambridge 1961)
7. J.W. Forrester: *Urban Dynamics* (MIT Press, Cambridge 1969)
8. J.W. Forrester: European Journal of Operational Research **30**, 104 (1987)
9. P. Glendinning: *Stability, Instability, and Chaos* (Cambridge University Press, Cambridge 1994)
10. G. Hardin: Science **16**, 1243 (1968)
11. Ch. Heij, H. Schumacher, B. Hanzon, K. Praagman (Eds): *System Dynamics in Economic and Financial Models* (John Wiley and Sons, Ltd., Chichester1997)
12. D.H. Kim: *Systems Archetypes, Diagnosing Systemic Issues and Designing High-Leverage Interventions* (Pegasus Communication, Cambridge MA 1993)
13. T. Kinnunen, H. Pastijn: Chaotic Growth. Attractors for the logistic model. In: *AIA, Revue-Tijdschrift* (Association of Engineers in Artillery, Belgian Royal Military Academy, Brussels 1986) pp 1–17
14. P.L. Kunsch, M. Theys, A. Chevalier, J.P. Iacopetta: A System Dynamics model of stock price movements. In: *Decision Making: Recent Developments and Worldwide Applications* ed by G. Doukidis, S.H. Zanakis, C. Zapounidis (Kluwer Academic Publisgers, 2000) pp 153–167
15. R.M.M. Mattheij, J. Molenaar: *Ordinary Differential Equations in Theory and Practice* (John Wiley and Sons, Chichester 1996)
16. R. May: Nature **261**, 459 (1976)
17. D.H. Meadows, D.L. Meadows, J. Randers: *The Limits to Growth: Report for the Club of Rome's Project on the Predicament of Mankind* (Universe Books, New York 1972)
18. G.P. Richardson, A.L. Pugh III: *Introduction to System Dynamics Modeling* (Productivity Press, Portland 1981)
19. P.M. Senge: *The Fifth Discipline. The Art and Practice of The Learning Organization* (Random House, Business Books, London 1990)
20. P.M. Senge, A. Kleiner, Ch. Roberts, R.B. Ross, B.J. Smith: *The Fifth Discipline Fieldbook, Strategies and Tools for Building a Learning Organization* (Nicholas Brealey Publishing, London 1994)
21. R.J. Shiller: *Market volatility* (MIT Press, Cambridge MA and London 1989)

22. J.D. Sterman: *Business Dynamics. Systems Thinking and Modeling for a Complex World* (Irwin McGraw-Hill, Boston 2000)
23. Ventana Systems: *VENSIM ®3.0 software* (Harvard 1997)
24. E.C. Zeeman *Catastrophe Theory – Selected Papers 1972-1977* (Addison-Wesley Publ. Co., Reading MA 1977)

Recurrent Generation of Verhulst Chaos Maps at Any Order and Their Stabilization Diagram by Anticipative Control

Daniel M. Dubois

asbl CHAOS, Centre for Hyperincursion and Anticipation in Ordered Systems,
Institute of Mathematics, B37, UNIVERSITY OF LIEGE,
12, Grande Traverse, B-4000 LIEGE 1, Belgium
Daniel.Dubois@ulg.ac.be
ulg.ac.be/mathgen/CHAOS

1 Introduction

In 1838, P.-F. Verhulst [1] proposed an extension of the Malthus equation of the continuous growth of a population. In the Malthus equation, the time derivative of the population is directly proportional to the population, and gives a solution as a temporal exponential curve without limitation and tending to infinity. Verhulst introduced, to the Malthus equation, a negative term, proportional to the square of the population, with a view to obtaining a stable stationary finite state of the population. The Verhulst nonlinear equation is called the logistic equation, for which an analytical solution exists.

For a non-continuous growth of a population, the differential equation is traditionally transformed with the Euler algorithm to a discrete equation, called the Verhulst map in the framework of chaos theory. This chaos map gives solutions of several types depending on the value of the growth parameter of the population: stable fixed points, bifurcations and then chaos. Simple analytical solutions of this chaos map exist for particular values of the growth parameter. In the chaos zone, this solution depends directly on the initial population, so the future evolution of the population is not predictable for long times if the initial condition is not known with precision. This phenomenon is called the sensitivity to initial conditions.

The purpose of this chapter is first to demonstrate that the Verhulst map is not the correct discrete equivalent to the Verhulst logistic equation. Due to the discrete time interval, the square of the population must be transformed to the product of the population at time t and the same population at the following discrete time $t + \Delta t$, giving a non-recursive equation called an incursive equation (for inclusive or implicit recursive equation) [4]. The solution of this incursive discrete equation is similar to the stable stationary solution of the differential Verhulst logistic equation. The same result can be obtained from an anticipative control to the Verhulst map.

Secondly, it will be shown that the Verhulst chaos map belongs to a class of chaos maps at any order n, where n is the degree of the map. In this class, the Verhulst map depends on the power two, $n = 2$, of the population. The concept of *canonical chaos map* will be introduced. The set of canonical chaos maps is given by the Tchebychev polynomials that verify a very simple relation of recurrence. Closed form solutions exist for these canonical chaos maps. This set of canonical chaos maps will be transformed to a set of chaos maps depending on a growth parameter.

Thirdly, this chapter will deal with the control of these chaos maps at any order, with the method of the *incursive predictive control* that belongs to the class of model predictive controls for which the model is the equation of the map itself, and for which no setpoint is defined [5]. Such a control transforms all the unstable states of the chaos maps to stable fixed points. A new type of diagram will be introduced, which we shall call the *stabilization diagram*. The resulting stabilization diagrams show a number of stable fixed points directly related to the order of the maps. Numerical simulations will be performed on these chaos maps for various orders: the first return map, its bifurcation diagram and its stabilization diagram will be displayed as a function of the growth parameter.

2 Analytical Solution of Chaos Maps

This section deals with the solution of the Verhulst logistic equation and the deduction of the so-called Verhulst chaos map. The closed form solution of the chaos map is then demonstrated and its relation to the Shift map is established.

2.1 From the Verhulst Differential Logistic Equation to the Verhulst Chaos Map

The original Verhulst [1] differential growth equation is given by

$$\frac{\mathrm{d}N(t)}{\mathrm{d}t} = rN(t)\left(1 - \frac{N(t)}{K}\right) \tag{1}$$

where $N(t)$ is the value of the population at the current time t, r is the growth rate and K is a limiting growth factor. The well-known analytical solution of this equation is given by

$$N(t) = \frac{\mathrm{e}^{rt}N(0)}{1 + (\mathrm{e}^{rt} - 1)N(0)/K} \tag{2}$$

where $N(0)$ is the initial condition of the population at time $t = 0$. This solution tends to $N(t) = K$, for $t \gg 1/r$. This means that the system loses its initial value and that its future is completely defined by the value of the parameter K, the value of which being fixed. So, such an equation is an anticipative system in the sense that the final future value, K, of the population is already known and completely fixed at the present time.

Let us now deduce the classical discrete version of this equation. In defining $P(t) = N(t)/K$, (1) becomes

$$\frac{\mathrm{d}P(t)}{\mathrm{d}t} = rP(t)[1 - P(t)] \; . \tag{3}$$

With the forward derivative, the discrete equation is given by the well-known Euler algorithm

$$P(t + \Delta t) = P(t) + \Delta t[rP(t) - rP^2(t)] \tag{4}$$

where Δt is the time step. The time step can be taken equal to one (in rescaling the growth rate r) without lack of generality. So, with $\Delta t = 1$, this equation can be rewritten as

$$\frac{P(t+1)}{1+1/r} = \frac{(1+r)P(t) - rP^2(t)}{1+1/r} \; . \tag{5}$$

With the change of variables $x(t) = P(t)/[1+1/r]$ and $a = 1+r$, the Verhulst chaos map is readily obtained:

$$x(t+1) = ax(t)[1 - x(t)] \tag{6}$$

with $a \in [0, 4]$ and $x(t) \in [0, 1]$.

Let us summarize some properties of this map [8].

The stationary states are given by $x(t+1) = x(t) = x_0$, with $x_0 = 0$ or $x_0 = 1 - 1/a$. The stability criterion of the stationary solutions is given by:

$$\left|\frac{\mathrm{d}x(t+1)}{\mathrm{d}x(t)}\right| = |a - 2ax_0| < 1 \; . \tag{7}$$

For $x_0 = 0$, $\mid \mathrm{d}x(t{+}1)/\mathrm{d}x(t) \mid = \mid a \mid < 1$, so this stationary solution is stable for $a < 1$. This means that the population disappears due to a negative growth rate $r = a-1 < 0$. The case $a = 1$ corresponds to a null growth rate $r = 0$. For $x_0 = 1 - 1/a$,$\mid \mathrm{d}x(t+1)/\mathrm{d}x(t) \mid = \mid -a + 2 \mid < 1$, so this stationary solution is stable for $1 < a < 3$, and unstable for $a \geq 3$. For $3 \leq a \leq 4$, bifurcations occur (period doubling: $2, 4, 8, 16, \ldots$), and then chaos (period $3, 5, 7, \ldots$ mixed with even periods $6, \ldots$). Chaos begins to occur for the value of a related to the universal constant of Feigenbaum [7, 6]: $a_c = 3.569945672\ldots$ So the chaos zone is defined for $3.569945672\ldots \leq a \leq 4$

2.2 Analytical Solution of the Verhulst Chaos Map

The analytical solution of the Verhulst chaos map (6), for $a = 4$,

$$x(t+1) = 4x(t)[1 - x(t)] \tag{8}$$

with $x(t) \in [0,1]$, is given by

$$x(t) = \frac{1 - \cos(2^t g)}{2} \tag{9}$$

with $g = \arccos[1 - 2x(0)]$ where $x(0)$ is the initial condition at $t = 0$.

This solution of the Verhulst chaos map is an exact closed form solution, because it is not necessary to compute all the successive iterates of the equation to obtain any iterate $x(t)$ for any t. From the fixed value of the initial condition $x(0)$ at time $t = 0$, the variable g is fixed to a constant, and the successive values of $x(t)$ as a function of time t, are given by the calculation of $\cos(2^t g)$ for the values of time $t = 1, 2, 3, \ldots$, that is $\cos(2^1 g), \cos(2^2 g), \cos(2^3 g), \ldots$, and, contrary to the solution of the original Verhulst differential equation, the chaos map never loses its initial condition $x(0)$, via the g function.

So a natural chaos system, if it is not perturbed by external effects, never loses its initial condition, and its future is completely written in its initial condition. But man, who tries to predict the future of this system, is limited by the exact knowledge of the initial condition. Indeed, one can only measure the initial condition with a certain number of decimals. Therefore, the future of this system is written in the successive values of the decimals, in theory, until infinity!

The next section explains the shift of the digits of the initial condition, written in binary.

2.3 The Verhulst Chaos Map Transformed to the Shift Map

In order to explain that the whole future of a chaotic system is written in the digits of its initial condition, let us consider the following change of variables

$$x(t) = \frac{1 - \cos[2\pi y(t)]}{2} = \sin^2[\pi y(t)] \tag{10}$$

in the Verhulst map (8) which becomes successively:

$$\begin{aligned}
&\frac{1 - \cos[2\pi y(t+1)]}{2} = 4\frac{1 - \cos[2\pi y(t)]}{2}\left\{1 - \frac{1 - \cos[2\pi y(t)]}{2}\right\}, \text{ or} \\
&1 - 2\cos[2\pi y(t+1)] = 2\{1 - \cos[2\pi y(t)]\}\{1 + \cos[2\pi y(t)]\}, \text{ or} \\
&1 - 2\cos[2\pi y(t+1)] = 2 - 2\cos^2[2\pi y(t)], \text{ or} \\
&\cos[2\pi y(t+1)] = -1 + \{1 + \cos[4\pi y(t)]\},
\end{aligned}$$

or

$$\cos[2\pi y(t+1)] = \cos[4\pi y(t)] \tag{11}$$

and the following Shift map is obtained

$$y(t+1) = [2y(t)] \bmod 1 \tag{12}$$

with $y(t) \in [0,1]$, where mod 1 is the modulo 1, which means that only the fractional part of $2y(t)$ is taken.

The exact closed form solution of (12) is given by

$$y(t) = [2^t y(0)] \bmod 1 \ . \tag{13}$$

For example, with $y(0) = 0.3203125\ldots$,
for which $x(0) = \sin^2[\pi y(0)] = 0.71377754671514104716048342844444\ldots$,
the Table 1 gives the successive values of $y(t)$ and $x(t)$.

Table 1. Numerical example of the Shift map and the Verhulst map

t	$y(t)$	$y(t+1)$	binary $y(t)$	$x(t) = \sin^2(\pi y(t))$	$x(t)$	$x(t+1)$
0	0.3203125	0.640625	0.0101001	0.7137775	0.7137775	0.8171966
1	0.640625	0.28125	0.101001	0.8171966	0.8171966	0.5975451
2	0.28125	0.5625	0.01001	0.5975451	0.5975451	0.9619397
3	0.5625	0.125	0.1001	0.9619397	0.9619397	0.1464466
4	0.125	...	0.001	0.1464466	0.1464466	...

In Table 1, the first column gives the time steps $t = 0, 1, 2, 3, 4$, the second column shows the $y(t)$ which are calculated from the shift map in the third column $y(t+1) = 2y(t) \bmod 1$, the fourth column shows the binary value of the decimal $y(t)$, the fifth column shows the $x(t) = \sin^2(\pi y(t))$ calculated from the shift map $y(t)$, the sixth column gives the $x(t)$ calculated from the Verhulst map in column seven $x(t+1) = 4x(t)[1 - x(t)]$.

It is clearly shown in Table 1 that, on one hand, the iterations shift to the left the digits of the binary $y(t)$, and on the other hand that the $x(t)$ calculated from the shift map and from the Verhulst map are identical.

But it is possible to obtain any iterate $y(t)$ at any time t, without computing all the preceding iterates, from the exact closed form solution.

Indeed, from the initial condition: $y(0) = 0.3203125\ldots$, (13) gives
$y(4) = [24y(0)] \bmod 1 = (5.125\ldots) \bmod 1 = 0.125\ldots$
and from (10), one obtains
$x(4) = \sin^2[\pi y(4)] = 0.14644660940672623779957781894758\ldots$

Let us now demonstrate that the chaos map does not represent a correct discrete Verhulst equation.

3 The Verhulst Incursive Map is the Correct Discrete Verhulst Equation

I proposed [4], several years ago, to transform the Verhulst chaos map (6) to the following Verhulst incursive map:

$$x(t+1) = ax(t)[1 - x(t+1)] \tag{14}$$

where the limiting factor $[1 - x(t)]$ is defined in the next time step as $[1 - x(t+1)]$.

This incursive equation (for inclusive or implicit equation) can be transformed (see the demonstration in the following section) to the following recursive equation

$$x(t+1) = ax(t)\left[1 - \frac{ax(t)}{1+ax(t)}\right] = \frac{ax(t)}{1+ax(t)} \tag{15}$$

where the limiting factor is equal to the whole equation.

This Verhulst incursive map (15) gives a stable solution for any value of a.

Indeed, the two stationary states of this incursive equation are the same as for the chaos map, $x_0 = 0$ and $x_0 = 1 - 1/a$. In applying the criterion of stability

$$\left|\frac{\mathrm{d}x(t+1)}{\mathrm{d}x(t)}\right| = \left|\frac{a}{[1+ax(t)]^2}\right| < 1 \,, \tag{16}$$

it is proved that, firstly, the stationary state $x_0 = 0$,

$$\left|\frac{\mathrm{d}x(t+1)}{\mathrm{d}x(t)}\right| = |a| < 1 \,, \tag{17}$$

is stable for $a < 1$, and, secondly, the stationary state $x_0 = (a-1)/a$,

$$\left|\frac{\mathrm{d}x(t+1)}{\mathrm{d}x(t)}\right| = 1/a < 1 \,, \tag{18}$$

is always stable for $a > 1$.

The analytical solution of this nonlinear (15) is given by

$$x(t) = \frac{(1-1/a)Ca^t}{1+Ca^t} \tag{19}$$

as a closed form solution, where C is a constant. The initial value $x(0)$ defines the value of C by, $x(0) = (1-1/a)C/(1+C)$, so the solution

$$x(t) = \frac{a^t x(0)}{1+(a^t-1)x(0)/(1-1/a)} \tag{20}$$

is similar to the solution of the original Verhulst differential logistic (2).

Let us demonstrate that the incursive (15) represents the correct discrete equivalent of the logistic (1). For that, let us write the solution (2) of the Verhulst logistic differential equation as

$$N(t_2) = \frac{\mathrm{e}^{r(t_2-t_1)}N(t_1)}{1+(\mathrm{e}^{r(t_2-t_1)}-1)N(t_1)/K} \tag{21}$$

which gives the growth of the population from the time t_1 to the time t_2. Indeed, in taking $t_1 = 0$ and $t_2 = t$, the continuous solution (2) is obtained.

Defining a time interval Δt between t_2 and t_1 as $t_2 = t_1 + \Delta t$, (21) becomes

$$N(t_1 + \Delta t) = \frac{e^{r\Delta t} N(t_1)}{1 + (e^{r\Delta t} - 1)N(t_1)/K} \tag{22}$$

Choosing $\Delta t = 1$, and $t_1 = t$, one obtains

$$N(t+1) = \frac{e^{r} N(t)}{1 + (e^{r} - 1)N(t)/K} \tag{23}$$

and with the change in variables, $e^r = a$ and $N(t)/K = x(t)/(1 - 1/a)$, (23) is written as

$$x(t+1) = \frac{ax(t)}{1 + ax(t)} \tag{24}$$

which is the discrete (15), and this is the correct discrete algorithm for the Verhulst logistic differential equation.

As result, we conclude that the chaos emerging from the so-called Verhulst chaos map is due to instabilities of the Euler algorithm and not from fundamental biological properties of the Verhulst logistic differential equation.

This (24) can also be obtained from an incursive control as shown in the next section.

4 Incursive Control for Stabilizing Chaos Maps

This section presents a survey of the role of incursive control with an example of control of the Verhulst chaos map, with simulations.

Let us begin by the definitions of recursive and incursive systems.

4.1 Recursive and Incursive Systems Applied to the Verhulst Chaos Map

A recursive system computes its vector current state $\boldsymbol{x}(t)$, at successive time steps $t = 0, 1, 2, \ldots$, from a vector function $\boldsymbol{R}$ of its past and present states as

$$\boldsymbol{x}(t+1) = \boldsymbol{R}(\ldots, \boldsymbol{x}(t-2), \boldsymbol{x}(t-1), \boldsymbol{x}(t); \boldsymbol{p}) \tag{25}$$

where the vector $\boldsymbol{p}$ is a set of parameters.

A *weak* incursive system [4] computes its current state at time t, as a function of its states at past times,..., $t-2, t-1$, present time, t, and even its *predicted* states at future times $t+1, t+2, \ldots$.

$$\boldsymbol{x}(t+1) = \boldsymbol{A}(\ldots, \boldsymbol{x}(t-2), \boldsymbol{x}(t-1), \boldsymbol{x}(t), \boldsymbol{x}^*(t+1), \boldsymbol{x}^*(t+2), \ldots; \boldsymbol{p}) \tag{26}$$

where the future states $\boldsymbol{x}^*(t{+}1), \boldsymbol{x}^*(t{+}2), \ldots$, are computed with a predictive model of the system.

A strong incursive system [4] computes its current state at time t, as a function of its states at past times,..., $t-3, t-2, t-1$, present time, t, and even its states at future times $t+1, t+2, t+3, \ldots$.

$$\boldsymbol{x}(t+1) = \boldsymbol{A}(\ldots, \boldsymbol{x}(t-2), \boldsymbol{x}(t-1), \boldsymbol{x}(t), \boldsymbol{x}(t+1), \boldsymbol{x}(t+2), \ldots; \boldsymbol{p}) \quad (27)$$

where the future states $\boldsymbol{x}(t+1), \boldsymbol{x}(t+2), \ldots$, are computed by the system itself.

The Verhulst chaos map (6):

$$x(t+1) = ax(t)(1 - x(t)) \quad (28)$$

is a **recursive system**, and the Verhulst incursive map (14):

$$x(t+1) = ax(t)[1 - x(t+1)] \quad (29)$$

is a **strong incursive system**, because the future value in the saturation factor, $[1 - x(t+1)]$, is computed by the system itself. Indeed, in replacing successively $x(t+1)$ by $ax(t)[1 - x(t+1)]$ in (29), one obtains the following equation

$$x(t+1) = ax(t)(1 - ax(t)(1 - ax(t)(1 - ax(t)(1 - ax(t)(1 - ...))))) \quad (30)$$

which is an infinite recursive equation that converges to

$$x(t+1) = \frac{ax(t)}{1 + ax(t)} \quad (31)$$

which is the (15), given at the preceding section.

This incursive map (29) can be obtained from the following recursive map (28) to be controlled

$$x(t+1) = ax(t)[1 - x(t)] + u(t) \quad (32)$$

by an incursive control $u(t)$

$$u(t) = ax(t)[x(t) - x(t+1)] \quad (33)$$

which can be transformed to a recursive control with (31) as follows

$$u(t) = ax(t)\left[x(t) - \frac{ax(t)}{1 + ax(t)}\right] . \quad (34)$$

Indeed, in including (34) to (32), (31) is obtained. As it was shown in the preceding section, this incursive map is always stable. This incursive controller, which is a powerful tool for stabilizing chaos maps, belongs to a special class of controller, because no setpoint is defined. Such a controller stabilizes by itself the unstable states to stable fixed points.

In the next section, this incursive controller will be compared to model predictive controllers through numerical simulations in order to indicate how to stabilize the chaos of the Verhulst map.

4.2 Incursive and Model Predictive Controls of the Verhulst Chaos Map

Predictive models were developed over the past two or three decades in the field of control, what is referred as the Model-based Predictive Control (MPC). These include Model Predictive Heuristic Control (MPHC) [9], Dynamic Matrix Control (DMC) [3], Internal Model Control (IMC) [10] and Generalized Predictive Control (GPC) [2].

The key difference between a conventional feedback control and a predictive control is that the control error $e = x - r$, which is the difference between the process output x and the setpoint r (the desired output), used by the predictive controller is based on future and/or predicted values of the setpoint $x(t + \tau)$, and also on future and/or predicted values of the process output $x(t + \tau)$, rather than their current values. See [1] for an overview of MPC.

The basic principle of the model predictive control of a discrete system, with a time step Δt,

$$x(t + \Delta t) = F[x(t)] + u(t) \tag{35}$$

where $u(t)$ is a control action, consists in minimizing a cost function J, given for example by a weighted least squares criterion:

$$J = E\left\{\sum_{i=1}^{N}[x(t + i\Delta t) - r(t + i\Delta t)]^2 + \sum_{i=1}^{Nu} w_i u(t - \Delta t + i\Delta t)^2\right\} \tag{36}$$

where $\{r(t + i\Delta t)\}$ is the setpoint sequence (target tracking), $\{w_i\}$ is the weight sequence, and N and Nu are fixed integers representing the time horizons of the predicted outputs, $x(t + i\Delta t)$, and control sequence, $u(t - \Delta t + i\Delta t)$. The anticipative outputs $x(t + i\Delta t)$, $i = 1$ to N, are computed from the model of the process to be controlled. The Verhulst map (28), to be controlled by a control action $u(t)$, in the chaos regime with $a = 4$, is written as

$$x(t + 1) = 4x(t)[1 - x(t)] + u(t) \tag{37}$$

with the cost function (36) given by

$$J = [x(t + 1) - x_0]^2 + w_1 u^2(t) \tag{38}$$

with $\Delta t = 1$, $N = Nu = 1$, with a constant setpoint $r(t + 1) = x_0 = 3/4$, that is the unstable equilibrium of the map [5]. The objective of the control is to stabilize the map at its unstable equilibrium given by $x_0 = 3/4$, obtained from $x_0 = 4x_0(1 - x_0)$, corresponding to the equilibrium condition $x(t+1) = x(t) = x_0$ in (37) [without the control $u(t)$]. Indeed, in applying the classical criterion of stability given by

$$\left|\frac{\mathrm{d}x(t + 1)}{\mathrm{d}x(t)}\right| < 1 \tag{39}$$

to the chaos map at the equilibrium $x_0 = 3/4$, one obtains

$$\left| \frac{\mathrm{d}x(t+1)}{\mathrm{d}x(t)} \right| = \mid 4 - 8x_0 \mid = 2 \tag{40}$$

which is greater then 1, so this equilibrium state x_0 is unstable.

Putting (37) into (38), one obtains the following cost function at the current time

$$J = [4x(t)(1 - x(t)) + u(t) - x_0]^2 + w_1 u^2(t) \,. \tag{41}$$

The minimum of J is obtained with the condition $\mathrm{d}J/\mathrm{d}u(t) = 0$, so

$$u(t) = \frac{x_0 - 4x(t)[1 - x(t)]}{1 + w_1} \,. \tag{42}$$

Now, let us put (42) into (37), to obtain

$$x(t+1) = 4x(t)[1 - x(t)] + \frac{x_0 - 4x(t)[1 - x(t)]}{1 + w_1} \,. \tag{43}$$

After elementary mathematical transformations, (43) becomes

$$x(t+1) = \frac{x_0 + 4w_1 x(t)[1 - x(t)]}{1 + w_1} \,. \tag{44}$$

The equilibrium conditions $x(t+1) = x(t) = x_0$ of (44) are $x_0 = 0$ and $x_0 = 3/4$, as desired.

Applying the stability criterion (39) to (44), for $x_0 = 3/4$, one obtains

$$\left| \frac{\mathrm{d}x(t+1)}{\mathrm{d}x(t)} \right| = \left| \frac{4w_1 - 8w_1 x_0}{1 + w_1} \right| = \left| - \frac{2w_1}{1 + w_1} \right| < 1 \tag{45}$$

and the chaotic map is stabilized to $x = x_0 = 3/4$ for a weight in the range $0 < w_1 < 1$.

Let us show the differences between the incursive control and the model predictive control by considering the control of the chaos map (37), with the three following control functions [5]:

a) The incursive control [Equation (34), in the chaos regime with $a = 4$]:

$$u(t) = 4x(t) \frac{x(t) - 4x(t)[1 - x(t)]}{1 + 4x(t)} \tag{46}$$

b) The model predictive control [Equation (42)]:

$$u(t) = \frac{x_0 - 4x(t)[1 - x(t)]}{1 + w_1} \tag{47}$$

c) The incursive predictive control:

$$u(t) = \frac{x(t) - 4x(t)[1 - x(t)]}{1 + w_1} \tag{48}$$

This third control is based on the incursive control applied to the model predictive control by replacing the setpoint x_0 by $x(t)$ in the cost function (38) as follows

$$J = [x(t+1) - x(t)]^2 + w_1 u^2(t) \tag{49}$$

This new controller (48) is obtained by minimizing this cost function (49).

Let us point out that the control (46) is similar to the control (48) by taking a variable weight $w_1 = 1/4x(t)$.

Figures 1, 2 and 3 show the simulations of the chaos map (37), with the three controllers (46), (47) and (48).

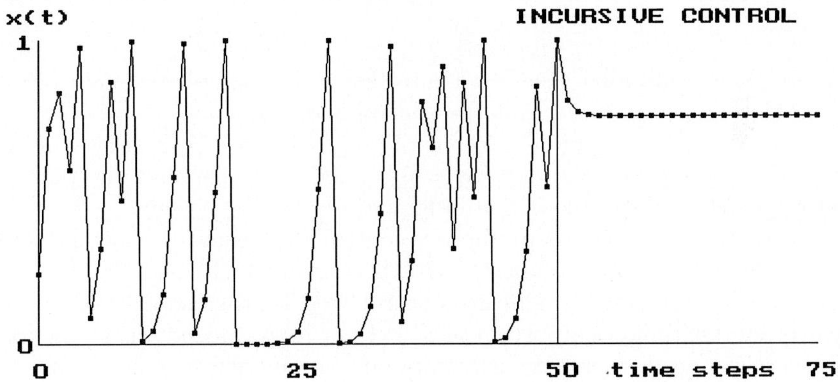

Fig. 1. Incursive Control (46), starting at step 50, of the chaos map

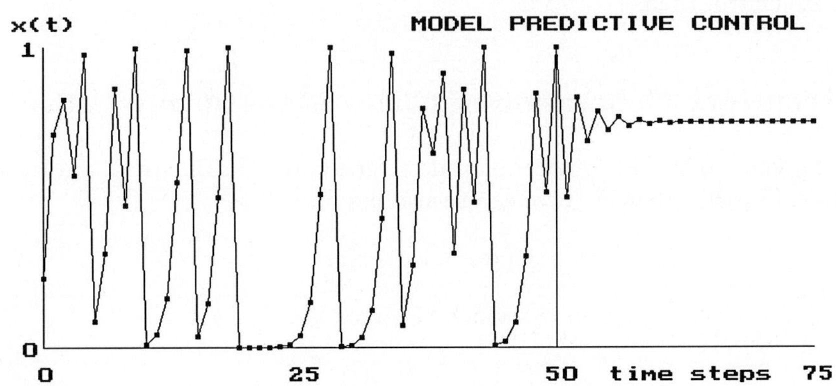

Fig. 2. Model Predictive Control (47), starting at step 50, of the chaos map

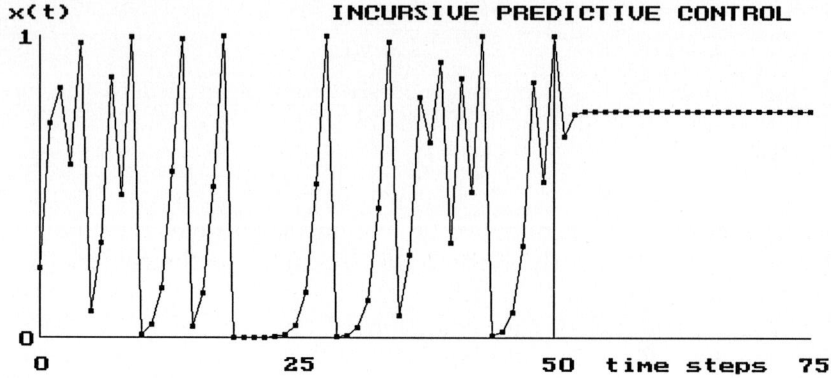

Fig. 3. Incursive Predictive Control (48), starting at step 50, of the chaos map

From the simulations, the best controller is the incursive control (46), in Fig. 1, where it is well seen that the control is optimal.

In Fig. 2, the model predictive control (47), with $w_1 = 1/2$, shows damped oscillations around the setpoint $x_0 = 3/4$. Let us notice that the damping effect depends on the value of the weight w_1.

In Fig. 3, the incursive predictive control (48) shows a better control than the model predictive control, with the same weight w_1. The incursive control and the incursive predictive control, find by themselves the setpoint which is the unstable equilibrium of the chaos map. Indeed, these incursive controllers do not use an explicit setpoint, but an implicit setpoint given by the unstable equilibrium state which is stabilized. The incursive control minimizes the distance between $x(t+1)$ and $x(t)$, instead of the distance between $x(t+1)$ and a setpoint x_0.

Let us now show how to generate Verhulst chaos maps at any order and how to control them.

5 Recurrent Generation of Chaos Maps at any Order

With a view to obtaining a simple rule to generate chaos maps at any order, let us make the following change of variables

$$X(t) = 1 - 2x(t) \tag{50}$$

in the Verhulst chaos map (8) which becomes

$$X(t+1) = 2X^2(t) - 1 \tag{51}$$

with $X(t) \in [1, -1]$, and the exact closed form solution (9) becomes

$$X(t) = \cos(2^t g) \tag{52}$$

with $g = \arccos[X(0)]$ or $X(0) = \cos g$.

The purpose of the next section is to define a set of canonical chaos maps with a general exact closed form solution given by

$$X(t) = \cos(n^t g) \tag{53}$$

for $n = 0, 1, 2, 3, 4, 5, \ldots$ and $g = \arccos[X(0)]$ or $X(0) = \cos g$.

5.1 Canonical Chaos Maps at any Order

The set of canonical chaos maps is defined by the following general equation

$$x(t+1) = \cos[n \arccos x(t)] \tag{54}$$

with $x(t) \in [1, -1]$ and $n \in N$. With the relation

$$\cos[n \arccos(x(t)] = x^n(t) + C_n^2 x^{n-2}(t)[x^2(t) - 1] + C_n^4 x^{n-4}[x^2(t) - 1]^2 + \cdots \tag{55}$$

where $C_n^m = n!/m!(n-m)!$, (54) becomes

$$x(t+1) = T_n[x(t)] = x^n(t) + C_n^2 x^{n-2}(t)[x^2(t) - 1] + C_n^4 x^{n-4}[x^2(t) - 1]^2 + \cdots \tag{56}$$

with $x(t) \in [1, -1]$ and $n \in N$. This equation (56) is the set of canonical chaos maps, for which the exact closed form solution is given by

$$x(t) = \cos(n^t g) \tag{57}$$

for $n = 0, 1, 2, 3, 4, 5, \ldots$ and $g = \arccos[x(0)]$ or $x(0) = \cos g$.

Let us remark that (55) corresponds to what is called the Tchebychev polynomials T_n given by

$$T_n(X) = X^n + C_n^2 X^{n-2}(X^2 - 1) + C_n^4 X^{n-4}(X^2 - 1)^2 + \cdots \tag{58}$$

which gives:

$$\begin{aligned}
T_0 &= 1 \,, \\
T_1 &= X \,, \\
T_2 &= 2X^2 - 1 \,, \\
T_3 &= 4X^3 - 3X \,, \\
T_4 &= 8X^4 - 8x^2 + 1 \,, \\
T_5 &= 16X^5 - 20X^3 + 5X \,, \\
T_6 &= 32X^6 - 48X^4 + 18X^2 - 1 \,, \\
&\ldots
\end{aligned} \tag{59}$$

where T_n is of degree n, and, for $n > 0$, the coefficient of the term of the highest degree of T_n is 2^n.

These Tchebychev polynomials verify a very simple relation of recurrence given by

$$T_{n+2} = 2XT_{n+1} - T_n \tag{60}$$

with the initial conditions $T_0 = 1$ and $T_1 = X$.

With the relation of recurrence (60), it is possible to generate the set of canonical chaos maps as follows:

$$\begin{array}{ll}
n=0 & x(t+1) = 1 \\
n=1 & x(t+1) = x(t) \\
n=2 & x(t+1) = 2x^2(t) - 1 \\
n=3 & x(t+1) = 4x^3(t) - 3x(t) \\
n=4 & x(t+1) = 8x^4(t) - 8x^2(t) + 1 \\
n=5 & x(t+1) = 16x^5(t) - 20x^3(t) + 5x(t) \\
n=6 & x(t+1) = 32x^6(t) - 48x^4(t) + 18x^2(t) - 1 \\
n=7 & x(t+1) = 64x^7(t) - 112x^5(t) + 56x^3(t) - 7x(t) \\
 & \ldots \\
n=10 & x(t+1) = 512x^{10}(t) - 128x^8(t) + 112x^6(t) - 400x^4(t) + 50x^2(t) - 1 \\
 & \ldots
\end{array} \tag{61}$$

Introducing a growth parameter a in (56), the set of chaos maps at any order is given by the following map

$$\begin{aligned} x(t+1) &= aT_n[x(t)] \\ &= a\{x^n(t) + C_n^2 x^{n-2}(t)[x^2(t) - 1] + C_n^4 x^{n-4}[x^2(t) - 1]^2 + \cdots\} \end{aligned} \tag{62}$$

with $a \in [-1, +1]$ and $x(t) \in [-1, +1]$. This set of chaos maps (62) can be controlled by an action control $u(t)$ as follows

$$x(t+1) = aT_n[x(t)] + u(t) \tag{63}$$

with the following general incursive predictive controller

$$u(t) = \frac{x(t) - aT_n[x(t)]}{1 + w_1} \tag{64}$$

where

$$T_n[x(t)] = x^n(t) + C_n^2 x^{n-2}(t)[x^2(t) - 1] + C_n^4 x^{n-4}[x^2(t) - 1]^2 + \cdots, \tag{65}$$

with $a \in [-1, +1]$ and $x(t) \in [-1, +1]$.

In Subsect. 5.2, some numerical simulations of these chaos maps and their stabilization will be given.

5.2 Simulation and Incursive Control of Chaos Maps

In this last subsection, we show the numerical simulation of the chaos maps (62) for $n = 2, 3, 5, 10, 25$, and their stabilization diagram with the incursive predictive control (63), (64) for $n = 2, 3, 5, 10, 25$.

The chaos maps, which are numerically computed, are given by

$$x(t+1) = a[2x^2(t) - 1] \quad \text{for } n = 2 , \tag{66}$$
$$x(t+1) = a[4x^3(t) - 3x(t)] \quad \text{for } n = 3 , \tag{67}$$
$$x(t+1) = a[16x^5(t) - 20x^3(t) + 5x(t)] \text{ for } n = 5 , \tag{68}$$

and by

$$x(t+1) = a[512x^{10}(t) - 128x^8(t) + 112x^6(t) - 400x^4(t) + 50x^2(t) - 1] \tag{69}$$

for $n = 10$.

Let us remark that it is also possible to obtain this chaos map $n = 10$ by inserting the canonical map $n = 5$ in the chaos map $n = 2$ as follows:

$$x(t+1) = a\{2T_5^2[x(t)] - 1\} . \tag{70}$$

In the same way, the chaos map $n = 25$ is obtained in inserting the canonical chaos map $n = 5$ in the chaos map $n = 5$ as follows
$n = 25$

$$x(t+1) = a\{16T_5^5[x(t)] - 20T_5^3[x(t)] + 5T_5[x(t)]\} \tag{71}$$

with $T_5[x(t)] = 16x^5(t) - 20x^3(t) + 5x(t)$.

Figures 4 and 5 give the first return diagram of the chaos maps $n = 2$ and $n = 10$ for $a = 1$.

Figures 6, 8, 10, 12 and 14 give the bifurcation diagrams of the chaos maps $n = 2, 3, 5, 10, 25$, with $a \in [-1, +1]$.

Figures 7, 9, 11, 13 and 15 give the stabilization diagrams of the chaos maps $n = 2, 3, 5, 10, 25$, with $a \in [-1, +1]$, by using the incursive predictive control.

For example, the control for $n = 3$ is given by

$$x(t+1) = a[4x^3(t) - 3x(t)] + u(t) , \tag{72}$$
$$u(t) = \frac{x(t) - a[4x^3(t) - 3x(t)]}{1 + w_1} . \tag{73}$$

These stabilization diagrams of the chaos maps show that the number of fixed points is equal to $n = 2, 3, 5, 10, 25$, respectively. The proposition to create such stabilization diagrams could be of great interest in practice. Indeed, it is often impossible to detect, in real time, the unstable points of real chaotic systems. Instead of defining setpoints arbitrarily, it would be more accurate to let the controller self-stabilizes an unstable point.

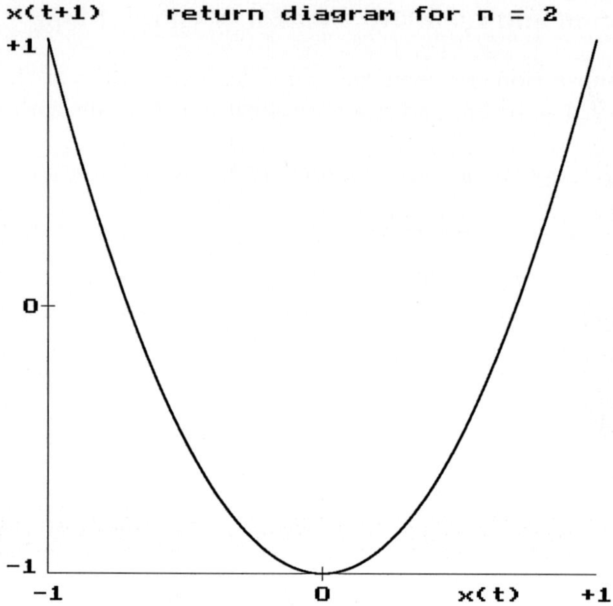

Fig. 4. First return diagram for $n = 2$

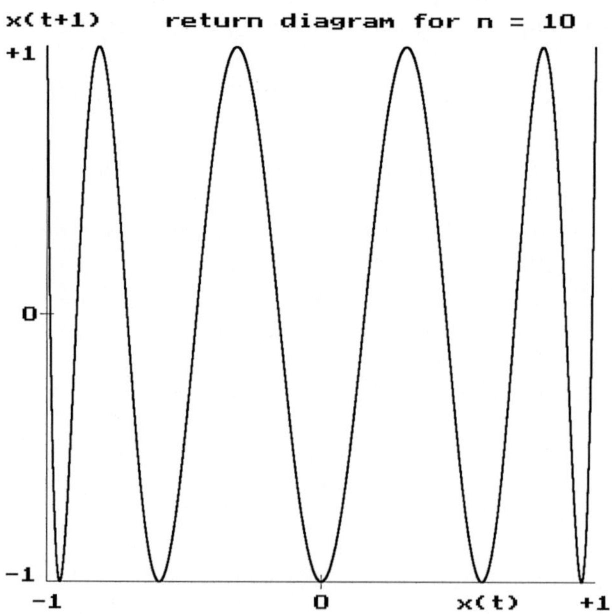

Fig. 5. First return diagram for $n = 10$

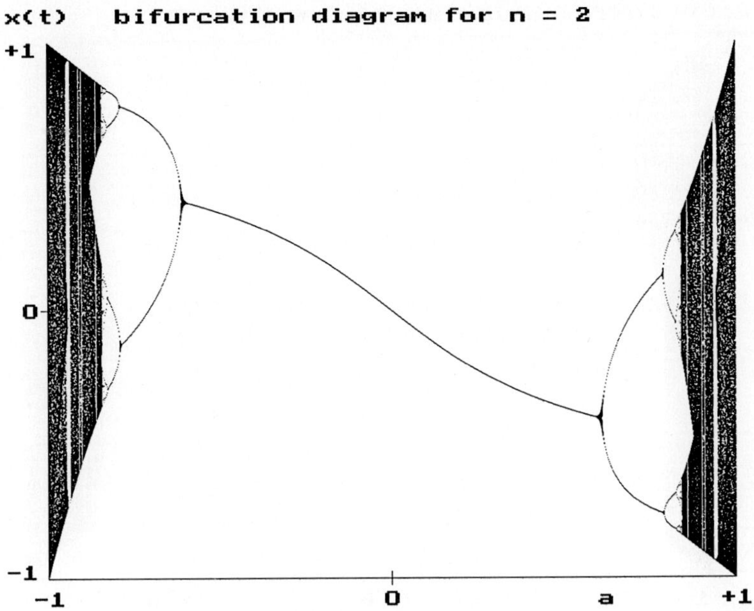

Fig. 6. Bifurcation diagram for $n = 2$

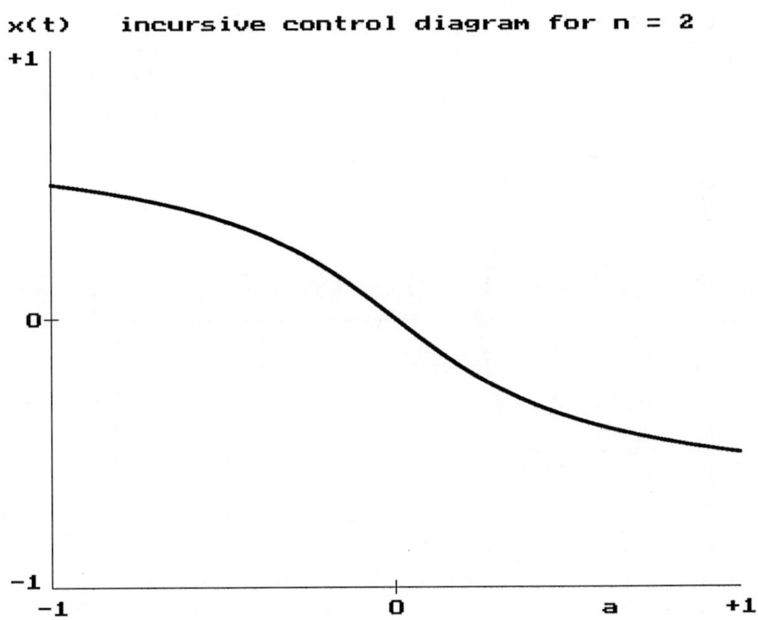

Fig. 7. Stabilization diagram for $n = 2$

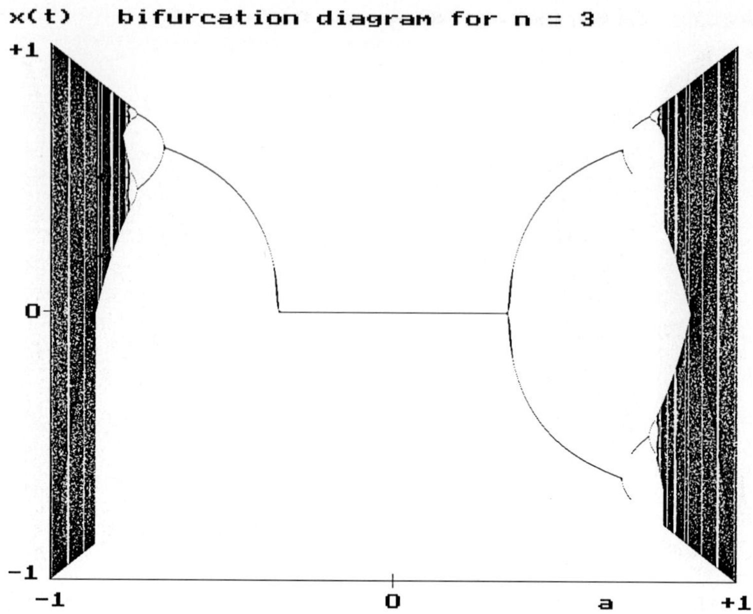

Fig. 8. Bifurcation diagram for $n = 3$

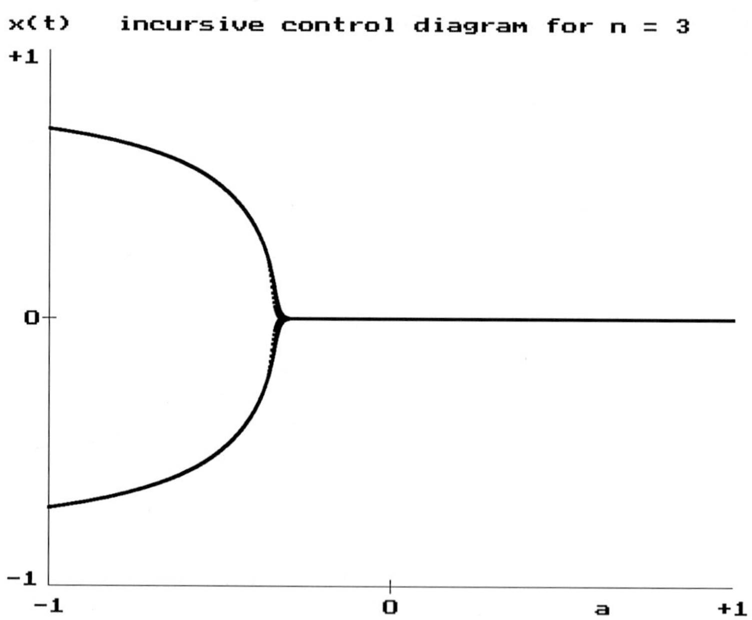

Fig. 9. Stabilization diagram for $n = 3$

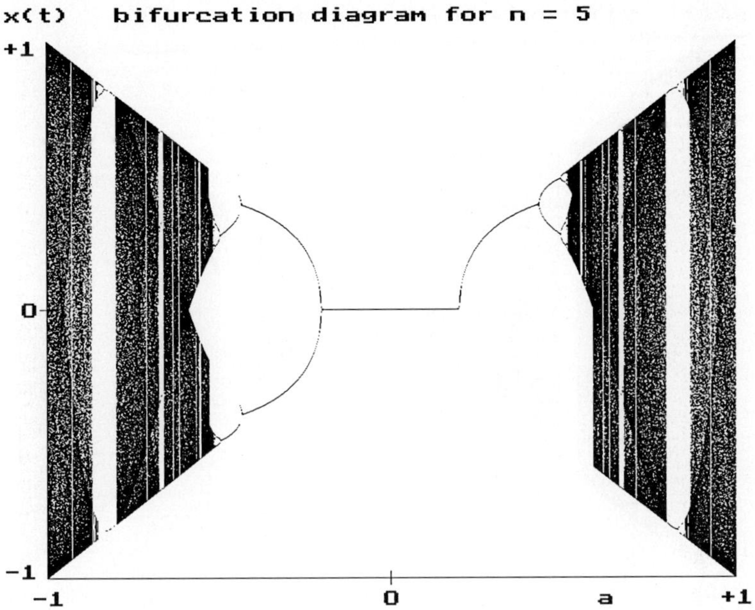

Fig. 10. Bifurcation diagram for $n = 5$

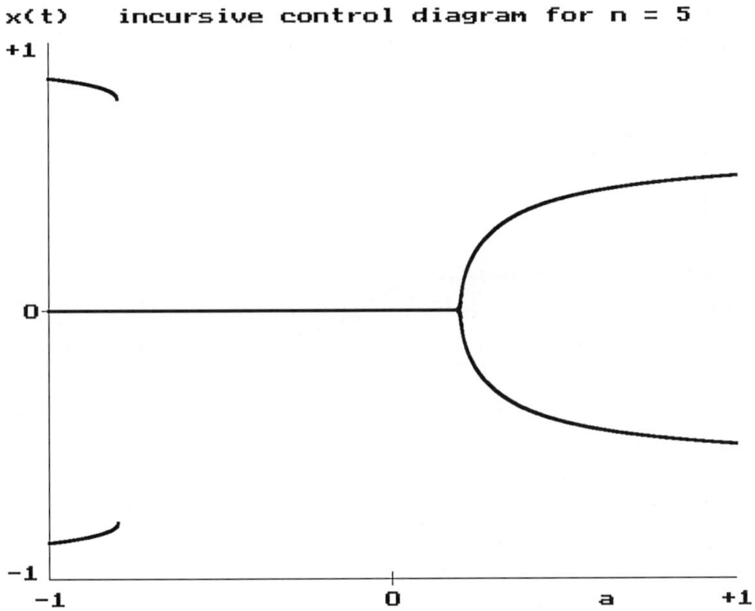

Fig. 11. Stabilization diagram for $n = 5$

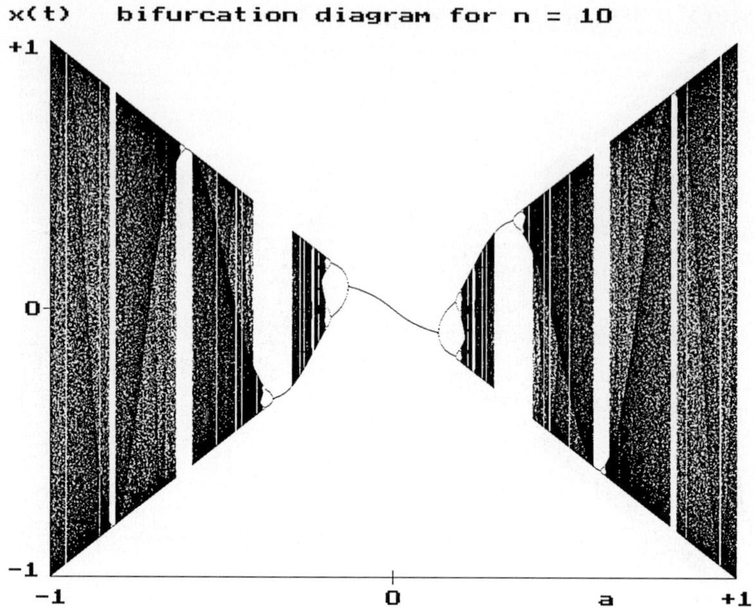

Fig. 12. Bifurcation diagram for $n = 10$

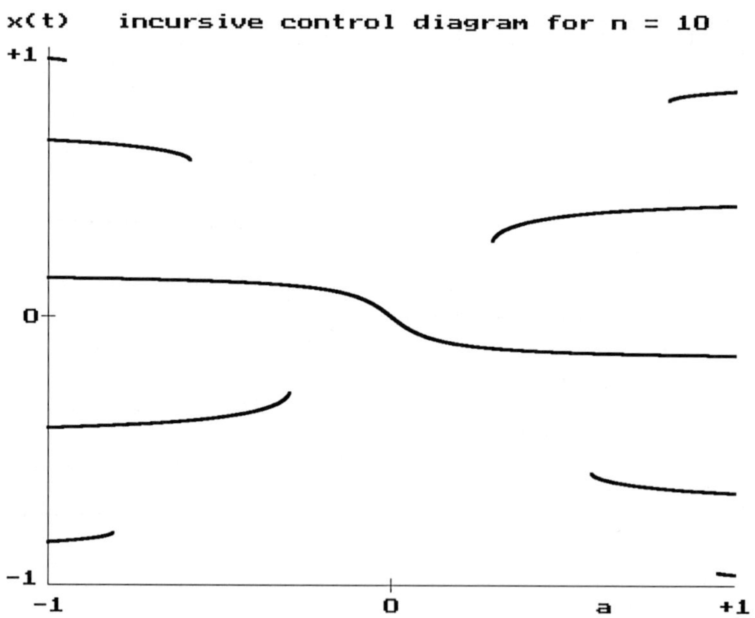

Fig. 13. Stabilization diagram for $n = 10$

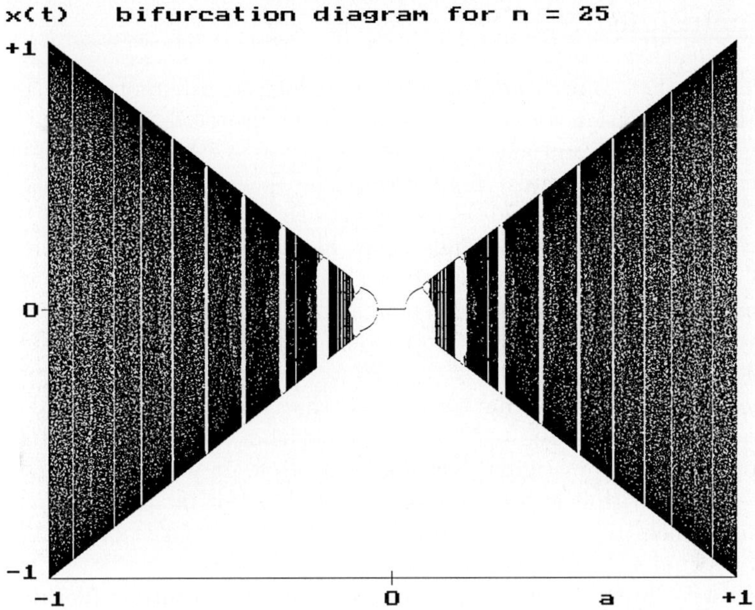

Fig. 14. Bifurcation diagram for $n = 25$

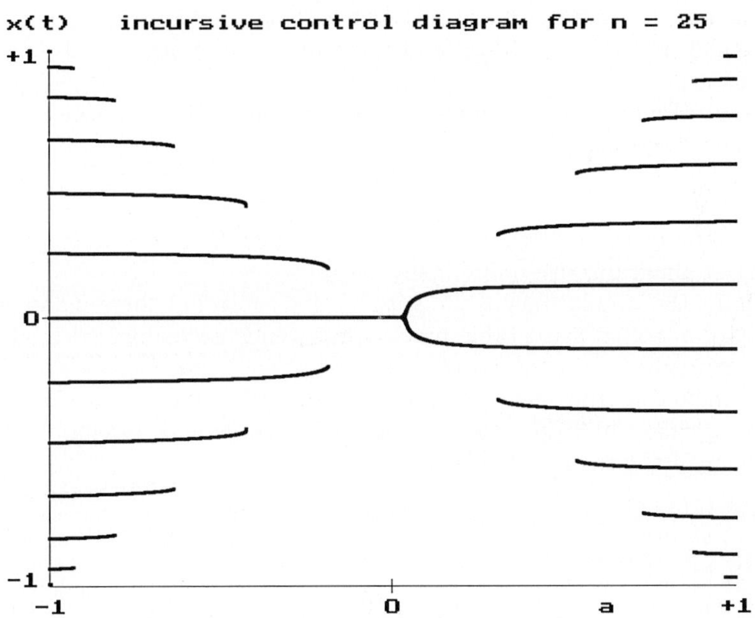

Fig. 15. Stabilization diagram for $n = 25$

6 Conclusions

This chapter firstly presented the Verhulst logistic differential equation and its analytical solution for various values of the growth parameter.

The classical discretization of this equation, with the Euler algorithm, gives the so-called Verhulst map, exhibiting fixed points, bifurcations and, then chaos for successive values of the growth parameter.

In the chaos zone, the Verhulst map has a closed form solution depending on the initial condition. It was recalled that the Verhulst map can be transformed to the Shift map for which the successive iterates are given by the fractional part of the shift, to the left, of the binary digits of the initial condition.

The first part of this chapter showed that the bifurcations and chaos in the Verhulst map are due to the Euler algorithm and not to the fundamental properties of the original Verhulst equation. It was demonstrated that the correct algorithm is the incursive algorithm (for inclusive or implicit algorithm) which gives fixed points similar to those of the original Verhulst equation, what we called the Verhulst incursive map.

Such a Verhulst incursive map can be obtained from the Verhulst chaos map with an incursive control, which is an anticipative controller.

The second part of this chapter demonstrated that the Verhulst chaos map belongs to a family of canonical chaos maps, given by the Tchebychev polynomials at any order (degree of the polynomials), for which closed form solutions exist.

These canonical chaos maps can be generated by a simple relation of recurrence.

With introduction of a growth parameter into these canonical chaos maps, a set of Verhulst chaos maps at any order is proposed.

For some of these Verhulst chaos maps, their first return diagrams and bifurcation diagrams are numerically computed.

Finally, the simulations of the anticipative control of these maps, with an incursive controller, give stable fixed points which are exhibited in stabilization diagrams.

In conclusion, the incursive tool is a powerful anticipative method for stabilizing chaos, and open new avenues of research and development in the framework of the theory of chaos.

Acknowledgements

I would like to thank Professor Mitchell J. Feigenbaum for very fascinating discussions and for having pointed out that the canonical chaos maps are related to the Tchebychev polynomials.

References

1. E.F. Camacho, C. Bordons: *Model Predictive Control* (Springer, Berlin 1999)
2. D.W. Clarke, C. Mohtadi: Automatica **23**, 137 (1987)
3. C.R. Cutler, B.L. Ramaker: Dynamic Matrix Control – A Computer Control Algorithm. In: *Proceedings of the 1980 Joint Automatic Control Conference* (San Francisco 1980) pp WP5-B,
4. D.M. Dubois: Review of Incursive, Hyperincursive and Anticipatory Systems – Foundation of Anticipation in Electromagnetism. In: *Computing Anticipatory Systems: CASYS'99 – Third International Conference*, ed by D.M. Dubois (The American Institute of Physics, AIP Conference Proceedings 517, 2000) pp 3–30
5. D.M. Dubois: Mathematical foundations of discrete and functional systems with strong and weak anticipations. In: *Anticipatory Behavior in Adaptive Learning Systems, State-of-the-Art Survey*, ed by M. Butz et al., Lecture Notes in Artificial Intelligence, Springer, LNAI 2684, pp 110–132, 2003
6. M.J. Feigenbaum: J. Stat. Phys **19**, 25 (1978)
7. M.J. Feigenbaum: Physica D **7**, 16 (1983)
8. H.-O. Peitgen , J. Hartmut, D. Saupe: *Chaos and Fractals, New Frontiers of Science* (Springer, Berlin 1992)
9. J. Richalet, A. Rault, J.L. Testud, J. Papon: Automatica, **14**, 413 (1978)
10. D.E. Rivera, M. Morari: Ind. Eng. Chem. Process Des. Dev. **25**, 252 (1986)
11. P.-F. Verhulst: Correspondance Mathématique et Physique **10**, 113 (1838)

Coherence in Complex Networks of Oscillators

Pedro G. Lind[1,2], Jason A.C. Gallas[1,3], and Hans J. Herrmann[1,4]

[1] ICP, Universität Stuttgart, Pfaffenwaldring 27, D-70569 Stuttgart, Germany
[2] Centro de Física Teórica e Computacional, Av. Prof. Gama Pinto 2, 1649-003 Lisbon, Portugal
[3] Instituto de Física, Universidade Federal do Rio Grande do Sul, 91501-970 Porto Alegre, Brazil
[4] Departamento de Física, Universidade Federal do Ceará, 60451-970 Fortaleza, Brazil

Summary. We study fully synchronized (coherent) states in complex networks of chaotic oscillators, reviewing the analytical approach of determining the stability conditions for synchronizability and comparing them with numerical criteria. As an example, we present detailed results for networks of chaotic logistic maps having three different scale-free topologies: random scale-free topology, deterministic pseudo-fractal scale-free network and Apollonian network. For random scale-free topology we find that the lower boundary of the synchronizability region scales approximately as $k^{-\mu}$, where k is the outgoing connectivity and μ depends on the local nonlinearity. For deterministic scale-free networks coherence is observed only when the coupling is heterogeneous, namely when it is proportional to some power of the neighbour connectivity. In all cases, stability conditions are determined from the eigenvalue spectrum of the Laplacian matrix and agree well with numerical results based on histograms of coherent states in parameter space. Additionally, we show that almost everywhere in the synchronizability region the basin of attraction of the coherent states fills the entire phase space, and that the transition to coherence is of first-order.

1 The Interplay Between Dynamics and Topology

The structure and dynamics underlying complex networks have been widely investigated, providing insight for many systems where they arise naturally [1–3]. Complex networks appear in a wide variety of fields ranging from lasers [4], granular media [5, 6], quantum transport [7], colloidal suspensions [8], electrical circuits [9], and time series analysis [10], to heart rhythms [11], epidemics [12, 13], protein folding [14], and locomotion [15] among others [1–3].

From the mathematical point of view, a network is a graph, composed by nodes or vertices and by their connections or edges [2]. Sometimes, each node is characterized by some dynamical state (amplitude), which evolves according to some local contribution and to the interaction with the neighbourhood. In other words, the complexity of the system underlying the network may be introduced either in the way nodes are interconnected (topology) or in the way nodes evolve in time (dynamics).

When studying network dynamics one frequently assumes a regular topology (lattice) where each node evolves according to some more or less complicated dynamics, typically fixed points [16], limit cycles [17] or chaotic attractors [18, 19]. One main goal of this approach is to study the so-called spatio-temporal chaos which appears in many different spatially extended systems out of equilibrium, such as hydrodynamical flows, chemical reactions and biological systems [20, 21]. Two main topics in this context concern the study of mechanisms underlying pattern formation and pattern selection [20–24] and also the study of chaotic synchronization behaviour [16, 25].

Spatially extended systems are fundamentally modelled by (i) sets of coupled differential equations [20] with nonlinear terms, where both time and amplitude are continuous, (ii) cellular automata [22], where both time and amplitude are discrete or (iii) coupled map lattices [21], where time is discrete as in cellular automata, but where the space of states is continuous. In all these models the underlying networks have connections whose range assumes all values from 1 (nearest neighbours) up to some maximum range, in particular the size of the system (global coupling regime). In other words, neglecting boundary conditions, these network systems assume translational symmetry and therefore the underlying network is called a regular network.

To study more complicated network structures, one usually neglects node dynamics and all complexity is introduced by the network topology, i.e. by the way nodes are connected to each other. This can be done in three different ways [2]: by randomly connecting the nodes (random networks [26, 27]), by considering some random long-range connections in a regular network with some small range of couplings (small-world networks [28, 29]), or by considering the introduction of new nodes which are connected to the previous ones following some rule of preferential attachment (scale-free networks [30]). For all these cases there is no translational symmetry and no typical range connection: connections do not have any 'spatial' interpretation. Therefore, one uses some general topological quantities to characterize each particular network, namely the average path length $\langle \ell \rangle$, i.e. the average minimum number of connections linking two nodes, the clustering coefficient C, defined as the average fraction of neighbours which are connected to each other, and the distribution of connections $P(k)$, representing the number of nodes having k connections. Table 1 shows the values of these three quantities for all three topologies.

Random networks were introduced by Erdös and Rényi in the late fifties [26] to study organizing principles underlying some real networks [27]. In random networks one defines some probability $p(N)$ for any two nodes to be connected in a total of N nodes. Consequently, the connections are typically long-range connections having a completely irregular structure. One main goal in studying random networks is to determine the critical probability $p_c(N)$, beyond which some specific property is very likely to be observed, e.g. the critical probability marking a transition to percolation [31]. One

Table 1. Characterizing complex topologies with the topological quantities: average path length $\langle \ell \rangle$, clustering coefficient C and distributions $P(k)$ of connections k. Here N is the total number of nodes, p is the probability for two nodes to be connected, $\bar{k}$ is the average number of connections per node, C_0 is the clustering coefficient of the regular network from which the small-world network is constructed, and m is the number of initial connections of each new node in a scale-free network

	Random	Small-world	Scale-free
$\langle \ell \rangle$	$\ln N / \ln (pN)$	N for small p $\ln N$ for large p	$\ln N / \ln \ln N$
C	$\bar{k}/N$	$C_0(1-p)^3$	$\sim N^{-3/4}$
$P(k)$	$\mathrm{e}^{-\bar{k}}\bar{k}^k/k!$	$\mathrm{e}^{-\bar{k}}\bar{k}^k/k!$	$2m^2/k^3$

important feature of random networks, which also appears in real networks, is their small average path length $\langle \ell \rangle$, i.e. the average distance between any two nodes increases slowly with the system size. However, unlike random networks, real networks also have large cluster coefficients C.

Small-world networks were introduced recently by Watts and Strogatz [28] in order to implement both short $\langle \ell \rangle$ and large C features. Small-world networks have short-range connections between neighbours, as in regular networks, but they also have long-range connections similar to random networks, without middle range ones. There are mainly two procedures to construct a small-world network: starting from the same regular network, where each site is coupled to some number of first neighbours, one either *rewires* each regular connection with probability p (Watts–Strogatz model [28]) or *adds* a random connection to each node with probability p (Newman–Watts model [32]). The second procedure is more appropriate for most purposes, since it avoids the possibility of generating disconnected clusters [32].

Both random and small-world topologies produce typically networks where connections obey a Poisson distribution (see Table 1). However, there are real systems which are scale-free, i.e. where the connection distribution obeys a power law.

Scale-free networks were introduced by Barabási and Albert [30] using growing and preferential attachment: one starts with a small amount of fully interconnected nodes, and adds iteratively one node with m connections to the previous nodes, chosen from a probability function proportional to their number of connections. With this construction one obtains analytically [33–36] a distribution of the connections $P(k) \propto k^{-\gamma}$, where $\gamma \to 3$ as $N \to \infty$, independently of the initial number of fully interconnected nodes and of m. It is also possible to generate scale-free networks, either by imposing *a priori* a power-law distribution of all randomly distributed connections, or by following a deterministic iterative rule for new nodes. The first procedure generates what is usually called a generalized random graph, while the latter was recently referred as deterministic scale-free network [37].

Deterministic scale-free networks are hierarchical structures composed by some succession of generations of nodes, i.e. the set of new nodes appearing simultaneously at a given iteration during the 'construction' of the network, whose connections follow a particular power-law distribution [37–40], being more easier to handle. The main difference between random and deterministic scale-free networks arises from the local connectivity character of the latter: random constructions generate irregular long-range connections, while deterministic networks impose a succession of generations of new nodes which are, in some way, organized in 'space'. Therefore, deterministic networks are applied for instance in spin systems [39], and geographical and social networks [39, 41].

After considering separately dynamical and topological complexity, the next logical step toward real network dynamics is to consider them together. One important question addressed in this context is to know whether full synchronization between oscillators in such complex topologies would appear and under which conditions the full synchronized state would be stable. By full synchronization we mean the convergence of the amplitudes of all oscillators to the same value, evolving *coherently* from then on. Therefore we call henceforth these fully synchronized states *coherent states*, to distinguish them from *partially* synchronized configurations, when several different clusters of nodes with the same amplitude are observed [16]. Synchronization and coherent behaviour of oscillator networks with complex topologies have been studied for the random topology [42–45] and the small-world topology [46–49], and also scale-free networks [45, 50–53]. In random networks, it is already known [42] that with high coupling strengths it is possible to fully synchronize oscillators and the corresponding stability condition may be computed [43] from the matrix of connections characterizing the network. In small-world networks, synchronizability is observed [47] only at the end of the small-world regime (high values of p), and recently [46] it was found that heterogeneity in the coupling may destroy coherence. These findings for small-world networks are somehow contradictory to the ones of other studies [45, 48] and other topological quantifiers have been proposed [49]. In scale-free networks some recent studies indicate that synchronizability among oscillators depends on the average connectivity [52] and is robust to a delayed flow of information [50] and to the removal of low-connected nodes [52].

In this chapter we describe the general approach to study coherent states in any general complex network of oscillators, and apply it to the particular case of a scale-free network of discrete-time oscillators, which is studied in great detail. We start in Sect. 2 by describing the stability analysis approach to the model introduced in (1) and deduce the corresponding conditions for synchronizability. In Sect. 3 we apply this stability analysis procedure to the particular case of scale-free networks, comparing our results with numerical simulations. The random scale-free case is treated in Sect. 3.1, where we show that the threshold value of such a transition as a function of coupling strength and outgoing connectivity obeys a power law with an exponent that depends on the nonlinearity, while deterministic scale-free networks are studied in Sect. 3.2, namely a pseudo-fractal network [38] and an Apollonian network [39, 40]. Discussion and conclusions are given in Sect. 4.

2 General Approach to Analyse Coherent States

For all the network topologies described above, if one considers discrete-time oscillators, namely maps of the interval, the equation of evolution for their amplitudes reads

$$\boldsymbol{x}_{t+1} = \boldsymbol{f}(\boldsymbol{x}_t) - \varepsilon \mathbb{G} \boldsymbol{g}(\boldsymbol{x}_t) , \tag{1}$$

where ε is the coupling parameter, t labels time, $\boldsymbol{x}_t = (x_{t,1}, \ldots, x_{t,N})$ with $x_{t,i}$ representing the amplitude at time-step t of node $i = 1, \ldots, N$, where N is the total number of nodes, $\boldsymbol{f} = (f(x_1), \ldots, f(x_N))$ and $\boldsymbol{g} = (g(x_1), \ldots, g(x_N))$ with f and g being real nonlinear functions, and $\mathbb{G}$ is the coupling (Laplacian) matrix, whose element G_{ij} represents the relative strength with which node i is coupled to node j, and satisfies the conditions $\sum_{j=1}^{N} G_{ij} = 0$ and $G_{ii} = 1$ for all $i = 1, \ldots, N$. In general $\mathbb{G}$ is a non symmetric matrix.

Usually, one chooses $\boldsymbol{g}(\boldsymbol{x}) \equiv \boldsymbol{x}$ when studying linear coupling, and $\boldsymbol{g}(\boldsymbol{x}) \equiv \boldsymbol{f}(\boldsymbol{x})$ when studying nonlinear coupling. Here we consider the nonlinear case. Apart from this choice, all the information about the dynamics is introduced in the function $\boldsymbol{f}(\boldsymbol{x})$, while all the information about the coupling topology (regular, random, small-world or scale-free) and the coupling regime (either homogeneous or heterogeneous) is included in the coupling matrix $\mathbb{G}$.

From (1) one easily sees that the coherent state $x_{t,1} = x_{t,2} = \cdots = x_{t,N} = X_t$ evolves in time according to the local map $X_{t+1} = f(X_t)$. There are two ways to study these coherent states: either by studying the stability of small perturbations of the coherent states or by making statistics over significant large samples of initial configurations, counting how many converge to a coherent state. Some attention to the parameter ranges must be taken, since the basin of attraction of the coherent states may be bounded by regions of phase space where amplitudes diverge. In particular, for maps of the interval one has $0 \le \varepsilon \le 1$ in order to guarantee convergence of any initial configuration.

In this chapter we shall illustrate both analytical and numerical approaches for the particular case of scale-free networks. To this end, we define the coupling matrix as $G_{ii} = 1$ and

$$G_{ij} = -\frac{k_j^{\alpha}}{\sum_{k \in \mathcal{K}_i} k_k^{\alpha}} \tag{2}$$

if node i is coupled to node j, with k_j representing the number of neighbours of node j and $\mathcal{K}_i$ is the set of labels of all neighbours of node i. If nodes i and j are not coupled $G_{ij} = 0$. The parameter α is a real number controlling the heterogeneity in the coupling: positive values of α enhance the coupling strength with sites having a large number of neighbours, while negative values favour sites having less neighbours. For $\alpha = 0$ the coupling between each site and its neighbourhood is homogeneous.

For local dynamics we choose the well-known quadratic map $f(x) = 1 - ax^2$, where the free parameter a is restricted to the interval $-0.25 \leq a \leq 2$ and contains all possible dynamical regimes from a fixed point (e.g. $a = 0$) to fully developed chaotic orbits (e.g. $a = 2$).

When determining the stability of coherent states, various criteria are possible. For instance, one could compute the maximum Lyapunov exponent and obtain the conditions where it is negative. However, such a Lyapunov exponent does not indicate the existence of a local instability in the synchronous state, which may pull the trajectories apart from the coherent manifold.

The correct approach is based on the variational version of (1) proposed by Pecora and Carroll [54], which is valid for any network of identical oscillators as far as their local dynamics (quadratic map, Lorenz system, etc.) and their coupling regime (linear, nonlinear, etc.) are concerned. For the nonlinear coupling regime, the diagonal form of these variational equations reads [54–56]

$$\xi_{t+1,i} = \exp\left[\Lambda(\varepsilon\lambda_i)\right]\xi_{t,i} = \left[Df(X) - \varepsilon\lambda_i Df(X)\right]\xi_{t,i} \ , \qquad (3)$$

for the coherent states $x_{t,i} = X$, where $\Lambda(\varepsilon\lambda_i)$ is the Lyapunov exponent, $Df(X)$ represents the identity matrix multiplied by the derivative of $f(x)$ computed at $x = X$ and λ_i are the eigenvalues of the coupling matrix $\mathbb{G}$. If $\mathbb{G}$ has zero-sum rows, i.e. $\sum_{j=1}^{N} G_{ij} = 0\,\forall i$, and all its eigenvalues $\lambda_1 \leq \lambda_2 \leq \cdots \leq \lambda_N$ are real and nonnegative, then $\lambda_1 = 0$ corresponds to the mode parallel to the synchronization manifold and the largest Lyapunov exponent defines a master stability function [54]. The coherent state is stable whenever $\Lambda(\varepsilon\lambda_i) < 0$ for $i = 2, \ldots, N$ [54–56].

In our case, it is easy to check from (2) that $\mathbb{G}$ has indeed zero-row sum, yielding $\lambda_1 = 0$. Moreover, all the eigenvalues of the matrix $\mathbb{G}$ are real and nonnegative, since $\det(\mathbb{G} - \lambda\mathbb{I}) = \det(\bar{\mathbb{G}} - \lambda\mathbb{I})$ where $\bar{\mathbb{G}}$ is a positive semidefinite symmetric matrix, namely $\bar{\mathbb{G}} = \mathbb{H}^{1/2}\mathbb{K}^{1/2}\mathbb{A}\mathbb{K}^{1/2}\mathbb{H}^{1/2}$ with $\mathbb{A}$ being the adjacency matrix of the network [52], and matrices $\mathbb{H}$ and $\mathbb{K}$ being the diagonal matrices with elements $H_{ii} = 1/(\sum_{k\in\mathcal{K}_i} k_k^{\alpha})$ and $K_{ii} = k_i^{\alpha}$ respectively.

From (3) and regarding the ordering of the eigenvalues λ_i one easily concludes that the stability condition reads

$$\varepsilon_L \equiv \frac{1 - \exp\left(-\bar{\lambda}\right)}{\lambda_2} < \varepsilon < \frac{1 + \exp\left(-\bar{\lambda}\right)}{\lambda_N} \equiv \varepsilon_U \ , \qquad (4)$$

where $\bar{\lambda}$ is the Lyapunov exponent of the local single map. In particular there is a range of coupling strengths enabling synchronizability whenever $\lambda_N/\lambda_2 < (1+\mathrm{e}^{-\bar{\lambda}})/(1-\mathrm{e}^{-\bar{\lambda}})$ holds. Therefore, by computing the eigenvalues of the Laplacian matrix $\mathbb{G}$ one is able to find the range of couplings for which coherent states are stable. This approach can be applied for any system ruled by (1).

3 Scale-Free Networks of Coupled Logistic Maps: An Example

For the particular case of scale-free networks, recent results [45] show a transition to full synchronization for two particular values of the nonlinearity a in the homogeneous regime ($\alpha = 0$), when either the coupling strength or the number of outgoing connections is varied. However, as far as we know there is no detailed study showing how these coherent states depend on all the model parameters. Therefore, we present in this section detailed numerical results concerning synchronization in oscillator networks with scale-free topologies. Our purpose is to give a complete example of how to study coherent solutions in complex networks of oscillators, comparing both the stability analysis and the numerical approaches.

The stability analysis is carried out just by computing the boundary values ε_L and ε_U in (4) as functions of the model parameters. The ranges of values where $\varepsilon_L < \varepsilon_U$ are the ones where coherent solutions appear. As stated above, for discrete oscillators, ruled by maps of the interval, the condition $0 \le \varepsilon \le 1$ must be added.

Numerically, to detect coherent solutions from a given sample of initial configurations, we compute the standard mean square deviation [45]

$$\sigma_t^2 = \tfrac{1}{N} \sum_{i=1}^{N} (x_{t,i} - \bar{x}_t)^2 \ , \tag{5}$$

where $\bar{x}_t$ is the average amplitude at a given time step t. Whenever σ^2 is zero within numerical precision, i.e. $\sigma^2 \sim 10^{-30}$, all the nodes are synchronized at the same amplitude.

We divide our approach into two parts, one concerning random scale-free networks (Sect. 3.1) and the other one concerning deterministic scale-free networks (Sect. 3.2).

3.1 Random Scale-Free Networks

In this section, we use the algorithm of Barabási and Albert [2, 30] to construct the random scale-free network (see Sect. 1), where at each node one places a chaotic logistic map. In a previous work [45] a transition to coherence between chaotic logistic maps was found for random scale-free networks, occurring for particularly high coupling strengths, typically of the order of $\varepsilon_c \sim 0.9$. Our simulations have shown that these transitions occur after discarding transients of $\sim 10^4$ time steps and they do not change significantly

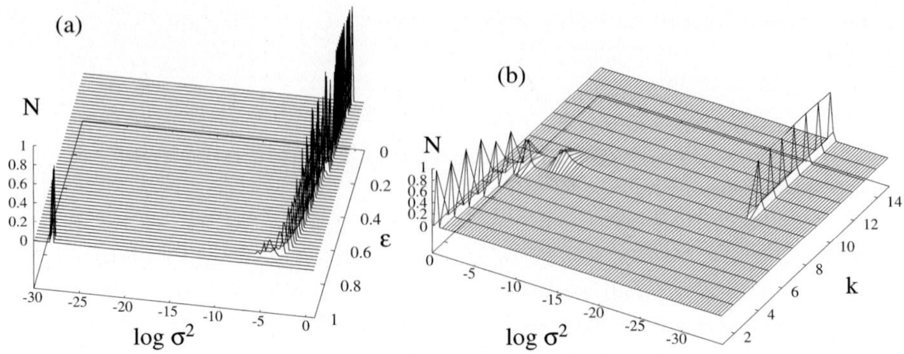

Fig. 1. Typical histograms of the standard mean square amplitude deviation σ^2 showing the sharp transition to coherence in function of **(a)** the coupling strength ε with $k = m_0 = 8$, and **(b)** the outgoing connectivity k with $\varepsilon = 0.95$. Values of N represent the fraction of configurations which converge to a coherent state from a total of 500 initial random configurations, after discarding transients of 10^4 time steps. Here $a = 2$, $N = 1000$, and $\alpha = 0$

with the network size. Moreover, as shown in Fig. 1, this transition to coherence is robust with respect to initial configurations, by varying either the coupling strength ε (Fig. 1a) or the outgoing connectivity k (Fig. 1b). In particular, above the threshold $\varepsilon_c \sim 0.9$, all initial configurations converge to a coherent state, indicating that in this parameter region the basin of attraction of coherent states fills almost the entire phase space.

From stability analysis, we find that in the fully chaotic regime ($a = 2$) the transition to coherence occurs for gradually smaller coupling strength if the connectivity k is increased. Figure 2a shows the boundaries ε_L and ε_U as a function of k for $a = 2$ (solid lines) and $a = 1.9$ (dashed lines) with the same parameter values as in Fig. 1. As one sees, in both cases the lower boundary ε_L decreases when k increases, while the upper boundary ε_U increases beyond $\varepsilon = 1$. Therefore, one expects that the region of synchronizability increases for larger values of connectivity k. Figure 2a also shows clearly that for $a = 2$ the intersection between both boundaries, $\varepsilon_L = \varepsilon_U$, occurs just above $k = 7$, which explains why the transition to coherence in Fig. 1b occurs at this value. For $a = 1.9$ this transition should occur near $k = 5$. Decreasing even more the nonlinearity coherent solutions are observed for even smaller connectivities and synchronizability regions increase, as shown in Fig. 2b. To see this feature more clearly we magnify in Fig. 2c the rectangle of Fig. 2b. As one sees, one exception occurs for $a = 1.8$, where the lower boundary is *below* the one for $a = 1.7$, due to the fact that for $a = 1.8$ the Lyapunov exponent of the logistic map is smaller than the one for $a = 1.7$, as illustrated below in Fig. 4. For all these values of a, the single uncoupled map shows chaotic orbits. Moreover, for any other network size N, the same curves are obtained.

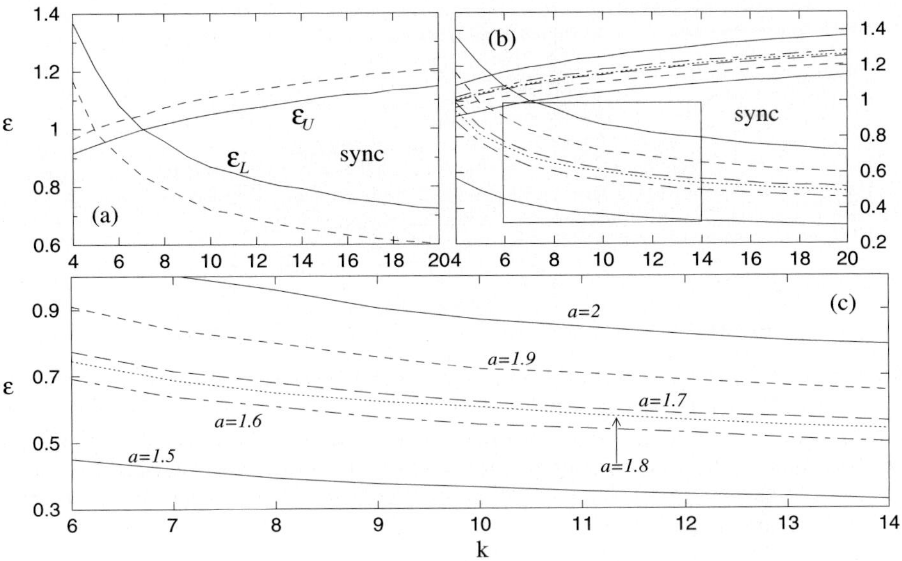

Fig. 2. Boundary values ε_L and ε_U in (4) in function of the connectivity k **(a)** for $a = 1.9$ (*dashed lines*) and $a = 2$ (*solid lines*), and **(b)** for $a = 1.5, 1.6, 1.7, 1.8, 1.9$ and 2. The data in the rectangle in **(b)** is magnified in **(c)**. The regions labelled with 'sync' are the ones where coherent solutions are observed, i.e. $\varepsilon_L < \varepsilon_U$. Notice that in (c) the boundary for $a = 1.8$ is below the one for $a = 1.7$, contrary to other values (see text)

These analytical predictions extracted from the stability condition in (4) and shown in Fig. 2 are strongly corroborated with our numerical simulations as shown in Fig. 3. In Fig. 3a we plot the fraction F of initial configurations which converge to a coherent state for $a = 2$, while Fig. 3b shows the threshold values, ε_c and k_c, at the transition curves where the entire sample of initial configurations converge to a coherent state, for the same values of a as in Fig. 2c. Here, one clearly sees that there is a clear and sharp transition to coherence. Interestingly, the curves in Fig. 3b fit very well the ones in Fig. 2c, which means that whenever the synchronizability condition $\varepsilon_L < \varepsilon_U$ holds, coherent states fill almost entirely the phase space.

Moreover, as illustrated in Fig. 3c, all curves obey a power law, within our numerical precision,

$$\varepsilon_c \propto k_c^{-\mu} \,. \tag{6}$$

For the six above values of a, the exponents are respectively $\mu = 0.2345$, 0.2354, 0.2353, 0.2231, 0.2023 and 0.1804: the exponent is almost constant below $a \sim 1.7$ and decreases above this value, as illustrated in the inset of Fig. 3c.

To determine the nature of the transition to coherence seen in Figs. 3a and 3b, we plot in Figs. 3d and 3e the average standard deviation in the

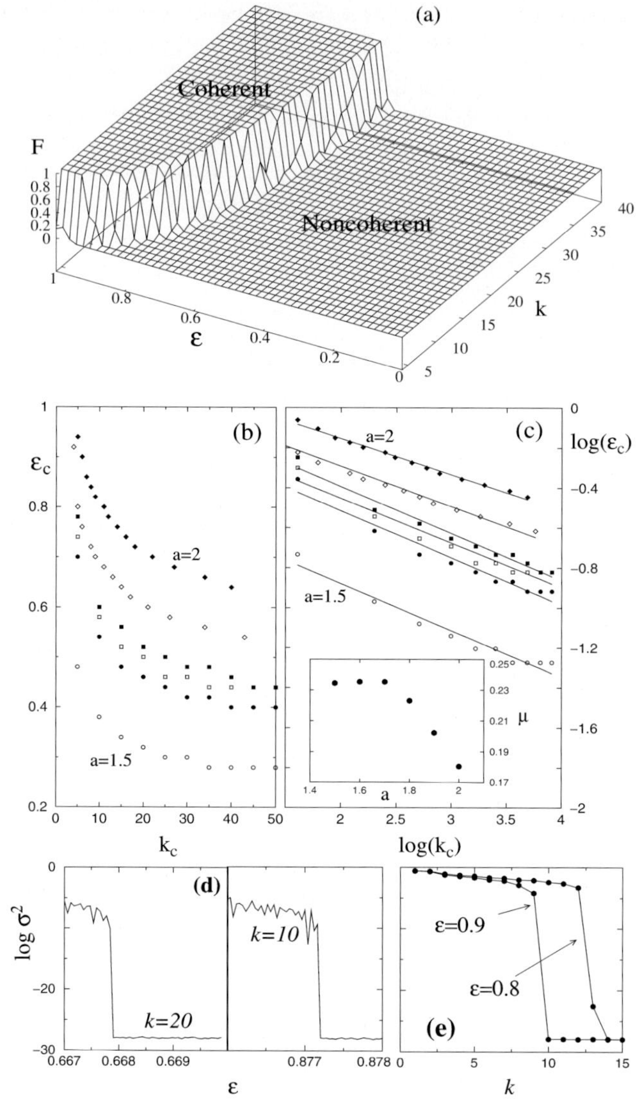

Fig. 3. Transition to coherence as a function of connectivity k and coupling strength ε. **(a)** Fraction $N_{\sigma=0}$ of coherent states from 500 random initial configurations for $a = 2$. **(b)** Coherence transition curves in the (ε, k) plane for (from bottom to top) $a = 1.5, 1.6, 1.8, 1.7, 1.9$ and $a = 2$, and **(c)** the same transition in a log-log plot, showing power-law dependence between connectivity and coupling strength for the transition curves, with an exponent μ which depends on the value of a (see inset). Here $\alpha = 0$, $L = 1000$ and we used transients of 10^4 time steps. By increasing the transient size to $\sim 10^6$ one clearly sees that the transition to coherence is of first-order either **(d)** when varying the coupling strength ε or **(e)** when varying the outgoing connectivity k

region where transition to coherence is observed, using much higher resolution. One clearly sees that the transition to coherence is of first-order, when varying ε or k. That the transitions are indeed of first order is easily recognised by the clear existence of hysteresis: when increasing either ε or k the configuration eventually falls into a coherent state, no longer spontaneously desynchronizing, no matter how far the parameters are tuned back.

All results till now, concern the case of homogeneous coupling ($\alpha = 0$). Next, we study the case of heterogeneous coupling. Figure 4 shows the boundaries ε_L and ε_U in (4) as functions of the outgoing connectivity k, the nonlinearity a and the heterogeneity α, covering both the homogeneous and heterogeneous regimes. Figure 4a shows the two boundaries as a function of k for $a = 2$ and $\alpha = 0$ (solid lines), $\alpha = 1$ (dashed lines) and $\alpha = 2$ (dotted lines). As one sees for nonzero values of α the boundaries are no longer smooth curves, but instead they show fluctuations as k is increased, enlarging and shrinking alternately the region of synchronizability, labelled as 'sync'. When varying a (Fig. 4b) the boundaries are mainly controlled by the Lyapunov exponent of the local map (see (4)), where ε_L (resp. ε_U) decreases (resp. increases) whenever a periodic window occurs [57]. The fluctuations observed in Fig. 4a are clearly seen in Fig. 4c, where the stability boundaries are plotted in function of α fixing $k = 8$ and $a = 2$ (solid lines) and $a = 1.5$ (dashed lines). The fluctuations are much higher for $\alpha > 1$ and for the fully chaotic regime both boundaries may even cross each other suppressing synchroniz-

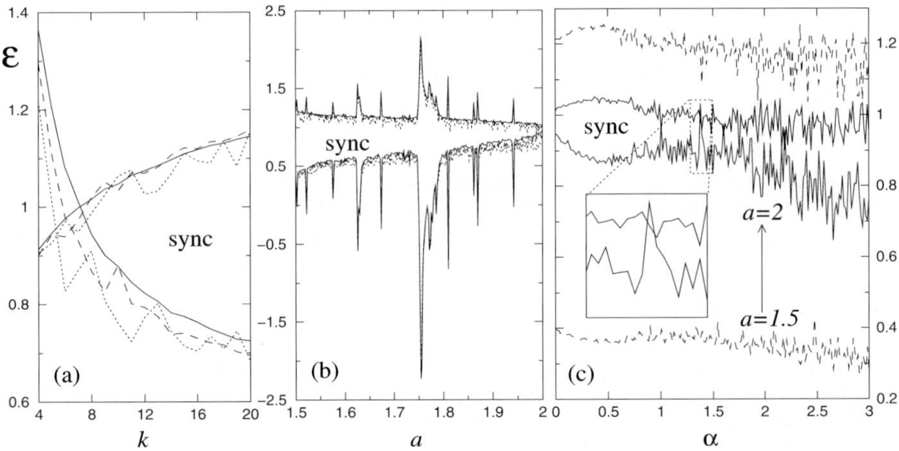

Fig. 4. Boundary values ε_L and ε_U in (4) in function of **(a)** connectivity k with $a = 2$ and $\alpha = 0$ (*solid lines*) $\alpha = 1$ (*dashed lines*) and $\alpha = 2$ (*dotted lines*), **(b)** nonlinearity a for $k = 8$ and $\alpha = 0, 1$ and 2, and **(c)** heterogeneity α with $k = 8$ and $a = 2$ (*solid lines*) and $a = 1.5$ (*dashed lines*). The inset in (c) emphasises one small region where synchronizability is not observed, $\varepsilon_L > \varepsilon_U$ (see text). Here $N = 1000$

ability (see inset of Fig. 4c). Moreover, the lower boundary ε_L decreases till $\alpha \sim 0.5$, then increases till $\alpha \sim 1$ and decreases in average from there on.

All these analytical results computed from (4) and matrix $\mathbb{G}$ in (1) are corroborated by our numerical simulations. In particular, the boundaries ε_L and ε_U seen in Fig. 4c are obtained also when plotting the contour of Fig. 5a, where we plot the average standard deviation from a sample of 500 initial configurations and vary the coupling strength and heterogeneity for $a = 2$ and $L = 1000$. While Fig. 5a shows the numerical results for $\alpha > 0$, i.e. in the case where nodes are more strongly coupled to the neighbours with higher connectivities, Fig. 5b shows the transition to coherence when $\alpha < 0$. Here synchronizability is observed only for $\alpha \gtrsim -0.15$ and for very high coupling strengths $\varepsilon \gtrsim 0.95$.

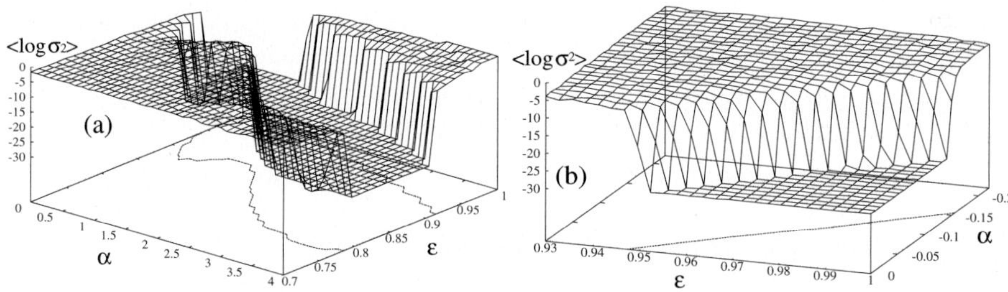

Fig. 5. Transition to coherence in function of coupling strength ε and heterogeneity α **(a)** when the most connected nodes dominate the dynamics ($\alpha > 0$) and **(b)** when the coupling to nodes with the least neighbours is strengthened ($\alpha < 0$). Here, we compute the average standard deviation from a sample of 500 initial configurations and fix $a = 2$, $k = m_0 = 8$ and $N = 1000$

We end our study of coherent solutions in random scale-free networks by investigating briefly the role of hubs in the lattice. Instead of strengthening the coupling to the most connected nodes by increasing $\alpha > 0$, we now fix $\alpha = 0$ and impose synchronization between all the nodes with more than a certain threshold k_t of neighbours and observe which fraction of the initial configurations converges to a coherent state. In this case the transition to coherence converges asymptotically to a limit of the coupling strength, as shown in Fig. 6a. The same occurs when synchronization is imposed to all nodes with *less* than k_t neighbours, as shown in Fig. 6b.

3.2 Deterministic Scale-Free Networks

In the previous section we focused on random scale-free networks, i.e. growing networks where new nodes are connected following probabilistic rules. In this section we study deterministic scale-free networks [37–39], using two different deterministic topologies: the pseudo-fractal scale-free network introduced by

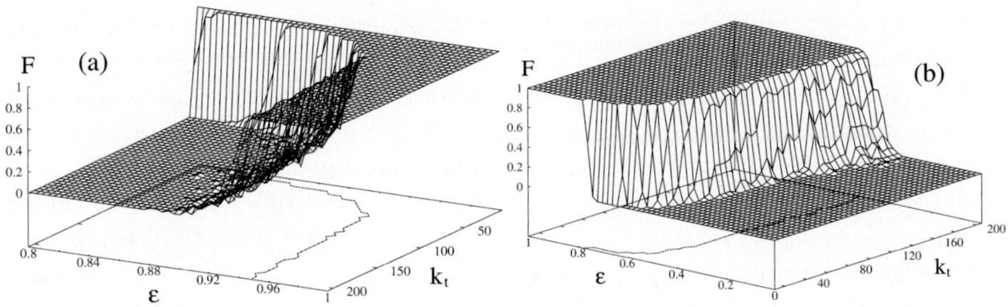

Fig. 6. Transition to coherence when synchronization is imposed to all nodes having a number of neighbours **(a)** larger than a threshold k_t, and **(b)** smaller than k_t (see text). Here $a = 2$, $k = 8$, $\alpha = 0$ and $N = 1000$

Dorogovtsev et al [38] and the Apollonian network introduced by Andrade et al. [39] and also studied in [40]. Both networks are illustrated in Fig. 7.

The pseudo-fractal network of Dorogovtsev is obtained, starting from three interconnected nodes, and at each iteration each edge generates a new node, attached to its two vertices. Figure 7a illustrates this network after three iterations, i.e. with three generations of nodes. The number of nodes N_n and the number of connections V_n increase with the number of generations as [38]

$$N_n = \tfrac{3}{2}(3^n + 1) \, , \tag{7a}$$

$$V_n = 3^{n+1} \, . \tag{7b}$$

From Fig. 7a one easily sees that this network has indeed a scale-free topology, since the number of nodes with degree $k = 2, 2^2, \ldots, 2^{n-1}, 2^n$ and 2^{n+1} is equal to $3^n, 3^{n-1}, \ldots, 3^2, 3$ and 3 respectively. In particular, the exponent of this power-law distribution is $\gamma = 1 + \ln 3/\ln 2$. Moreover, the cluster

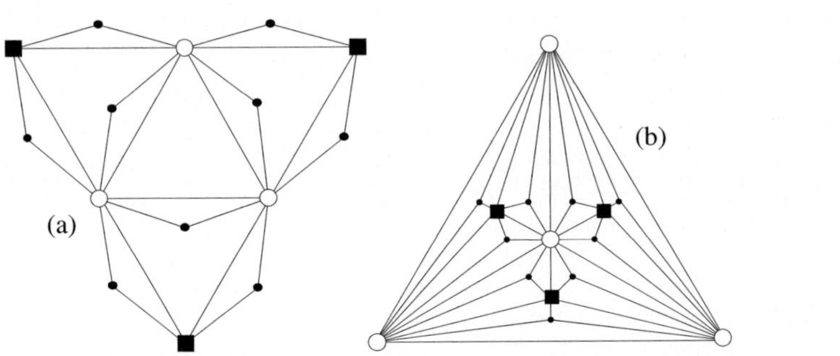

Fig. 7. Illustrations of two deterministic scale-free networks: **(a)** the pseudo-fractal network [41], and **(b)** the Apollonian network [39]. Identical symbols label nodes belonging to the same generation n (see text), namely ○ for $n = 0$, ■ for $n = 1$ and • for $n = 2$

coefficient of a node with degree k is $C = 2/k$, and the average path length is approximately $\langle \ell \rangle \simeq 4 \ln N_n / (9 \ln 3)$.

The Apollonian network is constructed in a different way: one starts with three interconnected nodes, defining a triangle; at $n = 0$ one puts a new node at the center of the triangle and joins it to the three other nodes, thus defining three new smaller triangles; at iteration $n = 1$ one adds at the center of each of these three triangles a new node, connected to the three vertices of the triangle, defining nine new triangles and so on (see Fig. 7b). The number of nodes and the number of connections are respectively given by

$$N_n = \tfrac{1}{2}(3^{n+1} + 5) , \tag{8a}$$
$$V_n = \tfrac{3}{2}(3^{n+1} + 1) . \tag{8b}$$

The distribution of connections obeys a power law, since the number of nodes with degree $k = 3, 3 \times 2, 3 \times 2^2, \ldots, 3 \times 2^{n-1}, 3 \times 2^n$ and 2^{n+1} is equal to $3^n, 3^{n-1}, 3^{n-2}, \ldots, 3^2, 3, 1$ and 3 respectively, and the exponent γ is the same as for the pseudo-fractal network. Moreover, a node with k neighbours has a cluster coefficient of $C \simeq 4/k$ as reported in [40], converging on average to $C_\infty = 0.828$, and the average path length grows weaker than $\ln N_n$ [39].

Although both networks have similar values for the topological quantities, they are quite different from the geometrical point of view: the pseudo-fractal network has no metric, while the Apollonian network is embedded in Euclidean space and fills it densely as $n \to \infty$, being particularly suitable to describe geographical situations [39].

For stability analysis purposes (see Sect. 2), the Laplacian matrix $\mathbb{G}$ of deterministic networks can be analytically determined from the adjacency matrix $\mathbb{A} = \{a_{ij}\}$, since they are related by

$$\mathbb{G} = \mathbb{I} + \mathbb{A}\mathbb{T} , \tag{9}$$

where $\mathbb{I}$ is the identity matrix and the values of matrix $\mathbb{T} = \{T_{ij}\}$ are defined by

$$T_{ij} = - \frac{a_{ji} \left[\sum_{k=1}^{N} a_{ik} \right]^{\alpha}}{\sum_{p=1}^{N} a_{jp} \left[\sum_{k=1}^{N} a_{pk} \right]^{\alpha}} . \tag{10}$$

A simple way to write the adjacency matrix of the pseudo-fractal network is

$$\mathbb{A}_n = \begin{bmatrix} \mathbb{A}_{n-1} & \mathbb{M}_{n-1} \\ \mathbb{M}_{n-1}^T & \emptyset \end{bmatrix}_{N_n \times N_n} , \tag{11}$$

where N_n is given by (7a), $\mathbb{M}^T$ represents the transposed matrix of $\mathbb{M}$ and for each generation $n = 1, 2, \ldots$ the matrix $\mathbb{M}_n$ reads

$$\mathbb{M}_n = \begin{bmatrix} \mathbb{M}_{n-1} & \mathbb{M}_{n-1} & \emptyset \\ \emptyset & \emptyset & \mathbb{B}_{n-1} \end{bmatrix}_{(2 \times 3^{n-1}) \times 3^n} , \tag{12}$$

with

$$\mathbb{B}_{n-1} = \begin{bmatrix} \mathbb{A}_0 & \emptyset & \dots & \emptyset \\ \emptyset & \mathbb{A}_0 & \dots & \emptyset \\ \vdots & \vdots & \ddots & \vdots \\ \emptyset & \emptyset & \dots & \mathbb{A}_0 \end{bmatrix}_{3^{n-1}\times 3^{n-1}} \tag{13}$$

and whose starting form is

$$\mathbb{M}_0 = \mathbb{A}_0 = \begin{bmatrix} 0\,1\,1 \\ 1\,0\,1 \\ 1\,1\,0 \end{bmatrix}_{3\times 3} . \tag{14}$$

For the Apollonian network, the adjacency matrix is given by the same recurrence of (11), but this time with

$$\mathbb{A}_0 = \begin{bmatrix} 0\,1\,1\,1 \\ 1\,0\,1\,1 \\ 1\,1\,0\,1 \\ 1\,1\,1\,0 \end{bmatrix} , \tag{15}$$

and $\mathbb{M}_n$ being a matrix with $(3^n + 5)/2$ rows and 3^n columns and having in each column only three nonzero elements.

Figure 8 shows the eigenspectra of the Laplacian matrices for both the pseudo-fractal (Fig. 8a) and the Apollonian (Fig. 8b) networks, in function of the heterogeneity. As one sees for $a = 2$ (solid lines) synchronizability is observed only above $\alpha \gtrsim 1.5$, and in particular there is no synchronizability for the homogeneous coupling regime ($\alpha = 0$). Figure 9a shows the distribution of the average standard deviation over a sample of 500 initial configurations, from which one clearly sees that there are no coherent solutions. Here the standard mean deviation is characterized by some large value which is almost constant beyond the weak coupling regime ($\varepsilon \gtrsim 0.2$). In the weak coupling regime ($\varepsilon \lesssim 0.2$) the standard mean square deviation is even larger, since the coupling is not strong enough to compensate the highly chaotic local dynamics ($a = 2$).

From Fig. 8 one also sees that, for the pseudo-fractal and $\alpha > 1.5$, the upper threshold ε_U increases monotonically with the heterogeneity, while for the Apollonian network the upper threshold decreases. This particular difference between both networks should be due to their geometrical differences, in particular the fact that Apollonian networks are embedded in Euclidean space could explain in some way that stronger dominance in the coupling to the most connected nodes *destroys* coherence.

Choosing other values of a for which the local dynamics is chaotic, one finds that the form of de curves $\varepsilon_L(\alpha)$ and $\varepsilon_U(\alpha)$ does not change. These curves are only shifted: ε_L gets smaller, while ε_U increases. Figure 8 illustrates this for the particular case of $a = 1.5$. Decreasing even further the nonlinearity

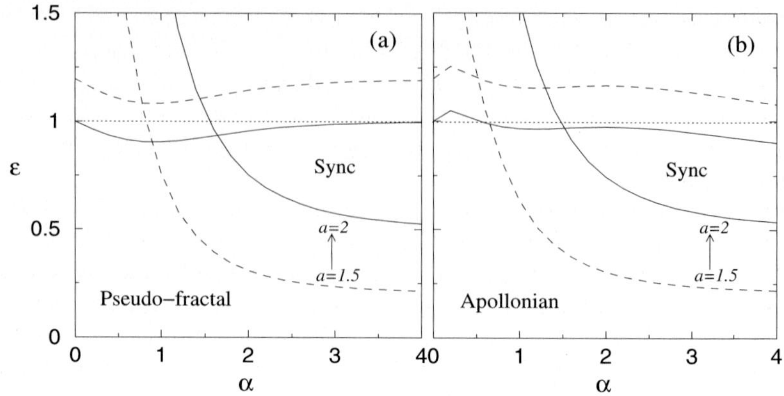

Fig. 8. Boundary ε_L and ε_U for synchronizability in function of the heterogeneity α for **(a)** the pseudo-fractal network and **(b)** the Apollonian network, with $a = 2$ (*solid lines*) and $a = 1.5$ (*dashed lines*). For each network we use 6 generations of nodes (see text). These eigenspectra are the same for any number of generations

below the accumulation point $a = 1.411\ldots$ synchronizability is attained for any positive value of α, whenever the coupling strength is sufficiently strong.

Moreover there is a complicated dependence of the average standard deviation on the coupling strength and nonlinearity. As shown in Fig. 9b for deterministic scale-free networks one finds two main regions in the (a, ε) plane: (I) a region where the standard mean square deviation is large and varies

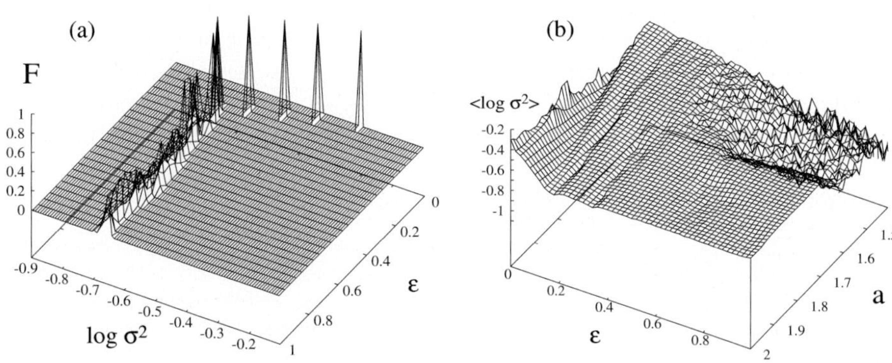

Fig. 9. **(a)** Typical histogram of the standard mean square deviation σ^2 for the pseudo-fractal network as a function of the coupling strength ε, with $a = 2$ and $\alpha = 0$. A similar result is obtained for the Apollonian network. **(b)** Histogram of the standard mean square deviation σ^2 as a function of nonlinearity a and coupling strength ε, for deterministic scale-free networks with $\alpha = 0$. The mean square deviation is averaged over a sample of 500 initial configurations and during 100 time steps, after discarding transients of 10^4 time steps

smoothly with the parameters and (II) a region where the mean square deviation is smaller but has larger fluctuations. This second region, observed for $a \lesssim 1.7$, is somehow surprising, since irregular variations of the standard mean square deviation occur for low nonlinearity and high coupling strengths, precisely where one would expect the most regular behaviour of the node dynamics.

As for the heterogeneous coupling regime ($\alpha \neq 0$), Fig. 10 illustrates the transition to coherence by varying the heterogeneity α for the pseudo-fractal (Fig. 10a) and the Apollonian network (Fig. 10b). For both networks, one sees that coherence sets in for $\alpha \gtrsim 1.5$, and the contour of the histograms marking the transition to coherence fits well the regions in Fig. 8 labelled as 'sync'. Moreover, from Figs. 10c and 10d one observes that all these transitions to coherence are of first-order.

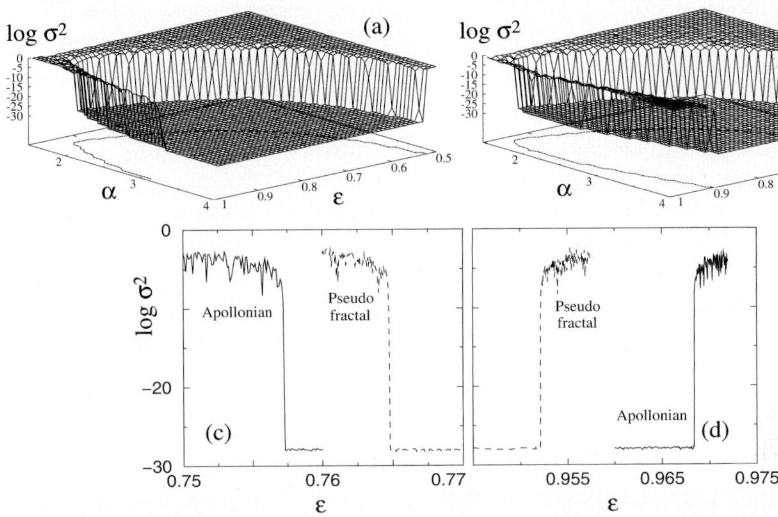

Fig. 10. Inducing transition to coherence by varying the heterogeneity α (see (1)) in scale-free networks. **(a)** Pseudo-fractal network and **(b)** Apollonian network. For strong heterogeneity coherence appears beyond a relatively high coupling strength, and disappears again for very large couplings (see text). For each network, we use $\ell = 6$ generations of nodes and fix $a = 2$. **(c)** and **(d)** show high-resolution plots of σ^2 as a function of ε for $\alpha = 2$, emphasizing the first-order phase transition to coherence

Finally, we study the role of hubs in deterministic scale-free networks, as we did in the previous section for random networks. To this end, we impose synchronization among $g = 1, \ldots, \ell$ generations, with ℓ being the total number of generations, and observed in what conditions coherent states are observed. In the pseudo-fractal network the first generation has $N_1 - N_0 = 3$ nodes, the second one has $N_2 - N_1 = 9$ nodes, and the nth generation has

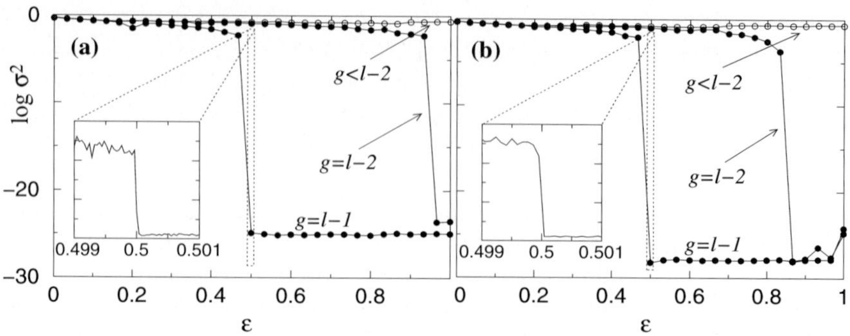

Fig. 11. Transitions to coherence in deterministic scale-free networks, when synchronizing the first g generations of nodes out of ℓ generations (see text). **(a)** Pseudo-fractal network and **(b)** Apollonian network. The collective dynamical behaviour is quite insensitive to hubs (see text). Insets show that transitions to coherence are of first-order. For each network, we use $\ell = 9$ generations of nodes and $a = 2$ fixed

$N_n - N_{n-1} = 3^n$ nodes. In the Apollonian network the number of nodes appearing at each generation is precisely the same.

Figure 11 shows the standard mean square deviation as a function of coupling strength for pseudo-fractal (Fig. 11a) and Apollonian networks (Fig. 11b). In each case we choose the fully chaotic map ($a = 2$) and impose synchronization among the nodes of the first g generations by setting them to their mean amplitude at each time-step. In both cases, one sees that the standard mean square deviation remains large when synchronization is imposed to all $g < \ell - 2$ generations. Coherent solutions are only observed for $g = \ell - 2$ and $g = \ell - 1$, beyond a coupling threshold which is smaller for the latter case. Surprisingly, for $g = \ell - 1$ the transition to coherence occurs for the same coupling strength in both networks. This may be due to the fact that the fraction $N_g/N_\ell = 3^{g-\ell} = 1/3$ of nodes on which one imposes synchronization is the same for both networks and is high enough to suppress the influence of local connectivities.

For $g = \ell - 2$ the pseudo-fractal network shows coherence only above very high coupling strengths, near $\varepsilon \sim 1$, while for Apollonian networks the threshold is much lower. This difference in the coupling strength threshold is due to the fact that here the fraction of nodes $N_g/N_\ell = 1/9$ to which synchronization is imposed is small enough not to suppress the influence of local connectivities. Therefore, since the hubs in the pseudo-fractal network are less connected than the hubs in Apollonian networks, one needs higher coupling strength to observe coherence. For any higher value ℓ of generations the same results are obtained, since one has for the quotient of the number of nodes between two successive generations $N_n/N_{n-1} \to 3$ as n increases.

As a general remark, one observes from Fig. 11 that one needs to synchronize a rather high fraction of nodes ($\gtrsim 1/9$) to induce coherence. Therefore, it seems that, dynamical collective behaviour on scale-free networks is quite insensitive to hubs. As shown in the insets of Figs. 11a and 11b, the transition to coherence is also of first-order.

4 Discussion and Conclusions

In this chapter we studied fully synchronized solutions for three scale-free network topologies. The main conclusion is the following: in random scale-free networks synchronization of chaotic maps not only depends on the coupling strength but is mainly controlled by the outgoing connectivity k, which is a measure of the cohesion in the networks. Because of that, one finds coherent solutions in random scale-free networks of fully chaotic logistic maps ($a = 2$) with outgoing connectivity $k = 8$ and homogeneous coupling, but not in deterministic scale-free networks, since they have rather small effective outgoing connectivity, namely $k = 2$ for the pseudo-fractal network and $k = 3$ for the Apollonian network. Therefore, although the exponent γ of connection distributions in scale-free networks does not depend on the outgoing connectivity [2], we have shown that, in general, synchronization of chaotic maps in such coupling topologies is quite sensitive to it.

Our results were obtained both numerically, from histogram of significantly large samples of initial configurations with a criterion for full synchronization based on the mean standard deviation of amplitudes, (5), and analytically from the eigenvalue spectra of the diagonalized variational equations computed at the coherent states, (3).

In particular, for random scale-free networks, the threshold values of the coupling strength obey a power law, (6), as function of the outgoing connectivity. The exponent of this power law depends on the nonlinearity a of the chaotic map, being almost constant below $a_c \sim 1.7$ and decreasing linearly above it. Interestingly this value of a_c is in the vicinity of the bifurcation of the quadratic map where the period-3 window appears, and coincides with the appearance of other nontrivial behaviours in coupled map lattices with regular topologies, namely in the velocity distribution of travelling wave solutions [24].

For deterministic scale-free networks with homogeneous coupling, the same value a_c indicates the threshold above which no coherent solutions are observed, independently of the coupling strength. Above a_c, coherence is observed only for heterogeneous coupling, namely for $\alpha \gtrsim 1.5$. However, for this range of values, we have also shown that coherence is also absent either for very small or for very large coupling strengths, due to spatial instabilities. Another particularly interesting result that still needs to be explained is that, for Apollonian networks, the coupling threshold beyond which coherence disappears gets smaller when the heterogeneity is further increased.

This point is not observed for the pseudo-fractal network and may be due to the geometrical differences between both deterministic networks.

As a general property, we have shown that all transitions to coherence are of first-order, indicating a similarity with other complex networks [29]. Furthermore, all results are robust not only against changes of the initial configurations of node amplitude but also, in random scale-free networks, against changes of the connection network. We also presented results indicating that in scale-free networks hubs play apparently no fundamental role in the dynamical collective behaviour.

Acknowledgments

The authors thank A.O. Sousa and C. Zhou for useful discussions. P.G.L. thanks Fundação para a Ciência e a Tecnologia, Portugal, for financial support. J.A.C.G. thanks Conselho Nacional de Desenvolvimento Científico e Tecnológico, Brazil and Sonderforschungsbereich 404 of DFG for financial support.

References

1. S. Bornholdt, H.G. Schuster (eds.): *Handbook of Graphs and Networks* (Wiley-VCH, Weinheim 2003)
2. R. Albert, A.-L. Barabási: Rev. Mod. Phys. **74**, 47 (2002).
3. S.N. Dorogovtsev, J.F.F. Mendes: Adv. Phys. **51**, 1079 (2002)
4. R. Meucci, R. McAllister, R. Roy: Phys. Rev. E **66**, 026216 (2002)
5. J.H. Snoeijer, T.J.H. Vlugt, M. van Hecke, W. van Saarloos: Phys. Rev. Lett. **92**, 054302 (2004)
6. M. Otto, J.-P. Bouchaud, P. Claudin, J.E.S. Socolar: Phys. Rev. E **67**, 031302 (2003)
7. C. Texier, G. Montambaux: Phys. Rev. Lett. **92**, 186801 (2004)
8. H. Tanaka, J. Meunier, D. Bonn: Phys. Rev. E **69**, 031404 (2004)
9. D.P. Almond, C.R. Bowen: Phys. Rev. Lett. **92**, 157601 (2004)
10. M. Small, C.K. Tse, Phys. Rev. E **66**, 066701 (2002)
11. I. Stewart: Nature **427**, 601 (2004)
12. Y. Moreno, M. Nekovee, A. Vespignani: Phys. Rev. E **69**, 055101(R) (2004)
13. Z. Dezso, A.L. Barabási: Phys. Rev. E **65**, 055103 (2002)
14. M. Compiani, E. Capriotti, R. Casadio: Phys. Rev. E **69**, 051905 (2004)
15. L. Zhaoping, A. Lewis, S. Scarpetta: Phys. Rev. Lett. **92**, 198106 (2004)
16. P.G. Lind, J. Corte-Real, J.A.C. Gallas: Phys. Rev. E **69**, 026209 (2004)
17. S.H. Strogatz: Physica D **143**, 1 (2000)
18. L.M. Pecora, T.L. Carroll, G.A. Johnson, D.J. Mar, J.F. Heagy: Chaos **7**, 520 (1997)
19. C. Anteneodo, A.M. Batista, R.L. Viana: Phys. Lett. A **326**, 227 (2004)
20. M. Cross, P. Hohenberg: Rev. Mod. Phys. **65**, 851 (1993)

21. K. Kaneko, I. Tsuda: *Chaos and Beyond* (Springer-Verlag, Berlin Heidelberg New York 2000)
22. S. Wolfram: *A New Kind of Science* (Wolfram Media Inc., New York 2002)
23. K. Kaneko (ed.): *Theory and Applications of Coupled Map Lattices* (John Wiley & Sons Inc., New York Chichester 1993)
24. P.G. Lind, J. Corte-Real, J.A.C. Gallas: Phys. Rev. E **69**, 066206 (2004)
25. S. Boccaletti, J. Kurths, G. Osipov, D.L. Valladares, C.S. Zhou: Phys. Rep. **366**, 1 (2002)
26. P. Erdös, A. Rényi: Publ. Math. Debrecen **6**, 290 (1959)
27. B. Bollobás: *Random Graphs* (Academic, London 1985)
28. D.J. Watts, S.H. Strogatz: Nature **393**, 440 (1998)
29. S.H. Strogatz: Nature **410**, 268 (2001)
30. A.-L. Barabási, R. Albert: Science **286**, 509 (1999)
31. K. Christensen, R. Donangelo, B. Koiler, K. Sneppen: Phys. Rev. Lett. **81**, 2380 (1998)
32. M.E.J. Newman, D.J. Watts: Phys. Rev. E **60**, 7332 (1999)
33. A.-L. Barabási, R. Albert, H. Jeong: Physica A **272**, 173 (1999)
34. S.N. Dorgovtsev, J.F.F. Mendes, A.N. Samukhin: Phys. Rev. Lett. **85**, 4633 (2000)
35. L. Kullmann, J. Kertész: Phys. Rev. E **63**, 051112 (2001)
36. P.L. Krapivsky, S. Redner, F. Leyvraz: Phys. Rev. Lett. **85**, 4629 (2001)
37. A.-L. Barabási, E. Ravasz, T. Vicsek: Physica A **299**, 559 (2001)
38. S.N. Dorogovtsev, A.V. Goltsev, J.F.F. Mendes: Phys. Rev. E **65**, 066122 (2002)
39. J.S. Andrade Jr, H.J. Herrmann, R.F.S. Andrade, L.R. da Silva: Phys. Rev. Lett. **94**, 018702 (2005)
40. J.P.K. Doye, C.P. Massen: Phys. Rev. E **71**, 016128 (2005)
41. M.C. González, A.O. Sousa, H.J. Herrmann: Int. J. Mod. Phys. C **15**, 45 (2004)
42. H. Chaté, P. Manneville: Chaos **2**, 307 (1992)
43. P.M. Gade: Phys. Rev. E **54**, 64 (1996)
44. S.C. Manrubia, A.S. Mikhailov: Phy. Rev. E **60**, 1579 (1999)
45. J. Jost, M.P. Joy: Phys. Rev. E **65**, 016201 (2001)
46. T. Nishikawa, A.E. Motter, Y.-C. Lai, F.C. Hoppensteadt: Phys. Rev. Lett. **91**, 014101 (2003)
47. M. Barahona, L.M. Pecora: Phys. Rev. Lett. **89**, 054101 (2002)
48. L.F. Lago-Fernández, R. Huerta, F. Corbacho, J.A. Siguenza: Phys. Rev. Lett. **84**, 2758 (2000)
49. H. Hong, B.J. Kim, M.Y. Choi, H. Park: Phys. Rev. E **69**, 067105 (2004)
50. F.M. Atay, J. Jost, A. Wende: Phys. Rev. Lett. **92**, 144101 (2004)
51. X.F. Wang, G. Chen: IEEE Trans. Circ. Sys. I **49**, 54 (2002)
52. A.E. Motter, C. Zhou, J. Kurths: Europhys. Lett. **69**, 334 (2005)
53. P.G. Lind, J.A.C. Gallas, H.J. Herrmann: Phys. Rev. E **70**, 056207 (2004)
54. L.M. Pecora, T.L. Carroll: Phys. Rev. Lett. **80**, 2109 (1998)
55. K.S. Fink, G. Johnson, T. Carroll, D. Mar, L. Pecora: Phys. Rev. E **61**, 5080 (2000)
56. S.C. Manrubia, A.S. Mikhailov, D.H. Zanette: *Emergence of Dynamical Order Synchronization Phenomena in Complex Systems, vol. 2* (World Scientific, Singapore 2004)
57. E. Ott: *Chaos in Dynamical Systems* (Cambridge University Press, New York 1993)
58. A. Pikovsky, O. Popovych, Yu. Maistrenko: Phys. Rev. Lett. **87**, 044102 (2001)

Growth of Random Sequences

K. Austin and G. J. Rodgers

Department of Mathematical Sciences, Brunel University, Uxbridge, Middlesex UB8 3PH, U.K.

Summary. We study the behaviour of two classes of random sequences. In these sequences each new element is generated by adding together two or more of the previous elements, or multiples of previous elements in the sequence. At least one of the elements is chosen randomly from a probability distribution of the previous elements. We find that a wide range of different types of behaviour emerge from linear to exponential growth and that the sequences exhibit a remarkably diverse phase space. Interestingly, new transitions in phase space are observed when the generating equations correspond to the backward difference equations.

Key words: Random sequences, Fibonacci sequence, Disordered systems, Anderson model

1 Introduction

Integer sequences have attracted widespread interest in many fields of science since (in the 13th century) Fibonacci discovered a sequence whose numbers are found frequently both in nature and in man-made systems.

1.1 The Fibonacci Sequence

The numbers of the Fibonacci sequence,

$$F_{n+1} = F_n + F_{n-1} \ , \tag{1}$$

are the same as the numbers of petals found on flowers, the numbers of seeds on the spiral arms of a sunflower, the numbers that give the spacial patterns of leaves growing up the stems of plants and the spiral pattern of a snail's shell. More recently, the Fibonacci numbers have also been found to have applications in man-made systems. For example, in computer science the Fibonacci numbers appear in database structures and random number generation. An important property of the Fibonacci sequence is the "golden ratio". The Fibonacci sequence grows exponentially, with the nth element of the sequence given by $F_n \sim [(1+\sqrt{5})/2]^n$. The ratio $(1+\sqrt{5})/2 \simeq 1.61803\ldots$ is found everywhere in nature and was even known about in the time of the ancient Greeks, who used it for its aesthetically pleasing proportions.

Interest in integer sequences has more recently focussed on the introduction of random properties to incorporate stochasticity into the sequences. In [1], Furstenberg and Kesten showed that for a general class of random-sequence generating processes, the absolute value of the nth member of the sequence will, with probability 1, get closer to the nth power of some fixed number, as with the Fibonacci sequence. Furstenberg and Kesten's work has been found to have many applications. It was used in the first rigorous explanation of Anderson localization, which came long after the mathematics had been worked out.

1.2 Random Fibonacci Sequences

Viswanath [2] made use of Furstenberg and Kesten's work in the study of a stochastic version of the Fibonacci sequence,

$$x_{n+1} = \pm x_n \pm x_{n-1} , \tag{2}$$

where the signs are chosen independently with equal probability. Unlike the deterministic rules of the original Fibonacci sequence, Viswanath's [2] rules are stochastic and generate not one but many different sequences. Viswanath [2] showed that when the absolute value of the numbers of any sequence generated by these rules is taken, the nth number in the sequence will, on average, get closer to the nth power of the number 1.1398... Therefore the sequences generated by these stochastic rules grow exponentially, with $y_n \equiv \langle |x_n| \rangle \sim (1.1398...)^n$, just like the Fibonacci sequence, but with a new "stochastic golden ratio" of 1.1398...

1.3 Generalised Random Fibonacci Sequences

In [3], Embree and Trefethen generalised this result to show that for the sequence generating equation

$$x_{n+1} = x_n \pm \beta x_{n-1} , \tag{3}$$

there is a critical value of

$$\beta = \beta^* \in (0.702582, 0.702585) , \tag{4}$$

above which the sequences grow exponentially, but below which they decay exponentially. At the critical value $\beta = \beta^*$, no exponential growth or decay occurs. When two constant parameters are used rather than one, $x_{n+1} = \alpha x_n \pm \beta x_{n-1}$, the same results are obtained by simple rescaling. This is because the random signs ensure that the terms x_n and x_{n-1} are uncorrelated.

1.4 Alternative Random Sequences

An alternative stochastic version of the Fibonacci sequence is introduced in [4]. Rather than adding or subtracting the two preceding elements with equal probability as in [2] and [3], random elements are added according to

$$\text{(i)} \quad x_n = x_{n-1} + x_q \quad \text{and} \quad \text{(ii)} \quad x_n = x_p + x_q \,, \tag{5}$$

where p and q are chosen randomly between 0 and $n-1$. The sequences produced by (i) grow more slowly than the sequences in [2] and [3], with stretched exponential growth. The value of the average n^{th} element of the sequence is given by $< x_n >= A_n \propto n^{-1/4} \exp(2\sqrt{n})$. The sequences produced by (ii) grow linearly according to $A_n \simeq n+1$, or have power-law growth, with $A_n \sim n^{2c-1}$, if a multiplicative factor is introduced into the generating equation so that $x_n = c(x_p + x_q)$ [4].

1.5 Applications

Just as the Fibonacci sequence is linked to a variety of deterministic systems in physics, biology and computer science, so too are stochastic sequences linked with random or disordered systems. Stochastic sequences are present in the random walk that appears in models of many stochastic processes in a variety of fields. For example, Einstein's explanation of Brownian motion was one of the earliest to incorporate the random walk [5]. Other examples include models of evolutionary biology to model adaptive walks on a fitness landscape [6] and to financial markets [7] to model random elements of the markets such as investor speculation and consumer demand. Stochastic sequences also appear in a number of problems in one-dimensional disordered systems [8], such as the Anderson model describing the propagation of electrons in a metal with impurities. In this model the electrons interact with the inhomogeneous potential of the system. The rates of transition of the electrons from one site to another is a random function of position and therefore analogous to the random progression of the stochastic sequence.

1.6 More Random Sequences

In this chapter we investigate the effect of generalising some of the random sequences introduced so far. We firstly consider sequences where the random elements are chosen randomly from a separable probability distribution of the previous elements in the sequence. We then investigate the effect of introducing multiplicative constants to the individual terms in the generating equations of random sequences where the random elements are chosen from a uniform probability distribution of the previous elements in the sequence.

In Sect. 2, we consider the sequences generated by

$$x_n = x_{n-1} + \gamma x_q \,, \tag{6}$$

with $\gamma > 0$ and q chosen randomly from a probability distribution $P_n(q)$, [9]. This is a generalisation of sequences considered in [4], where the case for $P_n(q) = 1/n$ and $\gamma = 1$ was considered. The mean and variance of the sequences generated by $P_n(q) = 1/n$ and the sequences generated by power-law and exponential $P_n(q)$ are obtained, as well as non-random $P_n(q) = \delta_{q,\beta(n-1)}$.

In Sect. 3, we investigate the effect of introducing multiplicative constants to the individual terms in the generating equations of the sequences studied in [4] and introduce a new sequence as an extension of these studies. In particular, we investigate the sequences generated by

$$x_n = \begin{cases} \gamma x_p + \delta x_q; & \text{(Section 3.1)} \\ \alpha x_{n-1} + \gamma x_q; & \text{(Section 3.2)} \\ \alpha x_{n-1} + \beta x_{n-2} + \gamma x_q\ . & \text{(Section 3.2)} \end{cases}$$

Since the random signs in [2] and [3] are removed, the terms become correlated so that the growth of the resulting sequences depends on the values of the multiplicative constants.

We find that for each type of generalisation, new types of behaviour emerge. In particular, the sequences are found to exhibit linear and exponential growth at the extremes of the free parameters, with intermediate types of growth in-between. We summarise and discuss these results in Sect. 4.

2 Sequences with Random Elements Chosen from a Probability Distribution

In this section we consider the sequences generated by

$$x_n = x_{n-1} + \gamma x_q\ , \tag{7}$$

where q is chosen randomly from the separable probability distribution $P_n(q) = P(q)/b_n$, with $b_n = \sum_{q=0}^{n-1} P(q)$, [9]. The average n^{th} element, $A_n = < x_n >$ is given by

$$A_n = A_{n-1} + \frac{\gamma}{b_n} \sum_{q=0}^{n-1} P(q) A_q\ . \tag{8}$$

We consider uniform, power-law and exponential forms of $P(q)$ and then go on to look at non-random $P_n(q) = \delta_{q,\beta(n-1)}$.

2.1 Uniform $P(q)$

First we consider the case of $P(q) = 1$ and hence $b_n = n$. Substituting this into (8), multiplying through by n and subtracting the equivalent expression for $n + 1$, we obtain

$$A_{n+1} - \left(2 + \frac{\gamma - 1}{n+1}\right) A_n + \frac{n}{n+1} A_{n-1} = 0 \,. \tag{9}$$

This is equal to the nth Laguerre polynomial, $L_n(-\gamma)$, [10] and for large n yields the asymptotic solution

$$A_n \sim k_\gamma n^{-1/4} \exp(2\sqrt{\gamma n}) \,, \tag{10}$$

[9]. This is the result obtained in [4] for $\gamma = 1$. We can calculate the average growth of the second moment, $V_n = \langle x_n^2 \rangle$ for large n, by introducing two averages

$$V_n = \langle x_n^2 \rangle \quad \text{and} \quad M_n = \sum_{r=0}^{n-1} < x_n x_r > \,. \tag{11}$$

Using (7), it is simple to show that V_n and M_n obey the coupled iterations

$$(n+1)V_{n+1} - [2n + (\gamma+1)^2]V_n + (n + 2\gamma)V_{n-1} = 2\gamma(M_n - M_{n-1}) \tag{12}$$

and

$$(n+1)M_{n+1} - (2n + 2\gamma + 1)M_n + nM_{n-1} = (n + \gamma + 1)V_n - nV_{n-1} \,. \tag{13}$$

We can then take the continuum limit for continuous n to give the second-order differential equations

$$[nV(n)]'' - (2\gamma + 1)V'(n) - \gamma^2 V(n) = 2\gamma M'(n) \tag{14}$$

and

$$[nM(n)]'' - M'(n) - 2\gamma M(n) = [nV(n)]' + V'(n) + \gamma V(n) \,. \tag{15}$$

In the limit $n \to \infty$, we can equate leading order terms to give the solutions $V(n) \sim n^\phi \exp(\delta\sqrt{n})$ and $M(n) \sim n^{\phi+1/2} \exp(\delta\sqrt{n})$, where

$$\delta = \sqrt{2\gamma \left(4 + \gamma + \sqrt{16 + \gamma^2}\right)} \tag{16}$$

for $\gamma > 0$, [9]. This is in good agreement with [4], where $\delta \simeq 4.3$ was obtained numerically for $\gamma = 1$.

2.2 Power-Law $P(q)$

If $P(q)$ takes the form of a power-law, such that

$$P(q) = (q+1)^\alpha \tag{17}$$

and $\gamma = 1$, five different classes of behaviour emerge for the cases i) $\alpha > -1$, ii) $\alpha = -1$, iii) $-2 < \alpha < -1$, iv) $\alpha = 2$ and v) $\alpha < -2$. We consider each of these in turn.

i) $\alpha > -1$

For $\alpha > -1$, $b_n/b_{n+1} = 1 - (\alpha+1)/n + O(1/n^2)$ for large n. From (8) we obtain

$$A_{n+1} - 2A_n + A_{n-1} = \frac{\alpha+1}{n} A_{n-1} \, . \tag{18}$$

This equation can be solved exactly in terms of generalised Laguerre polynomials [10]. In this case, $A_n \sim L_n^{(-\alpha)}(-(\alpha+1))$, and we obtain the asymptotic solution

$$A_n \sim c_\alpha \frac{1}{n^{(2\alpha+1)/4}} \exp[2\sqrt{(\alpha+1)n}] \tag{19}$$

as $n \to \infty$, with

$$c_\alpha = \frac{(\alpha+1)^{(2\alpha-1)/4} e^{-(\alpha+1)/2}}{2\sqrt{\pi}} \, , \tag{20}$$

[9].

ii) $\alpha = -1$

When $\alpha = -1$, $b_n/b_{n+1} = 1 - 1/(n+1)\log n + O(1/n(\log n)^2)$ and (8) gives

$$A_{n+1} - 2A_n + A_{n-1} = \frac{1}{(n+1)\log n} A_{n-1} \tag{21}$$

as $n \to \infty$. Using the WKB method [11], A_n is found to obey

$$A_n \sim \frac{1}{\sqrt{n}\log n} \exp\left[2\sqrt{\frac{n}{\log n}}\right] \tag{22}$$

for large n. The growth of the sequences for $\alpha = -1$ is slower than for $\alpha > -1$ as $P_n(q)$ is no longer as strongly dominated by $q \sim n$.

iii) $-2 < \alpha < -1$

For $-2 < \alpha < -1$, $b_n/b_{n+1} = 1 - (n+1)^\alpha/\zeta(-\alpha) + O(n^{\alpha-1})$ where ζ is the Riemann zeta function [12]. For large n (8) gives

$$A_{n+1} - 2A_n + A_{n-1} = \frac{(n+1)^\alpha}{\zeta(-\alpha)} A_{n-1} \tag{23}$$

and the WKB approximation [11] yields

$$A_n \sim \frac{1}{n^{\alpha/4}} \exp\left[\frac{2}{\alpha+2} \frac{n^{1+\alpha/2}}{\sqrt{\zeta(-\alpha)}}\right] . \tag{24}$$

Again, the growth of the sequences is found to be slower than for $\alpha \geq -1$ as $P_n(q)$ is not as strongly dominated by $q \sim n$ as for $\alpha > -1$ or for $\alpha = -1$.

iv) $\alpha = -2$

For $\alpha = -2$ the large n behaviour of A_n is purely power-law with

$$A_n \sim n^{\frac{1}{2}+\frac{1}{\sqrt{\zeta(2)}}} \tag{25}$$

where the exponent $1/2 + 1/\sqrt{\zeta(2)} \approx 1.108$ is greater than 1.

v) $\alpha < -2$

Finally, for $\alpha < -2$, the right hand side of (23) can be neglected so that (23) yields

$$A_n \sim n \tag{26}$$

as $n \to \infty$. The sequences are now dominated by $P_n(q) \sim 1$, resulting in linear growth.

2.3 Exponential $P(q)$

Next we consider an exponential form of $P(q)$, $P(q) = a^q$. Again there are found to be different solutions, in this case for different values of a. For i) $a = 1$ the solution $A_n = L_n(-1)$ is recovered. When ii) $a < 1$, $b_n \to 1/(1-a)$ for large n and $A_n = n$. When iii) $a > 1$, $b_n \sim a^n/(a-1)$ as $n \to \infty$ and hence

$$A_n \sim \left[1 + \sqrt{1 - \frac{1}{a}}\right]^n . \tag{27}$$

We find that the average sequence growth is linear for $a < 1$ when $P_n(q)$ is dominated by $q \sim 1$, exponential for $a > 1$, when $P_n(q)$ is dominated by $q \sim n$ and intermediate for $a = 1$.

2.4 Non-Random $P(q)$

Next we look at the non-random sequence

$$x_n = x_{n-1} + x_{\beta(n-1)} \tag{28}$$

where $0 \leq \beta \leq 1$ is fixed and the index in the last term on the right hand side is rounded to the nearest integer. If $\beta = 0$, the sequence grows linearly, with $x_n = n$ and for $\beta = 1$, the growth of the sequence is exponential, with $x_n = 2^n$. For $0 < \beta < 1$ we can solve (28) for large n by taking the continuum limit of continuous n to give

$$\frac{\mathrm{d}x}{\mathrm{d}n} = x(\beta n) . \tag{29}$$

By substituting a power series solution for $x(n)$ into (29) and solving for the coefficients, we obtain

$$x(n) = \sum_{r=0}^{\infty} \frac{n^r \beta^{\frac{1}{2}r(r-1)}}{r!} . \tag{30}$$

For large n this summation is dominated by the term $r \sim \log(n)/\log(1/\beta)$ and evaluating the summation for large n yields

$$x_n \sim \exp\left[\frac{(\log n)^2}{2\log(\frac{1}{\beta})}\right] . \tag{31}$$

Hence for $0 < \beta < 1$ an intermediate type of growth occurs.

3 Random Sequences with Multiplicative Constants

In this section we investigate the effect of introducing multiplicative constants to the generating equations of the random sequences studied in [4]

Table 1. Summary of results for Sect. 2. Here, ζ is the Riemann zeta function [12] and c_α and k_γ are coefficients for the Laguerre polynomials $L_n{}^{(-\alpha)}(-(\alpha+1))$ and $L_n(-\gamma)$ respectively, [13]

$P_n(q)$		A_n
$(q+1)^\alpha$	$\alpha < -2$	n
	$\alpha = -2$	$n^{\frac{1}{2}+\frac{1}{\sqrt{\zeta(2)}}}$
	$-2 < \alpha < -1$	$\frac{1}{n^{\frac{\alpha}{4}}} \exp\left[\frac{2}{\alpha+2} \frac{n^{1+\alpha/2}}{\sqrt{\zeta(-\alpha)}}\right]$
	$\alpha = -1$	$\frac{1}{\sqrt{n}\log n} \exp\left[2\sqrt{\frac{n}{\log n}}\right]$
	$\alpha > -1$	$c_\alpha \frac{1}{n^{\frac{2\alpha+1}{4}}} \exp[2\sqrt{(\alpha+1)n}]$
a^q	$\alpha < 1$	n
	$\alpha = 1$	$k_\gamma n^{-1/4} \exp(2\sqrt{\gamma n})$
	$\alpha > 1$	$\left[1+\sqrt{1-\frac{1}{a}}\right]^n$
$\delta_{\beta(n-1)}$	$\beta = 0$	n
	$0 < \beta < 1$	$\exp\left[\frac{(\log n)^2}{2\log(\frac{1}{\beta})}\right]$
	$\beta = 1$	2^n

and also introduce a new random sequence as an extension of these studies. We consider the sequences generated by $x_n = \gamma x_p + \delta x_q$, $x_n = \alpha x_{n-1} + \gamma x_q$ and $x_n = \alpha x_{n-1} + \beta x_{n-2} + \gamma x_q$ in turn.

3.1 Sequences Generated by Randomly Chosen Previous Elements

Introducing multiplicative constants to the generating equation $x_n = x_p + x_q$ studied in [4] gives

$$x_n = \gamma x_p + \delta x_q \tag{32}$$

where γ and δ are constants and p and q are chosen randomly between 0 and $n-1$. The average value of x_n, A_n, can be shown to obey

$$A_n = \frac{\gamma+\delta}{n} \sum_{r=0}^{n-1} A_r \tag{33}$$

and can be simplified to give the recursion relation

$$A_n = \frac{n+\gamma+\delta-1}{n} A_{n-1} \,. \tag{34}$$

For large n, the solution to this equation is $A_n \sim n^{\gamma+\delta-1}$. This is the result obtained in [14] for $\delta = 0$. It is a simple matter to show that adding extra random terms with multiplicative constants to (32) simply multiplies this result by extra constant powers of n. Following the method of [9], in Sect. 2, we next introduce the averages

$$V_n = \langle x_n^2 \rangle \qquad \text{and} \qquad M_n = \sum_{r=0}^{n-1} < x_n x_r > \tag{35}$$

to give the two coupled iterations

$$\begin{aligned}(n+1)^2 V_{n+1} - [(\delta^2+\gamma^2+2n)n + (\delta+\gamma)^2]V_n \\ +[(\delta^2+\gamma^2)n - (\delta-\gamma)^2 + (n-1)^2]V_{n-1} = 4\alpha\gamma(M_n - M_{n-1})\end{aligned} \tag{36}$$

and

$$(n+1)M_{n+1} - (2(\alpha+\gamma)+n)M_n = (\alpha+\gamma)V_n \,. \tag{37}$$

Since we are only interested in the large n behaviour of the sequences, we take the continuum limit of the iterations for continuous n to give the second order differential equations

$$\begin{aligned}(n+1)^2 V''(n) + [(4-\delta^2-\gamma^2)n+ \\ (\delta-\gamma)^2]V'(n) + 2(1-\delta^2-\gamma^2)V(n) = 4\delta\gamma M'(n)\end{aligned} \tag{38}$$

and

$$(n+1)M''(n) + (n+1)M'(n) + (1-2\alpha-2\gamma)M(n) = (\alpha+\gamma)V(n)\,. \quad (39)$$

In the limit $n \to \infty$, the solutions are $V(n) \sim n^{\alpha^2+\gamma^2-1}$ and $M(n) \sim n^{\alpha^2+\gamma^2-1}$ for $(\alpha-1)^2 + (\gamma-1)^2 = 1$ in the region $[\ \alpha \geq 0,\ \gamma \geq 0$ and $\alpha+\gamma \geq 1\ ]$. This was the result obtained in [14] for $\alpha = 2$ and $\gamma = 0$. For $1 < (\alpha-1)^2 + (\gamma-1)^2 < 2$ in the region $[\ \alpha \geq 0,\ \gamma \geq 0$ and $\alpha+\gamma \geq 1\ ]$, the solutions are $V(n) \sim n^{\alpha^2+\gamma^2-1}$ and $M(n) \sim n^{2\alpha+2\gamma-1}$, and for $(\alpha-1)^2 + (\gamma-1)^2 = 2$ in the region $[\ \alpha \geq 0,\ \gamma \geq 0$ and $\alpha+\gamma \geq 1$ $]$, the solutions are $V(n) \sim n^{2(\alpha+\gamma-1)}$ and $M(n) \sim n^{2\alpha+2\gamma-1}$. Thus for $1 \leq (\alpha-1)^2+(\gamma-1)^2 \leq 2$ in the region $[\alpha \geq 0,\ \gamma \geq 0$ and $\alpha+\gamma \geq 1]$ we were able to find the average growth of the second moment, $V_n = \langle {x_n}^2 \rangle$, and show that in this region it always diverges faster than the mean, $A_n =< x_n >$.

3.2 Sequences Generated by One Randomly Chosen and One Non-Random Previous Element

Adding multiplicative constants to $x_n = x_{n-1} + x_q$ [4] gives the more general equation

$$x_n = \alpha x_{n-1} + \gamma x_q\,. \quad (40)$$

A_n is then given by

$$A_n = \alpha A_{n-1} + \frac{\gamma}{n}\sum_{q=0}^{n-1} A_q\,. \quad (41)$$

Following the method that was used in Sect. 2 to obtain $V(n)$ and $M(n)$, we obtain the second order differential equation

$$(n+1)A''(n) + [(1-\alpha)n+1]A'(n) + (1-\alpha-\gamma)A(n) = 0 \quad (42)$$

for large n in the limit of continuous n. We find that the form of the solution to this equation is dependent upon the multiplicative constant α, with $\alpha = 1$ a special case. In particular, there are different solutions for i) $\alpha > 1$, ii) $\alpha = 1$ and iii) $\alpha > 1$. We consider these in turn.

i) $\alpha > 1$

For $\alpha > 1$, the solution is exponential, with

$$A(n) \sim \exp\left[(\alpha-1)n\right]. \quad (43)$$

In this region the growth of the average sequence is dominated by the term x_{n-1} in the sequence generating equation so that the sequence growth is exponentially fast.

ii) $\alpha = 1$

For $\alpha = 1$, there is intermediate growth of the average n^{th} element, given by

$$A_n \sim \exp\left[2\sqrt{\gamma n}\right] \tag{44}$$

as found in [4] and [9]. This 'special' case corresponds to the first backward difference equation with an added random term. Here the sequences are not so strongly dominated by the x_{n-1} term in the generating equation and the growth is slower than exponential.

iii) $\alpha < 1$

When $\alpha < 1$, the average sequence growth is purely power-law, with

$$A_n \sim n^{\frac{\alpha+\gamma-1}{1-\alpha}} \, . \tag{45}$$

In this region the sequences are no longer dominated by the x_{n-1} term and the solution has the same form as the solution to (32).

To summarise, on average the sequences are found to grow exponentially fast when $\alpha > 1$, as in this region the growth is dominated by the x_{n-1} term in the generating equation. For $\alpha < 1$, the average sequence grows according to a power law as the x_{n-1} term no longer dominates. For $\alpha = 1$, an intermediate type of growth occurs when the generating equation corresponds to the first backward difference equation. Results for Sect. 3 are summarised in Table 2.

3.3 Sequences Generated by One Randomly Chosen and Two Non-Random Previous Elements

The next sequences we study are those generated by the equation

$$x_n = \alpha x_{n-1} + \beta x_{n-2} + \gamma x_q \, . \tag{46}$$

This equation looks like the original Fibonacci sequence but with an extra random term and multiplicative constants. Following the method of Sect. 3.2, we find that the average of the nth element obeys the third order differential equation

$$\begin{aligned} &-\beta n A'''(n) + [(\beta+1)n + 1]A''(n) \\ &+[(1-\alpha-\beta)n + \beta + 1]A'(n) - (1-\alpha-\beta-\gamma)A(n) = 0 \end{aligned} \tag{47}$$

in the limit of continuous n as $n \to \infty$. The resulting sequences are dependent on both of the constants α and β, with four cases; i) $[\alpha + \beta > 1, \beta \neq 0]$ and $[\alpha + \beta = 1, \alpha > 2]$, ii) $\alpha = 2$ and $\beta = -1$, iii) $[\alpha + \beta = 1, \alpha < 2]$ and iv) $\alpha + \beta < 1$.

i) $[\alpha+\beta>1,\beta\neq 0]$ and $[\alpha+\beta=1,\alpha>2]$

We find that there are two regions in phase space, $[\alpha+\beta>1,\beta\neq 0]$ and $[\alpha+\beta=1,\alpha>2]$ where the average sequences grow exponentially according to

$$A_n \sim \exp\left[\frac{\beta+1+\sqrt{(\beta+1)^2+4\beta(1-\alpha-\beta)}}{2\beta}n\right] . \qquad (48)$$

In these regions the sequence growth is dominated by the x_{n-1} and x_{n-2} terms. For $\beta=0$ and $\alpha>1$, the average sequence still grows exponentially, but according to (43).

iii) $\alpha=2$ and $\beta=-1$

For $\alpha=2$ and $\beta=-1$, there is the 'special' solution

$$A_n \sim \exp\left[\frac{3}{2}(\gamma n^2)^{1/3}\right] , \qquad (49)$$

when the generating equation corresponds to the second backward difference equation with an added random term. This is a new intermediate type of sequence growth resulting from the extra term, x_{n-2}, in the sequence generating equation.

iv) $[\alpha+\beta=1,\alpha<2]$

For $[\alpha+\beta=1,\alpha<2]$ there is another intermediate type of growth given by

$$A_n \sim \exp\left[2\sqrt{\frac{\gamma n}{\beta+1}}\right] , \qquad (50)$$

corresponding to the solution for (40) when $\alpha=1$.

v) $\alpha+\beta<1$

Finally for $\alpha+\beta<1$, the solution is again purely power-law, with

$$A_n \sim n^{\frac{\alpha+\beta+\gamma-1}{1-\alpha-\beta}} . \qquad (51)$$

The sequences are no longer dominated by the x_{n-1} and x_{n-2} terms in the generating equation and the solution takes the same form as that for (40) when $\alpha<1$.

In summary, we find that there are two regions of phase space, $[\alpha+\beta>1]$ and $[\alpha+\beta=1,\ \alpha>2]$, where the average sequence grows exponentially fast. In these regions the sequence growth is dominated by the x_{n-1} and x_{n-2} terms in the sequence generating equation. For $\alpha+\beta<1$, the sequence growth is no longer dominated by the x_{n-1} and x_{n-2} terms and the sequences grow according to a power law. For $[\alpha+\beta=1,\ \alpha<2]$, an intermediate type of growth occurs, and interestingly a new intermediate type of growth occurs for $[\alpha+\beta=1,\ \alpha=2]$, when the generating equation corresponds to the second backward difference equation. Results for Sect. 3 are summarised in Table 2.

Table 2. Summary of results for Sect. 3

x_n		A_n
$\gamma x_p + \delta x_q$		$n^{\gamma+\delta-1}$
$\alpha x_{n-1} + \gamma x_q$	$\alpha < 1$	$n^{\frac{\alpha+\gamma-1}{1-\alpha}}$
	$\alpha = 1$	$\exp\left[2\sqrt{\gamma n}\right]$
	$\alpha > 1$	$\exp[(\alpha-1)n]$
$\alpha x_{n-1} + \beta x_{n-2} + \gamma x_q$	$\alpha+\beta<1$	$n^{\frac{\alpha+\beta+\gamma-1}{1-\alpha-\beta}}$
	$[\alpha+\beta=1,\alpha<2]$	$\exp\left[2\sqrt{\frac{\gamma n}{\beta+1}}\right]$
	$[\alpha=2 \text{ and } \beta=-1]$	$\exp\left[\frac{3}{2}(\gamma n^2)^{1/3}\right]$
	$[\alpha+\beta>1,\beta\neq 0]$ and $[\alpha+\beta=1,\alpha>2]$	$\exp\left[\frac{\beta+1+\sqrt{(\beta+1)^2+4\beta(1-\alpha-\beta)}}{2\beta}n\right]$

4 Discussion

We have generalised previously studied random sequences in two different ways to study the effect on their behaviour. In Sect. 2 we considered sequences of the form $x_n = x_{n-1} + \gamma x_q$, where $q = 0, 1, 2, ..., n-1$ is chosen randomly from the probability distribution $P_n(q)$. We found the exact solution for the mean of the sequence $< x_n >= A_n$ when $P_n(q) = 1/n$ and showed that the average of the second moment $\langle x_n^2 \rangle = V_n$ diverges faster than the mean. We also considered power-law, exponential and non-random forms of $P_n(q)$ and found that the sequences can grow exponentially, linearly, or with intermediate growth. The sequences were found to exhibit exponential growth when $P_n(q)$ is dominated by $q \sim n$ and linear growth when $q \sim 1$. Between these regimes the sequences exhibit an intermediate type of growth. The boundaries of the different types of behaviour are summarised in Table 1.

In Sect. 3, we investigated the effect of introducing multiplicative constants to the individual terms in the generating equations of both new and previously studied random sequences. We obtained analytical results for the mean of the random sequences generated by (i) $x_n = \gamma x_p + \delta x_q$, (ii) $x_n = \alpha x_{n-1} + \gamma x_q$ and (iii) $x_n = \alpha x_{n-1} + \beta x_{n-2} + \gamma x_q$, where p and q are chosen randomly between 0 and $n-1$, and α, β, γ and δ are constants, with $\gamma > 0$ and $\delta > 0$. We found that the solutions depend on all the multiplicative constants, while the form of the solution depends on α and β for $\gamma > 0$ and $\delta > 0$. For (i), each element in the sequence is made up from randomly chosen previous elements and the average sequence grows according to a power law. We were also able to find the average second moment, $V_n = \langle x_n^2 \rangle$, for $1 \leq (\alpha-1)^2 + (\gamma-1)^2 \leq 2$ in the region [$\alpha \geq 0, \gamma \geq 0$ and $\alpha + \gamma \geq 1$] for these sequences and to show that in this region, $V_n = \langle x_n^2 \rangle$ always diverges faster than $A_n =< x_n >$. For (ii) and (iii), we found that the type of growth that the average sequences exhibit is highly sensitive to the constant multiplicative parameters α and β. In (ii), when $\alpha > 1$, the next element in the sequence is dominated by the most previous elements and the average sequence grows exponentially. For $\alpha < 1$, the elements in the sequences are no longer dominated by the most previous elements and the average sequence grows according to a power law, as with (i). There is a transition at $\alpha = 1$, when the generating equation corresponds to the first backward difference equation with an added random term. In this case an intermediate type of growth occurs and we find that the average sequence grows as a stretched exponential. For (iii), a similar behaviour occurs, where in this case for $\alpha + \beta > 1$, the average sequence grows exponentially and for $\alpha + \beta < 1$, the average sequence grows according to a power law. However, for (iii), the intermediate behaviour becomes more complex, with three types of behaviour emerging for $\alpha + \beta = 1$. For $\alpha + \beta = 1$, $\alpha > 2$, the average sequence continues to grow exponentially. For $\alpha + \beta = 1$, $\alpha < 2$ the average sequence grows as a stretched exponential, but for $\alpha = 2$, $\beta = -1$, another intermediate type of behaviour is observed when the generating equation corresponds to the second backward difference equation with an added random term. This suggests that if we add more non-random previous elements with multiplicative constants to the generating equations, we obtain extra dimensions in phase space for the solutions to the average sequence growth. Also, new transitions are observed when the generating equations correspond to the backward difference equations.

Although these sequences are fairly straightforward generalisations of those previously studied, they are found to display a wide range of new and interesting behaviour with a remarkably diverse phase space.

Acknowledgments

KA would like to thank the EPSRC for their financial support.

References

1. H. Furstenberg, H. Kesten: Ann. Math. Stat. **31**, 457 (1960)
2. D. Viswanath: Math. Comput. **69**, 1131 (2000)
3. M. Embree, L.N. Trefethen: Proc. Roy. Soc. A **455**, 24 71 (1999)
4. E. Ben-Naim, P.L. Krapivsky: J. Phys. A **35**, L557 (2002)
5. A. Einstein, R. Furth, A.D. Cowper: *Investigations on the Theory of the Brownian Movement*, (Dover Publications Inc., New York 1956)
6. S.A. Kauffman: *The Origins of Order: Self-Organization and Selection in Evolution*, (Oxford University Press, Oxford 1993)
7. L. Bachelier: *Théorie de la Spéculation*, (Gauthier-Villars, Paris 1900)
8. J.M. Luck: *Systèmes Désordonnés Unidimensionnels*, (Alea-Saclay, 1992)
9. I. Krasikov, G.J. Rodgers, C.E. Tripp: J. Phys. A **37**, 2365 (2004)
10. M. Abramowitz, I.A. Stegun: *Handbook of Mathematical Functions*, (Dover Publications Inc., New York 1975)
11. E.J. Hinch: *Perturbation Methods*, (Cambridge Uiversity Press, Cambridge 1991)
12. I.S. Gradshteyn, I.M. Ryzhik: *Table of Integrals, Series and Products*, (Academic Press, 1980)
13. G. Szego: *Orthogonal Polynomials*, (American Mathematical Society, 1939)
14. E. Ben-Naim, P.L. Krapivsky: J. Phys. A **37**, 5949 (2004)

Part II

Life Relevant Physics

Logistic Population Growth and Beyond: The Influence of Advection and Nonlocal Effects

Emilio Hernández-García and Cristóbal López

Instituto Mediterráneo de Estudios Avanzados IMEDEA* (CSIC-UIB), Campus de la Universitat de les Illes Balears, E-07122 Palma de Mallorca, Spain
emilio@imedea.uib.es
clopez@imedea.uib.es

1 Introduction

By introducing the logistic equation in the context of demographic modelling [1], J.-F. Verhulst made seminal contributions to at least two important fields of research: On the one hand, to the quantitative approach to Population Dynamics, a subject in which substantial development beyond Verhulst's ideas only occurred 100 years after him. On the other, to the basics of Nonlinear Science through a detailed study of the predictions of a nonlinear model and its comparison with experimental data.

The dynamics of biological populations in aquatic environments [2–4] is an excellent framework to see recent developments in which these disciplines work together. Growth, limitation, competition, predation and all the other kinds of biological interactions appear in this context, and nonlinear processes are relevant both in the biological dynamics and in the motion of the turbulent fluid in which the population lives.

In this contribution we present two examples in the above field. In both cases a prominent role is played by the logistic growth process (i.e. population growth limited by finite resources), but other ingredients are also included that strongly change the phenomenology. In Sect. 1, a phytoplankton population experiencing logistic growth is studied, but in interaction with zooplankton predators that maintain it in a state below the carrying capacity of the supporting medium. In the appropriate parameter regime the system behaves in an excitable way, with perturbations inducing large excitation-deexcitation cycles of the phytoplankton population. The excitation cycles become strongly affected by the presence of chaotic motion of the fluid containing the populations.

In Sect. 2, an individual based model of interacting organisms is presented, for which logistic growth is again the main ingredient. Reproduction of a given individual is limited by the presence of others in a neighbourhood of finite size. This *nonlocal* character of the interaction is enough to produce an instability of the basic state of particles homogenously distributed, and clustering of

*http://www.imedea.uib.es/PhysDept

the individuals occurs, which form groups arranged in an hexagonal lattice (when the population lives in a two-dimensional space).

2 Plankton Dynamics Driven by an Environmental Open Flow

Plankton is the generic name given to a large number of aquatic organisms, most of them microscopic, living in the oceans, lakes or rivers, and characterized by the fact that they are transported by water currents in an approximately passive manner [2]. A major distinction in this group should be made between *phytoplankton*, the organisms with photosynthetic capabilities, that can grow from nutrients in the water and light, and *zooplankton*, the small predators that consume them. This distinction leads to the simplest models of aquatic population dynamics, which take into account just these two trophic levels.

Here we consider the phytoplankton-zooplankton competition model proposed in [5]. Phytoplankton growths logistically, and in addition it is limited by the zooplankton grazing, which itself dies grazed by upper trophic levels not explicitly modelled. The fundamental feature of the model is its excitable behaviour. In just a few words, this means that *activator* and *inhibitor* variables can be identified. The *activator* (phytoplankton in our case) displays some kind of autocatalytic growth behaviour, but the presence of the *inhibitor* (zooplankton) controls it so that the dynamical system has a stable fixed point of low phytoplankton population as the unique global attractor. The essence of the excitability phenomenon is the presence of a threshold, such that if the system is perturbed above it, variables reach the stable fixed point only after a large excursion in phase space in which phytoplankton population grows to values close to the carrying capacity of the medium. This behaviour usually appears when the activator has a temporal response much faster than the inhibitor, which then takes some time before stopping the growth of the activator. Observed *plankton bloom* phenomena have been interpreted in this dynamical context [4, 5].

Explicitly, the dynamics of the space- and time-dependent phytoplankton, $P = P(\boldsymbol{x}, t)$, and zooplankton, $Z = Z(\boldsymbol{x}, t)$, concentrations is ruled by

$$\frac{\partial}{\partial t}P + \boldsymbol{v}\cdot\nabla P - D\nabla^2 P = r\left[\beta P\left(1-\frac{P}{K}\right) - f\left(P\right)Z\right]$$
$$\frac{\partial}{\partial t}Z + \boldsymbol{v}\cdot\nabla Z - D\nabla^2 Z = r\epsilon\left[f\left(P\right)Z - \omega Z\right] \ . \tag{1}$$

The left-hand-side terms represent the transport processes: both species are advected by the same fluid flow characterized by the velocity field $\boldsymbol{v} = \boldsymbol{v}(\boldsymbol{x}, t)$, that we assume to be incompressible and not altered by the back-influence of the biological concentration fields. We choose $\boldsymbol{v}$ to be two-dimensional to emphasize the role of horizontal transport [3], and defined on a two-dimensional square box domain $\Omega = [0, 2L] \times [-L, L]$ with Cartesian coordinates $\boldsymbol{x} = (x, y)$. The diffusion operator $D\nabla^2$, assumed to be the same for both species, models small-scale complex turbulent motions not explicitly included in $\boldsymbol{v}$. The right-hand-side contains the biological interactions terms: r controls the ratio of the transport time scales to the biological activity time scales, ϵ sets the ratio of phytoplankton- to the much larger zooplankton-growth time scale, K is the phytoplankton carrying capacity, β the phytoplankton growth rate, ω a linear zooplankton mortality, and

$$f(P) = \frac{P^2}{P_0^2 + P^2} \tag{2}$$

is a Hollings type III response function, describing the zooplankton predation on phytoplankton. As demonstrated in [5], is the functional form in (2) the responsible for the excitable character.

Following [5] we take non-dimensional units such that $\beta = 0.43$, $K = 1$, $\epsilon = 0.01$, $P_0 = 0.053$, and $\omega = 0.34$. This corresponds to phytoplankton doubling times of the order of days, and zooplankton time scales in the range of months. Biological concentrations have been scaled so that the phytoplankton carrying capacity (of the order of 100 µg of Nitrogen equivalent per liter) is the unity. For these parameter values the biological dynamical system is in the excitable regime. Excitable behaviour generally appears for ϵ small enough, which biologically means that, as already stated, time scales for phytoplankton growth are much shorter than for zooplankton. This is the biologically relevant case. We study the influence of transport by varying its relative strength via the parameter r. D is fixed to 10^{-5} and $L = 9$, which means that the diffusive spatial scale corresponding to the phytoplankton doubling time is between three and four orders of magnitude smaller than system size.

Since the velocity field $\boldsymbol{v} = (v_x, v_y)$ is two-dimensional and incompressible it can be written in terms of a stream function $\Psi(x, y, t)$:

$$\begin{aligned} v_x &= \frac{\partial \Psi}{\partial y} \\ v_y &= -\frac{\partial \Psi}{\partial x} \; . \end{aligned} \tag{3}$$

We consider the following stream function [6]:

$$\Psi = \Psi_0 \tanh\left(\frac{y}{d}\right) + \mu \exp\left(-\frac{(x-L)^2 + y^2}{2\sigma^2}\right) \cos\left(k(y - vt)\right) \; . \tag{4}$$

It represents an oceanic jet perturbed by a localized wave-like feature, trapped by topography or some geographical accident. The first term is the main jet, of width d, flowing towards the positive x direction with maximum velocity

Ψ_0/d at its centre. The wave-like perturbation, of strength μ, is represented by the second term. It is localized in a region of size σ around the point $(x, y) = (L, 0)$, and the wavenumber and phase velocity (directed towards the positive y direction) are k and v, respectively. The complete velocity field is time-periodic with period $2\pi/kv$.

It is very important to note that the region Ω is open with respect to this flow, so that we have the situation of chaotic scattering [7]: particles enter Ω from the left, following essentially straight trajectories, experience the irregular motion called *transient chaos* [7] when reaching the wave region (which in consequence becomes a *mixing* region), and finally they leave the system. For μ large enough, recirculation in the mixing region gives birth to a chaotic saddle in Ω. The chaotic saddle is the (fractal) subset of the mixing region where particles are trapped forever. It is formed by an infinite number of bounded hyperbolic orbits in the mixing region. The stable manifold of the saddle contains orbits coming from the inflow region but never escaping from the mixing zone. Concerning the unstable manifold, if a droplet of dye (or any other passive tracer) is injected into the mixing region, most of it will be advected downstream in a short time. But part of the dye will remain close to the chaotic saddle for very long times, and continuously ejected along its unstable manifold. In this way passive tracers such as dye traces out the unstable manifold of the chaotic saddle, giving rise to the fractal patterns characteristic to open flows. We next study how these flow structures affect the plankton dynamics given by the right-hand-side of (1). Pictures of the stream function and of the chaotic saddle can be seen in [6].

Equations (1) are solved by a semilagrangian method. The fixed point representing stable phytoplankton-zooplankton coexistence in the absence of flow and diffusion is given by $P = P_e$ and $Z = Z_e$, with $P_e = P_0\sqrt{\sigma/(1-\sigma)} = 0.03827$ and $Z_e = \beta(1 - P_e/K)(P_0^2 + P_e^2)/P_e = 0.04603$. We choose these values to be imposed as Dirichlet boundary conditions on the boundary of Ω. In this way fluid particles enter in the system with a plankton content corresponding to the equilibrium concentrations, which is a rather natural condition from the biological point of view. During an excitation phase, the values of phytoplankton concentration rise to $P \approx 0.8 - 0.9$

Since (P_e, Z_e) is a stable equilibrium point, dynamics will be trivial without an initial seed to trigger the excitation dynamics. Our initial condition is a localized patch of high phytoplankton concentration close to the left part of Ω: $P(\boldsymbol{x}, t = 0) = P_e + Q\exp[-((x - x_0)^2 + y^2)/l^2]$, $Z(\boldsymbol{x}, t = 0) = Z_e$. We take $Q = 0.5$, $x_0 = 0.3L$, and $l = 0.11L$. The jet transports the patch towards the scattering or mixing region, where interesting dynamics occurs. The flow parameters are $d = 1$, $\Psi_0 = 2$, $\sigma = 2$, $k = 1$, and $v = 1$, giving a flow period $T = 2\pi/kv = 2\pi$.

We now consider $\mu = 3$ (a chaotic saddle is present in the system above $\mu \approx 2$). For small r, the biological dynamics is slow compared to the time scales of stirring by the flow. The phytoplankton patch is strongly deformed when reaching the scattering region. Plankton is stretched into long and thin filaments that become rapidly diluted into the surrounding unexcited fluid by the effect of diffusion. Thus excitation is destroyed by the fast stirring and by diffusion. Increasing r, i.e. by making the biological dynamics faster or the flow slower, a dramatic change occurs. The transition to the new regime occurs around $r \approx 1$. Plankton is again stretched in filaments but the width stabilizes and the excitation becomes distributed in the system, without leaving it (it remains forever) and oscillating in shape following the period of the flow. Some features of the distributions of both phytoplankton and zooplankton (see Fig. 1) seem to mimic the shape of the unstable manifold of the chaotic saddle, so that in these zones we can say that plankton is basically covering it with a finite width. Summing up, the relevant result we have shown is the following: *transient* chaos, characteristic of open flows, plus *transient* excitation, characteristic of excitable systems, give rise to a *permanent* pattern of high biological activity (excitation).

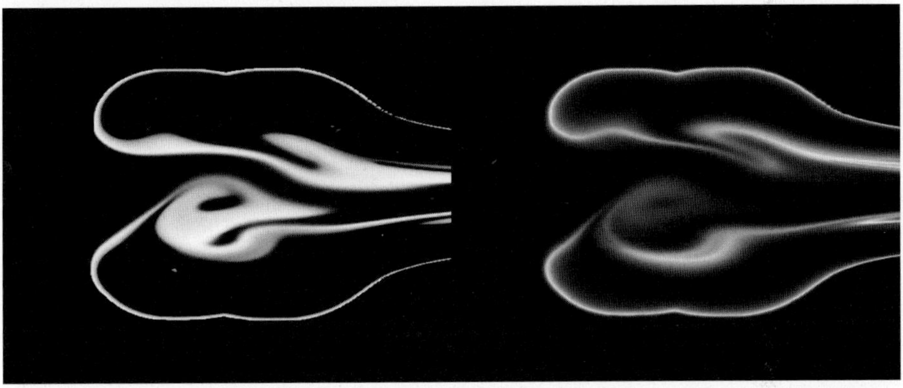

Fig. 1. Distribution of phytoplankton (left) and zooplankton (right) at time $t = 100$ and parameters $r = 10$ and $\mu = 3$. Dark grey corresponds to low concentration and lighter grey to higher concentration. A state of permanent excitation is sustained in the region close to the chaotic saddle and its unstable manifold (being the domain far from this region in the unexcited equilibrium state). The shape of the distributions changes in time with the period of the flow

The explanation for the observed behaviour can be elaborated along the lines of previous works [8, 9] as follows: The tendency of the chaotic flow to stretch fluid elements into long and thin filaments competes with the effect of diffusion and biological growth, which tends to expand excited regions, so that a compensation can be achieved in some parameter range. Above a given biological growth rate, steady filament solutions appear via a saddle-node bifurcation. This can be explicitly shown in simplified models capturing some of the features of the full system (1). When this steady solution does not exist, initial perturbations decay as in the usual excitation-deexcitation cycle, so that excitation disappears at long times. At sufficiently large biological growth rate, however, the steady filament solution exists, is stable, and the initial perturbation can be locally attracted by it. Chaotic flow deforms the simple filament solution obtained under simplified assumptions [8,9], but the results in [8] indicate that it still provides a useful description of the process. Chaotic stretching and folding of the excited filament in a closed system ends up when it fills the whole domain, after which an homogeneous deexcitation finishes the excitation cycle. In an open system, however, the continuous outflow of excited material inhibits the filling of the full domain, so that distributions related to relatively simple filament steady solutions can persist permanently.

3 Nonlocal Logistic Growth

In this section we discuss the interacting particle model with non-local interactions introduced in [10]. It is just a simple modification of the Brownian Bug model of [11] where there is an ensemble of diffusing particles (the *bugs*), each of them dying or duplicating with given probabilities per unit of time. As we shall see below the modification consists in the introduction of an interaction among the particles so that the birth rates for any of them diminish in regions of high particle density. This is precisely Verhulst logistic mechanism, here implemented in a particle model instead that in the original Verhulst equation for the global population.

The microscopic rules are enumerated in the following. Let $N(t)$ be the number of bugs in the system (a two-dimensional periodic box Ω of size $L \times L$; in all our computer simulations we shall take $L = 1$):

1. There is an initial population of $N(t = 0) = N_0$ bugs or particles, randomly located.
2. One particle, j, is selected at random and it reproduces (i.e., it is transformed into two particles) with a rate (probability per unit time) $\lambda(j)$ or dies with a rate $\beta(j)$. Both rates are not constant but depend on the number of particles surrounding the particle j. Explicitly we take:

$$\lambda(j) = \max\left(0, \lambda_0 - \frac{1}{N_s} N_R^j\right) , \tag{5}$$

and

$$\beta(j) = \max\left(0, \beta_0 - \frac{\alpha}{N_s} N_R^j\right) , \tag{6}$$

where λ_0 and β_0 are constants. N_R^j denotes the total number of particles which are at a distance smaller than R from particle j (excluding the particle j itself). R is thus a range of interaction, N_s is a saturation parameter, and α controls the asymmetry between its influence on death

and on reproduction (in the rest of this paper we take $\alpha = 0$ so that only birth rates are modified by the local density of particles). In the case of reproduction, the newborn particle is located at the same place as the parent particle. The process is repeated a number of times equal to $N(t)$, so that each particle is checked for birth or reproduction once in average.
3. Each particle moves in random direction a distance drawn from a Gaussian distribution of standard deviation σ (this models Brownian motion).
4. When advection is considered, the particles are transported by an external flow to be described later.
5. Time is incremented an amount $\tau = 1$, and the algorithm repeats.

Figure 2 shows typical spatial configurations observed at large times under this algorithm. When the maximum growth rate $\mu \equiv \lambda_0 - \beta_0$ is small ($\mu < \mu_c$), population dies at long times. Above this value of μ, an *active phase* with a persistent average number of individuals is attained. The nature of the spatial

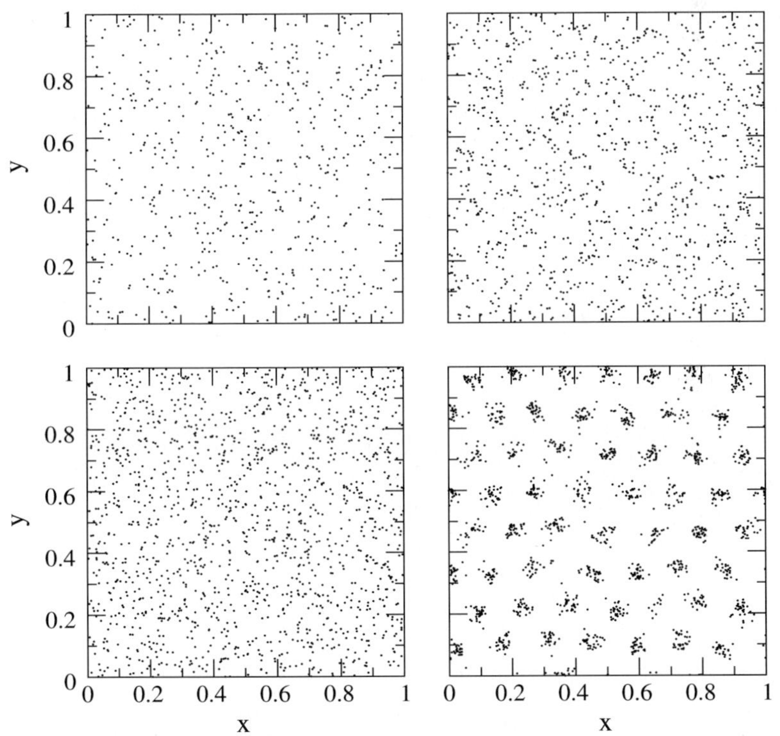

Fig. 2. Long-time spatial structures for the interacting particle model. Left column corresponds to two patterns with the same value of $D = 10^{-4}$, and two different values of $\mu = 0.5$ (*up*) and $\mu = 0.9$ (*bottom*). Right column corresponds to fixed $\mu = 0.7$, and $D = 10^{-4}$ (*upper*), and $D = 10^{-5}$ (*bottom*). In all the plots, $N_s = 50$ and $R = 0.1$

distribution in the active phase depends on the values of the parameters. For large enough value of the effective diffusion coefficient $D \equiv \sigma^2/\tau$, the spatial distribution of particles is homogeneous on average, whereas clear clustering occurs for small D. As in the Brownian Bug model [11], different clusters are coming from different families (i.e. each cluster is made of descendants of a different initial reproducing individual). But the most striking feature is that they *organize in a periodic pattern.* The periodicity of the pattern is of the order of R, the interaction range. In addition to decreasing D, this transition to a periodic organization occurs by increasing R and, for small enough D, by increasing μ.

Particle clustering seems to be a rather natural way to make compatible the high local growth at relatively large value of μ, with the reduction of this growth that a too crowded neighbourhood would imply: the empty space between the clusters acts as a buffer zone keeping the competition for resources less limiting than in a homogeneous distribution. We believe that this dispersion of the total population in small groups over a large spatial area is a general consequence of the logistic mechanism when applied to particle systems.

We try now to understand this pattern forming process. To this end we write down a mean-field-like description of the model, which completely neglects fluctuations, and check if the clustering instability appears there. The mean-field equation is written in terms of an *expected density* $\phi(\boldsymbol{x},t)$ as follows

$$\begin{aligned} \partial_t \phi(\boldsymbol{x},t) &= D\nabla^2\phi(\boldsymbol{x},t) + \\ (\lambda_0 - \beta_0)\phi(\boldsymbol{x},t) &- \frac{1}{N_s}\phi(\boldsymbol{x},t)\int_{|\boldsymbol{x}-\boldsymbol{r}|<R} \mathrm{d}\boldsymbol{r}\ \phi(\boldsymbol{r},t) \,. \end{aligned} \tag{7}$$

This expression can be understood as a nonlocal version of the Verhulst logistic equation, complemented also with the diffusion term arising from the Brownian motion of the bugs. With more generality, the nonlocal term may be written as $\phi(\boldsymbol{x},t)\int_A \mathrm{d}\boldsymbol{r} G(\boldsymbol{x}-\boldsymbol{r})\ \phi(\boldsymbol{r},t)$, where $A \subset \Omega$. Our model corresponds then to a kernel $G(\boldsymbol{x})$ given by

$$G(\boldsymbol{x}) = \begin{cases} 1 \text{ if } & |\boldsymbol{x}| \le R \\ 0 \text{ if } & |\boldsymbol{x}| \ge R \,. \end{cases} \tag{8}$$

Stationary homogenous solutions of equation (7) are the empty phase $\phi(\boldsymbol{x},t) = 0$, and the *active* or *survival* phase $\phi(\boldsymbol{x},t) = \phi_s = \mu N_s/\pi R^2$ (remember that $\mu = \lambda_0 - \beta_0$). For $\mu < 0$ the only stable solution is the absorbing one; the transition to the survival state is approached at $\mu = 0$, and this state is stable for a range of positive values of μ. At the deterministic level the transition is transcritical. We note that the critical value $\mu = 0$ is smaller than the one observed numerically in the particle model ($\mu_c > 0$), this deficiency of the mean-field description being a well known consequence of neglecting fluctuations.

We perform now a stability analysis of the ϕ_s solution by considering small harmonic perturbations around it, $\phi(\boldsymbol{x},t) = \phi_s + \delta\phi(\boldsymbol{x},t)$, with $\delta\phi(\boldsymbol{x},t) \propto \exp(\lambda t + i\mathbf{k}\cdot\boldsymbol{x})$. After simple calculations (see details in [10]) one arrives to the following dispersion relation

$$\lambda(K) = -DK^2 - \frac{2\mu}{KR} J_1(KR) , \tag{9}$$

where $K = |\mathbf{k}|$, and J_1 is the first-order Bessel function. It is clear that the relevant parameters in the problem are μ and D/R^2 (in fact the precise dimensionless combinations are $\mu\tau$ and $D\tau/R^2$, see [12]). The eigenvalue $\lambda(K)$ (which is in fact a function of KR, μ, and D/R^2) is real and can be positive for some values of the parameters. This is shown in Fig. 3 where we plot λ against K for fixed D/R^2 and different values of μ around the critical value $\mu_P = 185.192D/R^2$, which is the value of μ at which $\lambda(K)$ becomes positive [10]. Positive values of $\lambda(K)$ in a range of values of K imply instability of the homogeneous distribution against perturbations containing this range of wavenumbers (K around $K_c = 4.779/R$) and pattern formation with the corresponding wavelengths ($\approx 1.31R$).

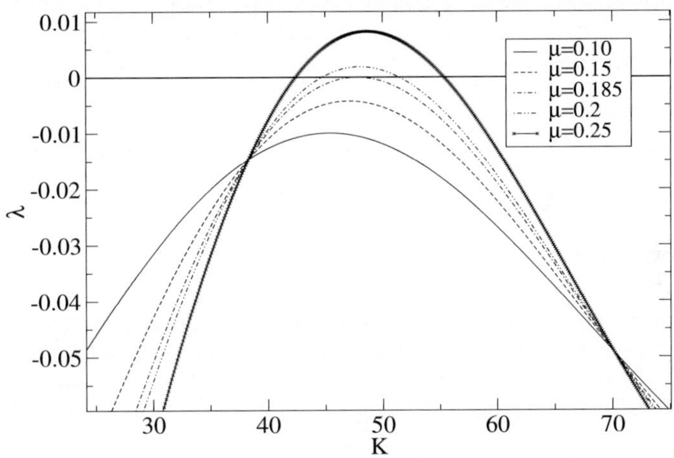

Fig. 3. Linear growth rate λ vs wavenumber K from (9) for different values of μ close to the onset of pattern formation. We take $R = 0.1$ and $D = 10^{-5}$ so that $\mu_P = 0.185$

The behaviour of the deterministic (7) is thus clear: for $\mu < 0$ the only stable solution is $\phi = 0$. Then there is an interval, $0 < \mu < \mu_P$, where one has the homogeneous density $\phi = \phi_s$, and for $\mu > \mu_P$ spatial patterns emerge. This last transition can also be crossed by decreasing D/R^2 at fixed $\mu > 0$. The details of this sequence of transitions and the critical parameter values do not coincide with the behaviour of the particle model, implying that fluctuations

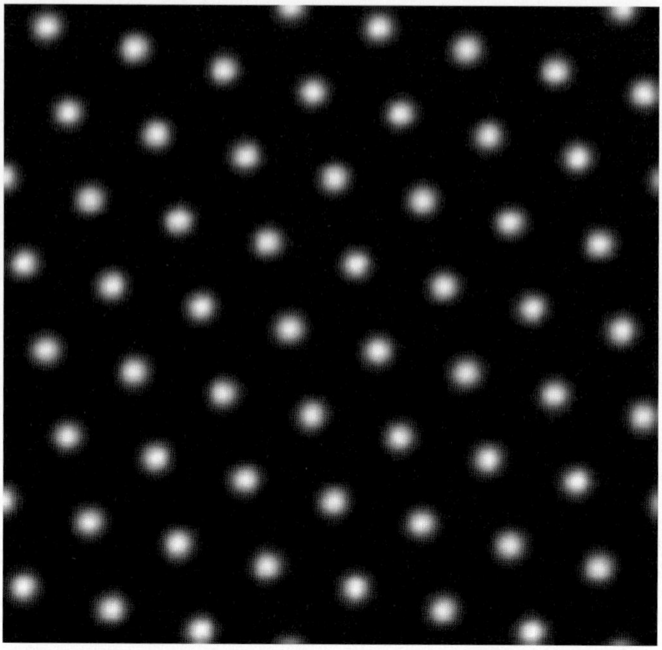

Fig. 4. Steady spatial pattern from the deterministic (7). $\mu = 0.70$, $R = 0.1$, $D = 10^{-5}$, and $N_s = 50$. Note the strong similarity with the pattern in the bottom-right plot in Fig. 2

are rather important. However, the presence of a pattern forming instability is well explained and even the selected pattern wavenumber ($K_c \approx 4.779/R$) is quantitatively reproduced by the mean-field approach [10]. Figure 4 shows a steady pattern of density which is the solution of (7) reached at long times. It is analogous to the one shown for the discrete model in the bottom-right panel of Fig. 2, confirming for the full nonlinear model (7) the behaviour identified from the linear stability analysis of the homogeneous solutions.

So far the only motion considered for the bugs has been Brownian motion. If they live in a turbulent aquatic medium, they will be also subjected to straining fields that will deform the clusters and, as in the previous Section, alter the population dynamics. We have implemented in the fourth step of the algorithm defining the particle model a simplified flow consisting in shear motions alternating in direction: if we denote by $(x_i(t), y_i(t))$ the coordinates of the particle i at time t, after one iteration of the map they become

$$x_i(t') = x_i(t) + A\cos(y_i(t)) \ , \tag{10}$$

$$y_i(t') = y_i(t) + A\cos(x_i(t')) \ , \tag{11}$$

where $t' = t+\tau$. The parameter A gives the strength of the flow and, depending of its value, particles can follow regular or chaotic trajectories. Next we

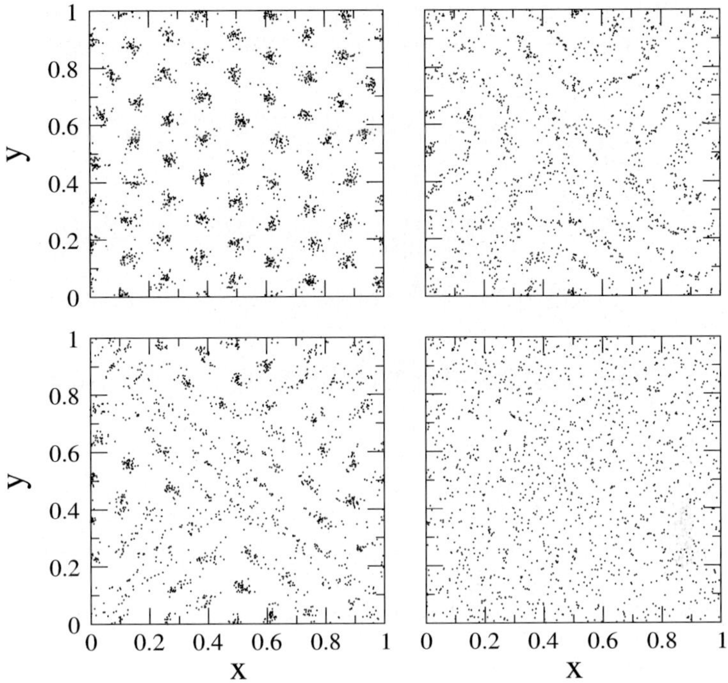

Fig. 5. Snapshots of the long-time spatial structure for the distribution of particles with an external flow. From *top to bottom* and *left to right*, $A = 0$, $A = 0.1$, $A = 0.5$, and $A = 1$. The other parameters: $\mu = 0.9$, $D = 5 \times 10^{-6}$, $N_s = 50$, and $R = 0.1$

analyze the behaviour of the model when the flow changes. In Fig. 5 we show instantaneous configurations of the particle distributions as they are being stirred by the flow at different values of A. Increasing A leads to increasingly chaotic trajectories and mixing behaviour in the flow. It is seen that the periodic array of clusters in the absence of flow becomes more filamental-like as the flow strength increases. The shape of the filamental structures reflects the known unstable and stable foliation of phase space for the map (11). Inhomogeneity persists for rather strong flow, but finally the distribution becomes homogenized. At this point, the particle distribution should be very close to Poissonian, with density given by the homogeneous solution of (7). This is indeed what is observed in Fig. 6 where we plot time evolution of the total number of particles. For large values of A the total number of particles, $N(t)$, fluctuates around the homogeneous deterministic solution value $\phi_s = \mu N_s/\pi R^2$. For smaller flow strength A the spatial structure in the neighbourhood of each particle becomes relevant, and the number of particles approaches the value in the absence of flow, corresponding to the pattern state of clustered particles. Thus, we see that the nonlocal interactions lead to

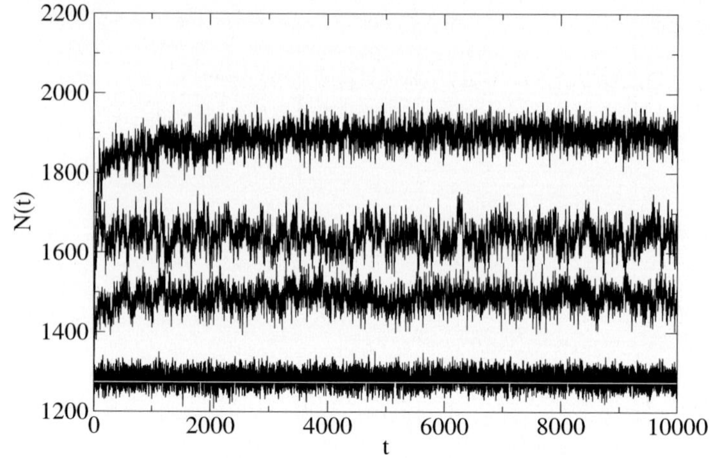

Fig. 6. $N(t)$ vs time for different values of the external flow strength, A. From *top to bottom*: $A = 0$, $A = 0.01$, $A = 0.05$, and, fluctuating around the value $\phi_s = \mu N_s/\pi R^2$ (*horizontal white line*), $A = 1$ and $A = 3$. The other parameters are $\mu = 0.8$, $N_s = 50$, $D = 10^{-5}$, and $R = 0.1$

a coupling between flow and population dynamics, mediated by the changes in local distribution geometry that the flow induces.

4 Summary

In this contribution we have presented results on two model systems coming from the context of population dynamics in aquatic flowing media. In both cases the logistic mechanism of P.-F. Verhulst is an essential ingredient, although the presence of nontrivial predation dynamics in one case, and of a finite range of interaction in the other lead to interesting new phenomena, namely excitability and pattern formation, respectively. In both model systems the consideration of fluid flow leading to chaotic trajectories (the process known as chaotic advection or Lagrangian turbulence [13]) has additional impact on the dynamics.

Acknowledgements

Financial support from MEC (Spain) and FEDER from grants CONOCE2 (FIS2004-00953) and IMAGEN (REN2001-0802-C02-01/MAR) is greatly acknowledged. C.L. is a *Ramón y Cajal* fellow.

References

1. P.-F. Verhulst: Corresp. Math. Phys. **10**, 113 (1838)
2. K.H. Mann, J.R.N. Lazier: *Dynamics of Marine Ecosystems. Biological-Physical Interactions in the Oceans* (Blackwell Scientific Publications, Boston 1991)
3. A.P. Martin: Progr. Oceanogr. **57**, 125 (2003)
4. E. Hernández-García, C. López, Z. Neufeld: Spatial Patterns in Chemically and Biologically Reacting Flows. In: *Chaos in Geophysical Flows*, ed by G. Bofetta, G. Lacorata, G. Visconti, A. Vulpiani (OTTO editore, Torino 2003) pp 35–61. Available from the publisher website: `http://www.otto.to.it/otto/index2_e.php`
5. J.E. Truscott, J Brindley: Bull. Math. Biol. **56**, 981 (1991)
6. E. Hernández-García, C. López: Ecological Complexity **1**, 253 (2004)
7. E. Ott, T. Tél: Chaos **3**, 417 (1993)
8. Z. Neufeld, C. López, E. Hernández-García, O. Piro: Phys. Rev. E **66**, 066208 (2002)
9. E. Hernández-García, C. López, Z. Neufeld: Physica A **327**, 59 (2003)
10. E. Hernández-García, C. López: Phys. Rev. E **70**, 016216 (2004)
11. W.R. Young, A.J. Roberts, G. Stuhne: Nature **412**, 328 (2001)
12. C. López, E. Hernández-García: Physica D, in press (2004)
13. H. Aref: Phys. Fluids **14**, 1315 (2002)

Predator-Prey Encounters Studied as Relative Particle Diffusion

J. Mann[1], S. Ott[2], H. L. Pécseli[3], and J. Trulsen[4]

[1] Risø National Laboratory, DK-4000 Roskilde, Denmark
jakob.mann@risoe.dk
[2] Risø National Laboratory, DK-4000 Roskilde, Denmark
soeren.ott@risoe.dk
[3] University of Oslo, Institute of Physics, Box 1048 Blindern, N-0316 Oslo, Norway
hans.pecseli@fys.uio.no
[4] University of Oslo, Institute of Theoretical Astrophysics, Box 1029 Blindern, N-0315 Oslo, Norway
jan.trulsen@astro.uio.no

Summary. The feasibility of an experimental method for investigations of the particle flux to an absorbing surface in turbulent flows is demonstrated in a Lagrangian as well as an Eulerian representation. A laboratory experiment is carried out, where an approximately homogeneous and isotropic turbulent flow is generated by two moving grids. The simultaneous trajectories of many small approximately neutrally buoyant polystyrene particles are followed in time. In a Lagrangian analysis, we select one of these as the centre of a "sphere of interception", and obtain estimates for the time variation of the statistical average of the inward particle flux through the surface of this moving sphere. The variation of the flux with the radius in the sphere of interception, as well as the variation with basic flow parameters is well described by a simple model, in particular for radii smaller than a characteristic length scale for the turbulence. Applications of the problem to, for instance, the question of the feeding rate of micro-organisms in turbulent marine environments are pointed out.

1 Introduction

Often the problem of turbulent diffusion in neutral turbulent flows is analysed in terms of an initial value problem [1, 2]. However, for many applications, a boundary value problem is more relevant. As such an example we here consider the turbulent particle flux to a perfectly absorbing spherical surface, which is a realistic physical model for many practical applications. This formulation of the problem serves, for instance, as a model for predator-prey encounters in turbulent waters, and seems to be the application of the problem that has received most attention recently [3, 4]. For small predators, fish larvae for instance [5], it can safely be assumed that their self-induced motion is small or negligible, and that they are passively convected by the local flow velocity, at least to a good approximation. Similarly, it can be assumed that their food (micro-zooplankton, for instance) is also passively convected by the same flow. The feeding process can be modelled by assuming that any individual prey entering a suitably defined "sphere of interception" is captured with certainty. The surface is thus "virtual" in the sense that it does not disturb the flow.

In turbulent waters, the prey flux to a passively convected predator is related to the problem of relative diffusion, but now considered as a boundary value problem, with the sphere of interception acting as a perfect absorber of prey. This is the standard model for this particular problem [6,7]. We use this as a terminology in the following, for simplicity and definiteness. The general interest in the problem arises essentially from the simple observation that the food concentration in the near region of a predator will rapidly be depleted, and without any self-induced motion a predator will be starving, unless the prey within its sphere of interception is replaced by turbulent mixing in the surrounding flow. Although the results presented in this work explicitly refer to spherical volumes, the scaling laws that are obtained will have a wider range of applications.

We propose and demonstrate the feasibility of an experimental method for a quantitative study of turbulent transport into an absorbing surface, and present results for varying parameters [8]. It is demonstrated that a relatively simple model equation is adequate for describing the basic features of our observations. The paper is organized as follows; in Sect. 2 we give a short summary of the experimental set-up, and the experimental conditions. In Sect. 3 we discuss experimental results for particle fluxes to an absorbing sphere where the centre is identified by a particle which is moving with the flow. Section 4 contains a discussion of a simple analytical model which gives results in fair agreement with observations. Finally, Sect. 5 contains our conclusions.

2 Experimental Set-up

The basic features of the present experiment are described elsewhere [9,10]. A short summary will suffice here. The turbulence is generated by the motion of two plastic grids, in the top and bottom of a tank with $320\times320\times450\,\mathrm{mm}^3$ inner dimensions, see Fig. 1 for a schematic illustration.

Typical Taylor micro-scale Reynolds numbers [11], $R_\lambda = \lambda^2/(\eta^2\sqrt{15})$, are ~ 100 for the present conditions, using the Taylor micro scale $\lambda = \sqrt{15\nu\sigma^2/\epsilon}$, where $\nu \simeq 0.89\,\mathrm{mm}^2/\mathrm{s}$ is the kinematic viscosity of the water, ϵ the specific energy dissipation rate, and σ^2 is the variance of one velocity component. The Kolmogorov length scale $\eta = (\nu^3/\epsilon)^{1/4}$ is less than $1/2\,\mathrm{mm}$ for the present conditions, while Kolmogorov time scales τ_η are in the range 0.05–0.12 s. The "micro-scale" η represents the length-scales, where the viscous effects become important. A characteristic Eulerian length scale, $\mathcal{L}_E$ as well as ϵ are determined by fitting a von Kármán type wavenumber spectrum [9,10] to the experimentally obtained data

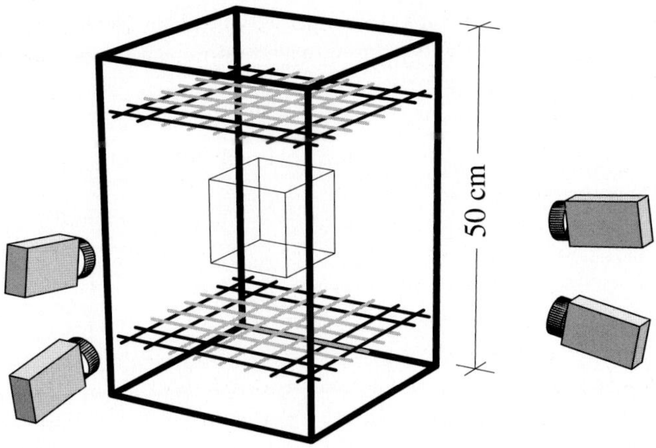

Fig. 1. Schematic illustration of the experimental set-up, showing the movable grids and the 4 video cameras. A restricted measuring volume of $140 \times 140 \times 120\,\mathrm{mm}^3$ is shown by thin lines

$$E(k) = \alpha \epsilon^{2/3} \mathcal{L}_E^{5/3} \frac{(\mathcal{L}_E k)^4}{[1 + (\mathcal{L}_E k)^2]^{17/6}} , \tag{1}$$

where α is the spectral Kolmogorov constant [12]. $\mathcal{L}_E$ is found to be in the range 20–25 mm. We can interpret $\mathcal{L}_E$ as the lower limit for separations between fixed frame detection points, where the velocities of fluid elements tend to become uncorrelated. As a working hypothesis we can assume that velocities are also statistically independent for separations larger than $\mathcal{L}_E$. An integral length scale can be defined by the integral of the parallel velocity component correlation function $R_{\|}(r)$ as $\mathcal{L}_{\text{int}} = \int_0^\infty R_{\|}(r)\mathrm{d}r$. A summary of parameters for 8 different conditions used in the present work is given in Table 1.

The motions of small polystyrene particles of size $a = 0.5 - 0.6$ mm are followed with 4 video cameras, and the simultaneous positions of typically 500–1000 particles recorded at time intervals of 1/25 s. The size of the effective measuring volume is approximately $140 \times 140 \times 120\,\mathrm{mm}^3$. It is ensured that the particles used in the experiment are approximately neutrally buoyant [9]. By a tracking procedure it is then possible to link the positions of particles [9], and thus to follow their individual motions in 3 spatial dimensions. In particular also their time varying velocity can be deduced. An illustrative sample trajectory is shown in Fig. 2. The figure shows a series of small spheres, centred at the particle positions, at individual sampling times. Since the time sequence used here is one of the longer ones obtained, the superposition of the spheres at subsequent sampling times gives rise to an appearance like a grey "band".

Experiments are carried out for different intensities of the turbulent velocity fluctuations, $\langle u^2 \rangle$. With the polystyrene particles acting as markers for the

Table 1. Summary of the parameters derived from the second order structure function and the spectra obtained from it, based on measurements in the restricted volume shown in Fig. 1

$\alpha\epsilon^{2/3}$ ($mm^{4/3}/s^2$)	$\mathcal{L}_E$ (mm)	σ (mm/s)	$\mathcal{L}_{int}$ (mm)	ϵ (mm^2/s^3)	τ_η (s)	η (mm)	λ (mm)	R_λ
45	31	18	23	160	0.075	0.26	5.1	100
41	27	16	20	140	0.080	0.27	4.9	88
40	29	16	22	135	0.081	0.27	5.1	93
45	28	17	21	160	0.075	0.26	4.9	91
24	29	12	22	62	0.120	0.33	5.8	81
65	29	21	22	279	0.056	0.22	4.5	104
56	28	19	21	225	0.063	0.24	4.6	97
25	27	12	20	65	0.117	0.32	5.6	78

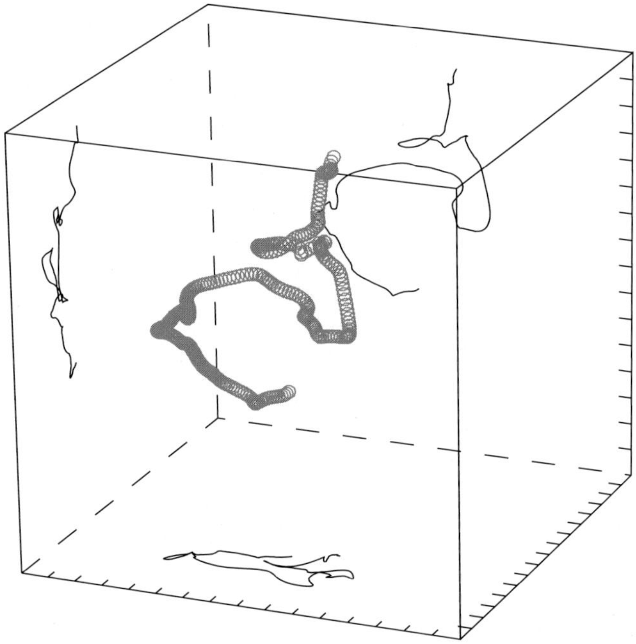

Fig. 2. Sample of particle trajectory obtained experimentally with 1/25 s time resolution. The small spheres give the particle position, and the continuous line a numerically interpolated particle trajectory, projected onto three of the bounding surfaces of the box. The distance between the tic-marks on axes is 10 mm. The spheres are here shown enlarged, for the sake of illustration

local flow velocities, experimental estimates can be obtained for the second order structure function, $\Psi_2(y) = \langle (u_i(\mathbf{r},t) - u_i(\mathbf{r}+\mathbf{y},t))^2 \rangle$ being independent of t for time stationary conditions. An example is shown in Fig. 3, including also a fit for small separations given by a dashed line, using the universal Kolmogorov $(\epsilon r)^{2/3}$ law. If we let the separation vector be along the y-axis, we have the longitudinal structure function $\Psi_{2\parallel}(y) \equiv \langle (u_y(0,t) - u_y(y,t))^2 \rangle$ given by the dotted line, with C_K being the Kolmogorov constant, related to the spectral constant α from (1) by $C_K \approx 1.315\,\alpha$ [12]. We note that C_K is known with some uncertainty, and a value of $C_K \approx 2.5$ can be justified as well as $C_K \approx 2.0$ [10,12]. The dash-dotted line in Fig. 3 gives the transverse structure function $\Psi_{2\perp}(y) \equiv \langle (u_x(0,t) - u_x(y,t))^2 \rangle$. By a general relation [11] we have

$$\Psi_{2\perp} = \frac{1}{2y}\frac{\mathrm{d}(y^2\Psi_{2\parallel})}{\mathrm{d}y} ,$$

for locally homogeneous and isotropic turbulence. With $\Psi_{2\parallel} \approx C_K(\epsilon y)^{2/3}$, we find $\Psi_{2\perp} \approx (4/3)\Psi_{2\parallel}$ in the universal subrange. The full line in Fig. 3 shows $\langle(\mathbf{u}(\mathbf{r},t) - \mathbf{u}(\mathbf{r}+\mathbf{y},t))^2\rangle = 2(\langle u^2\rangle - \langle \mathbf{u}(\mathbf{r},t)\cdot\mathbf{u}(\mathbf{r}+\mathbf{y},t)\rangle)$. The purpose of Fig. 3 is to demonstrate the existence of a universal range, and to indicate the range of its validity, here up to separations of the order of 20 – 25 mm.

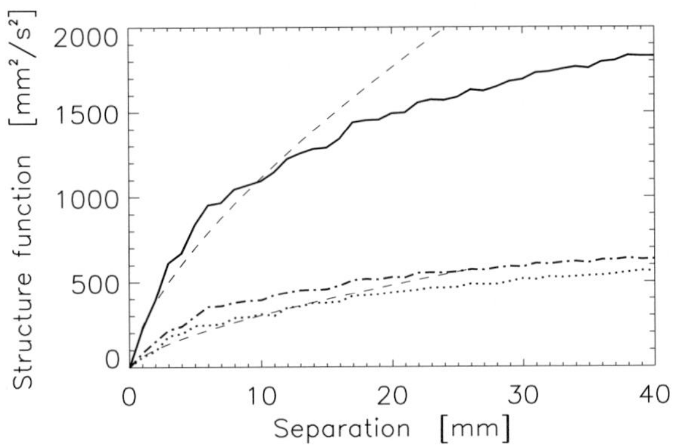

Fig. 3. Experimentally obtained second order structure function, as function of separation variable y. The heavy dashed line shows a $y^{2/3}$ fit. The *dotted line* refers to the longitudinal structure function, and the *dot-dashed line* to its transverse counterpart. The full line gives the structure function $\Psi_{2\parallel} + 2\Psi_{2\perp}$

The average distance between particles is much larger than their diameter, and particle interactions can be ignored. We estimate a Stokes number [13] as $St \equiv (2/9)(a/\mathcal{L}_E)^2 R_e$ with R_e being the Reynolds number based on $\sqrt{\langle u^2\rangle}$ and $\mathcal{L}_E$. For typical values [10] of $\mathcal{L}_E = 25$ mm and $R_e \approx 500$ we find $St \approx 0.05 \ll 1$. To the given accuracy, we assume that the particles follow the flow as passive tracers [14], and that the particle density is uniform, when interpreting fluxes to an absorbing surface. The assumption was substantiated by analysing the particle distributions and comparing the results to a model Poisson distribution.

3 Particle Flux into a Moving Sphere

With the records for simultaneous particle trajectories being available, we can select one of them to represent the "predator" and label all the others as "prey". We then select a predetermined radius $\mathcal{R}$ in the sphere of interception, and then remove all the particles which happen to be inside this sphere at the initial time. During the subsequent Lagrangian motion of the reference "predator", we count the number of prey entering its co-moving sphere of interception between successive time steps. Each time a particle enters, it is "eaten" in the sense that it is removed from the database [8]. Of course, if the data analysis is carried out for very long times, all particles representing prey will eventually be removed. Here we are only interested in the time evolution of the prey flux for times up to an eddy turn-over time. As long as $\mathcal{R}$ is much smaller than the size of the measuring volume, we can with negligible error assume the prey concentration to be constant at large distances, corresponding to an ideally infinite system. By choosing a large number of realizations, we can give an estimate for the ensemble averaged Lagrangian prey flux as a function of time after release.

In Fig. 4 we show, with solid lines, examples of the time varying particle flux to a self consistently moving sphere of interception with a given radius, $\mathcal{R}$. This flux is the result of a competition between, on one hand, the depletion of the density of polystyrene particles in the near vicinity of the reference sphere

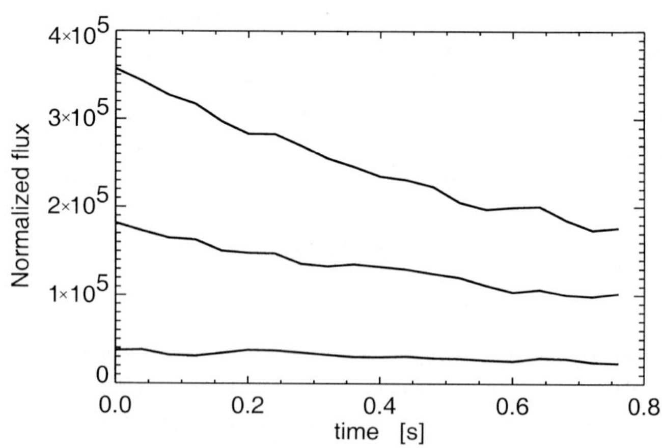

Fig. 4. Time variation of the estimate for the averaged particle flux for unit density $\langle J(t)\rangle/n_0$, to moving spheres, with radii, $\mathcal{R} = 10$, 20 and 30 mm

as they "absorbed", and, on the other hand, inward flux of such particles, due to the turbulent motions in the flow. In each realization, we divide the flux by the particle density for that particular realization. The result thus represents the particle flux for unit particle density, i.e. 1 particle per mm^3. For small radii, $\mathcal{R} < \mathcal{L}_E$, we find that the flux level is almost constant in time. A decreasing trend becomes more conspicuous as the radius is increased, and for $\mathcal{R} > \mathcal{L}_E$ we find a significant flux reduction for times approaching the eddy turn-over time, here estimated by $\tau_F \equiv \mathcal{L}_E/\sigma$. The flux is largest initially, when the concentration of "prey" in the surrounding is largest. At later times there will be a possibility for encountering fluid elements which *have* already been emptied, and the prey flux becomes smaller. The flux depletion due to this effect increases evidently for increasing radii in the reference sphere.

4 Analytical Results

The problem of turbulent particle flux to a perfect absorber moving with the flow can be studied analytically by allowing for some simplifying assumptions. Here, an absorbing spherical surface is assumed to have its centre defined by a particle, which is moving with the flow.

4.1 Dimensional Arguments

The present problem is characterized by a few dimensional quantities. With the viscosity, ν, being immaterial for the flow dynamics for scale lengths larger than the Kolmogorov length scale $\eta \equiv (\nu^3/\epsilon)^{1/4}$, we only have one quantity characterizing the turbulent flow, namely ϵ with dimension *length*^{2}/*time*^{3}, and the length scale $\mathcal{R}$ characteristic for the particular problem, here a moving sphere of interception. Out of these quantities the only combination giving a quantity with dimension *time* is $\mathcal{R}^{2/3}/\epsilon^{1/3}$, while $\epsilon^{1/3}\mathcal{R}^{7/3}$ gives *length*^{3}/*time*. The physical dimension of the averaged normalized particle flux $\langle J\rangle/n_0$ is *length*^{3}/*time*.

Quite generally it can then be argued, by purely dimensional reasoning, that the turbulent flux for given reference density n_0 must have the form

$$\frac{\langle J\rangle}{n_0} = \epsilon^{1/3}\mathcal{R}^{7/3} f\left(t\epsilon^{1/3}/\mathcal{R}^{2/3}\right) , \tag{2}$$

with f being a dimensionless function of a dimensionless variable. The actual form of f can only be determined by a more detailed model analysis. We can argue that we, in Fig. 4, have determined f experimentally, without reference to any explicit model equations. The arguments do not depend on any specific shape of the reference volume, and assume only that it scales self-similarly with one length scale, $\mathcal{R}$. The functional dependence f in (2) will, of course, be different for different shapes of the volume. Note that for $t > \mathcal{R}^{2/3}/\epsilon^{1/3}$, see Fig. 4, the variation of $f(\tau)$ is rather slow for parameters relevant here. The observations summarized in Fig. 4 seem to indicate that f approaches a constant value for large times. The constant is assumed to be universal, and we find it here to be in the range 5–10, as discussed in more detail later. The observation is not as trivial as it might seem [15].

4.2 A Model Diffusion Equation

The particle flux to a perfectly absorbing sphere, which is moving with the flow has been modelled by, for instance, a simple diffusion equation with a properly chosen diffusion coefficient which depends on the simultaneous mean square velocity differences obtained at given spatial separations, but independent of time [7]. Essentially, the argument is based on the second order structure function

$$\Psi_2(r) \equiv \left\langle (u_r(0,t) - u_r(r,t))^2 \right\rangle \approx C_K(\epsilon r)^{2/3} , \tag{3}$$

with the approximation being valid for separations r smaller than the length scale of the turbulence, see Fig. 3. A diffusion coefficient is constructed from a characteristic velocity and a characteristic length. The velocity is taken to be $\sqrt{\Psi_2(r)}$. For the limiting form expressed in (3), the only length characterizing the two particles is their separation r. The resulting diffusion coefficient is consequently $K(r) \sim r^{4/3}\epsilon^{1/3}$. The proposed diffusion equation for the density n is actually identical to the one suggested by Richardson in his study of distance-neighbour functions [16]

$$\frac{\partial}{\partial t} n(r,t) = C \frac{\epsilon^{1/3}}{r^2} \frac{\partial}{\partial r} r^{10/3} \frac{\partial}{\partial r} n(r,t) . \tag{4}$$

The result is written for spherically symmetrical geometry, with r being the radial coordinate, measured from the position of the centre of the reference sphere, and C is a numerical constant, assumed to be universal. While (4) was here argued by dimensional reasoning, it has also an analytical derivation [2]. As a consequence of (4) we have the well-known result for the mean-square separation of two initially close particles $\langle r^2 \rangle = C_R \epsilon t^3$, with the Richardson constant being $C_R \approx 0.5$ [10]. We have the relation $C = (3/2)(3C_R/143)^{1/3}$, giving $C \approx 0.33$. In the present model, the time varying diffusion flux of particles to a perfectly absorbing sphere is given by

$$J(t) = 4\pi C \epsilon^{1/3} \mathcal{R}^{10/3} \left. \frac{\mathrm{d}n(r,t)}{\mathrm{d}r} \right|_{r=\mathcal{R}} ,$$

with $n(r,t)$ obtained from (4).

The derivation of (4) assumes that ϵ is a deterministic constant, and thereby ignores intermittency corrections [17]. Although the relation (4) had some experimental support from the time when it was first proposed [16], and also supported more recently [10], its general validity has been criticized [1, 2], as also summarized recently [10]. The range of validity of (4) is thus not fully explored. For large separations, a simple diffusion equation, with constant diffusion coefficient, is expected to apply, as indicated for instance by experimental results [18], for initial conditions having scales larger than the integral length scale. These cases [10, 18] referred to particle releases considered as initial value problems. It seems that a diffusion equation as (4) can indeed be applied for analysing relative two-particle diffusion in certain variable ranges [10]. On the other hand, one cannot expect a diffusion coefficient depending solely on relative times or spatial separations to be universally applicable for this problem [2]. In general, a Fokker–Planck equation, with (4) being one special example, describes a Markov process, where the probabilities of future states depend solely on the present, and not past ones. Modelling of turbulence as a simple Markov process is known to be rather inaccurate, and a study of the limits of applicability of models like (4) is therefore worthwhile.

From (4) is easy to derive [7] a steady state flux to a sphere with radius $\mathcal{R}$ as

$$\frac{J_0}{n_0} = \frac{28\pi}{3} C\epsilon^{1/3}\mathcal{R}^{7/3} \,. \tag{5}$$

where n_0 is the constant particle density at $r \to \infty$.

4.3 Comparison Between Analytical and Experimental Results

In order to compare our observations with analytical results, we show by open circles in Fig. 5, the flux value at a time $t = \tau_F/2$, with τ_F being the eddy turn-over time. This time is sufficiently short to give a large number of

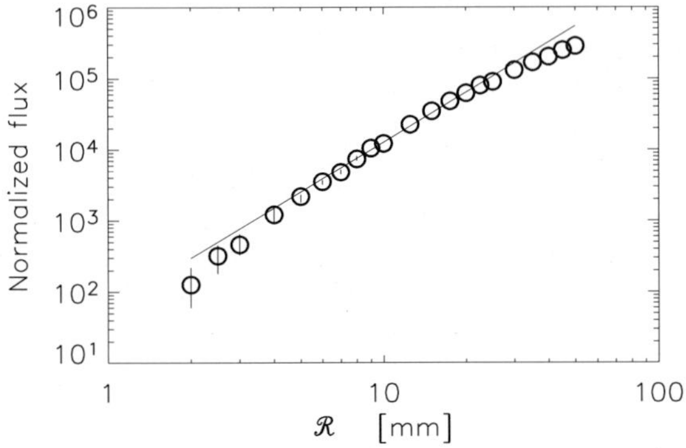

Fig. 5. The particle flux, $\langle J \rangle / n_0$, to a moving sphere of interception is shown with open circles for different radii, as is measured at 1/2 eddy turn over time, $t = \tau_F/2$. The full line gives the time-asymptotic result (5). Parameters are $\sigma = 19\,\mathrm{mm/s}$, $\tau_F = 1.6\,\mathrm{s}$, and $\epsilon = 225\,\mathrm{mm^2/s^3}$. The fluxes are normalized to unit density

particle traces for the averaging, and on the other hand, sufficiently long to give an estimate close to the asymptotic flux value of the particle flux. Vertical lines give the uncertainties on the experimental estimates. For small radii $\mathcal{R}$, this uncertainty is large because we only seldom find close particles. For $\mathcal{R} >$ 5 mm, on the other hand, this uncertainty is smaller than the size of the circles in Fig. 5. The analytical curve, given by a full line, is the asymptotic limit from (4), where we used $\epsilon = 225\,\mathrm{mm}^2/\mathrm{s}^3$. Taking into account that we have not introduced any free or adjustable parameters, we find the agreement between the analytical and experimental results to be satisfactory, although we note a slight, but systematic, reduction of the measured flux as compared with the analytical asymptotic result. The experimental results for the smallest radii give an underestimate, since in this limit a nontrivial fraction of the particles are "glancing", i.e. they manage to pass through the reference sphere within one sampling time, and are therefore not counted.

The model equations become inadequate for spatial separations larger than the largest eddies in the turbulence, $r \geq \mathcal{L}_E$, although we find that the $\mathcal{R}^{7/3}$-scaling seems to have a wider range of validity, in particular at early times, $t < \tau_F/2$. The analysis summarized here refers explicitly to spherical volumes. Qualitatively, the arguments will apply to different shapes as well, as long as they scale self-similarly with *one* characteristic length, $\mathcal{R}$.

We also present results for the flux variation for a fixed value of the radius of the moving sphere of interception, $\mathcal{R} = 20$ mm, and varying ϵ, see Fig. 6. In order to sample each dataset at a consistent time, we present results for a selected time $\tau_F/2$ used also in Fig. 5, using the proper value of σ. In this limit, we can in all cases assume that the particle flux is close to its asymptotic, or saturated, level. The circles show the result for ϵ obtained by fitting the second order structure function. Other methods for determining ϵ can be found, however [9], and these results are used to give the horizon-

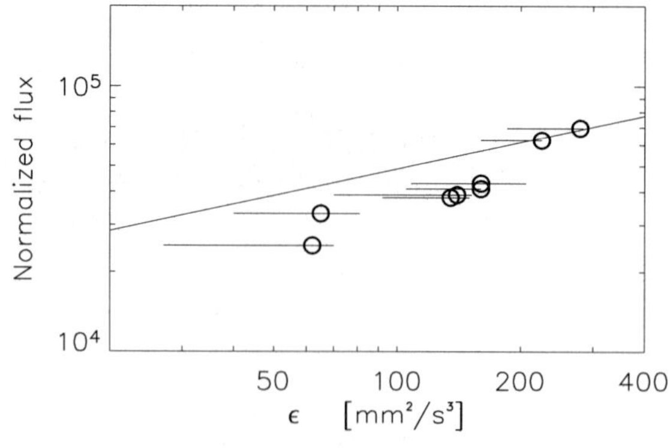

Fig. 6. Variation of the normalized flux, $\langle J \rangle / n_0$, with varying ϵ for a fixed value of $\mathcal{R} = 20$ mm. The full line gives the time-asymptotic result obtained from (4). See also Fig. 5

tal uncertainty bars. The theoretical full line is also in this case obtained as the asymptotic limit of the solution of Richardson's diffusion equation, using the most recent experimental value [10] of Richardson's constant. Within the range of variability, we find the scaling with ϵ to agree reasonably well with theoretical predictions based on (4). The numerical agreement between the measurements and analytical results is within a factor of 2, the analysis predicting a slightly larger flux than the observed value, also in agreement with the results shown in Fig. 5. The selected value $\mathcal{R} = 20$ mm can be taken as representative for the length scales smaller than or equal to $\mathcal{L}_E$ in the experiments.

The results summarized in Figs. 5 and 6 refer to fluxes obtained at fixed normalized times. We can also demonstrate a scaling law for the time variations of these fluxes, and compare the results to the results from a model like the one given by (4). In Fig. 7 we show the normalized fluxes for 8 different experimental conditions, see Table 1, and radii $\mathcal{R} = 5, 6, 7, 8, 9, 10, 12.5, 15,$ 17.5 and 20 mm. The figure demonstrates the experimental scatter, which is consistent with the uncertainties of the estimates for ϵ. Also here we note a "banded" structure in the figure. We find that the uppermost group of curves originates from the two datasets with the largest ϵ-values, see for instance also Fig. 6, where these two datasets also seem to be slightly distinct from the others.

Again, we note that the results have a wider range of applicability, and need not refer explicitly to spherical forms. A change in shape of the reference volume, will only imply a change in the numerical constant. Thus, the scaling law implied in Fig. 6 will apply, for instance, to the prey flux for any predator, independent of its range of vision, when it is exposed to different turbulence intensities.

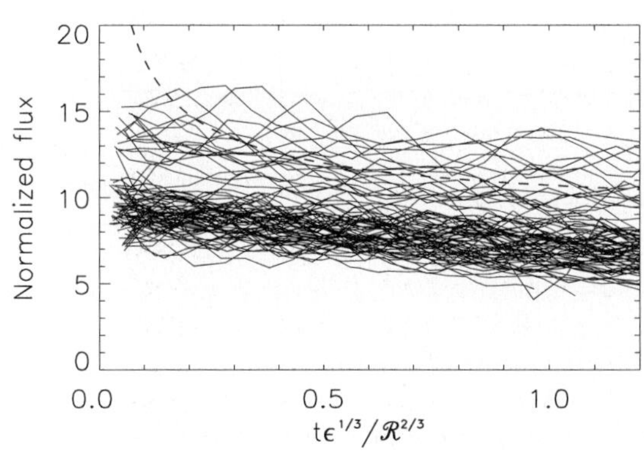

Fig. 7. Normalized flux, $\langle J\rangle/(n_0\epsilon^{1/3}\mathcal{R}^{7/3})$, as a function of the normalized temporal variable, $t\epsilon^{1/3}/\mathcal{R}^{2/3}$. Curves are shown for 8 realizations with different turbulence conditions, each with curves for different $\mathcal{R}$. The dashed line gives a theoretical result, obtained from (4). The (unphysical) singularity at $t = 0$ for the full line solution is due to the assumed infinite initial gradient at $r = \mathcal{R}$

Given the experimental uncertainties, the scaling relations obtained by dimensional reasoning are found to be well satisfied when analysing the data from the present experiment. The more specific diffusion equation model (4), is only giving qualitative agreement for the measured Lagrangian fluxes at early times. It seems, however, that the asymptotic limit is well accounted for by the model, in particular also the numerical coefficient obtained by use of the most recent value of the Richardson constant [10]. To some extent, the modest agreement between the model and the experimental results at early times might be surprising, since (4) has given better agreement with estimates of the distance-neighbour functions [10]. We note, however, that in the present case there is an ambiguity in the model for the diffusion coefficient: the result (4) uses characteristic eddies being of magnitude comparable to the predator-prey separation [7], which is the most obvious choice when modelling an equation for the distance-neighbour function. For the present case, it could as well be argued that the characteristic eddies should have a size comparable to the separation distance between prey and the surface of interception. Since such models can serve as guidelines only, we shall not pursue the problem any further here.

As particles are absorbed by the surface, with fluxes shown in Fig. 7, the particle density will be depleted in the flow surrounding the moving reference sphere. We can analyse also here the average particle density for $r > \mathcal{R}$, as a function of time, with results shown in Fig. 8. The radius $\mathcal{R}$ is chosen to be in the universal subrange. The first curve is shown at the first sampling time, i.e. $t = 1/25$ s. Variations with distance are obtained in "bins" of 1 mm, and the second bin from the surface at $r = \mathcal{R}$ is the first one shown. To reduce the noise level, we normalized also here the curves with the radial density variation found at $t = 0$. If we choose a smaller value for $\mathcal{R}$, the noise level increases, while larger $\mathcal{R}$ will fall outside the universal subrange.

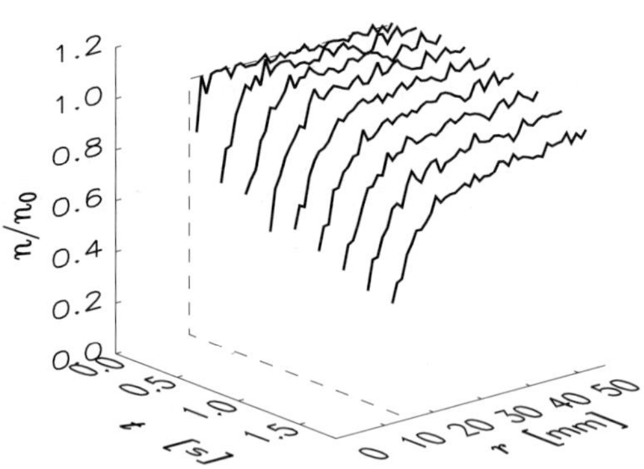

Fig. 8. Time evolution of the normalized density around an absorbing spherical surface moving with the flow, for $\mathcal{R} = 15$ mm

5 Conclusions

In this correspondence we investigated the turbulent flux to a perfectly absorbing surface, with particular attention to the problem of predator-prey encounters in turbulent flows. We summarized the basic elements of an experimental method for investigating the prey flux to a moving predator. In the limit of small $\mathcal{R}$, we found evidence for an $\mathcal{R}^{7/3}$ flux scaling (see Fig. 5) in terms of the radius of the sphere of interception. We also found indications of an $\epsilon^{1/3}$ scaling with the turbulent dissipation rate (see Fig. 6), in agreement with the predictions of the model (4). This model agrees to some extent also quantitatively with the observations. In the asymptotic time limit, to the extent it can be reached in the present experiment, the data gives a flux well approximated by (5), provided $\mathcal{R} < \mathcal{L}_E$. This will in general be the limit relevant for marine environments [19]. We suppose that the observations justify extrapolation to radii, $\mathcal{R}$, smaller than those experimentally accessible. In a general sense, our results provide experimental evidence also for the importance of turbulent motion for the feeding process in marine environments. We expect that in order to obtain a general analytical model, which can give results for extended time periods and all $\mathcal{R}$, we shall have to allow for a diffusion coefficient which depends on time as well as spatial separations, in particular including also memory effects [2].

The turbulent flux to a moving sphere can be significantly smaller than the flux to a stationary one. This can be argued simply by noting that the relative mean square velocity of a particle convected past a stationary sphere is $\langle u^2 \rangle$, while it is $\langle (\mathbf{u}(\mathbf{r},t) - \mathbf{u}(\mathbf{r}+\mathbf{y},t))^2 \rangle$, for a passively convected sphere-particle pair, with separation $\mathbf{y}$. For small separations, $y \ll \mathcal{L}_E$, we have [7, 20] the result (3), and the relative velocity is small, implying a small flux to the passively convected sphere. For large separations, $y \gg \mathcal{L}_E$, on the other hand, $\mathbf{u}(\mathbf{r},t)$ and $\mathbf{u}(\mathbf{r}+\mathbf{y},t)$ can be supposed to be statistically independent. The mean square relative velocity is then $2\langle u^2 \rangle$. The flux in this latter case is expected to be larger than that to the stationary sphere with a factor $\sqrt{2}$, although such large separations cannot be achieved for the present experimental conditions.

We can define a "gain factor" as the ratio between the flux to a stationary sphere divided by the flux to the passively convected sphere with the same radius, $\mathcal{R}$. In Fig. 9 we show this gain factor for various radii, $\mathcal{R}$. All points are obtained at the reference time $\tau_F/2$ used before. We find that the gain factor is considerable for small spheres of interception, using the length scale $\mathcal{L}_E$ as a measure. For $\mathcal{R} \approx \mathcal{L}_E$ this gain factor is close to 1, and the particle flux is the same for a stationary as for the moving sphere. For larger values, $\mathcal{R} > \mathcal{L}_E$, the flux to a moving sphere exceeds that to a stationary one. The variation of the initial value of the fluxes seen, for instance, on Fig. 4 are consistent with these observations. Heuristically, we can argue for a parameter variation of the gain factor as given by the ratio of the two scaling laws obtained by dimensional arguments, which here gives $\sigma/(\epsilon\mathcal{R})^{1/3}$. This ratio is shown by

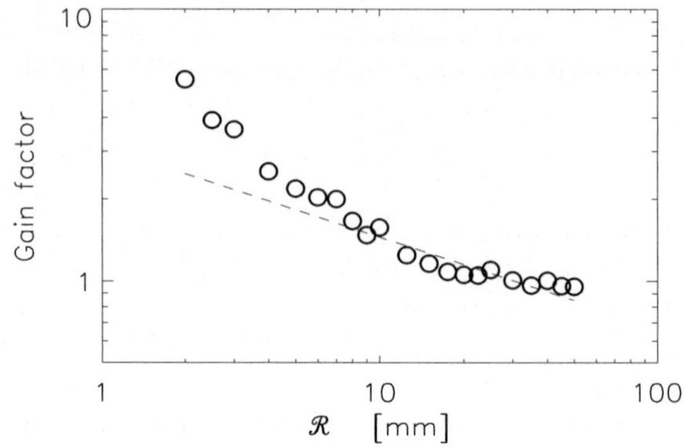

Fig. 9. Variation of the gain factor for a stationary sphere for various radii of the sphere of interception, $\mathcal{R}$. The figure refers to a time $t = \tau_F/2$.

a dashed line in Fig. 9, with a numerical constant not accounted for. We find, in this case, that this scaling law does not follow the data points in any convincing way, although the *trend* seems reasonable. After all, neither the Eulerian nor the Lagrangian data followed the scaling perfectly, the Eulerian data best at small $\mathcal{R}$, the Lagrangian data best at somewhat intermediate values. To expect a perfect agreement for the *ratio* of the two quantities may seem somewhat optimistic, in particular also because a small mean flow in the system gives a bias for the Eulerian fluxes. The gain factor shown in Fig. 9 can, for instance, be interpreted as the gain in prey flux for an imagined predator with possibilities for self-induced motion, which it uses to exactly compensate the motions in the surrounding water.

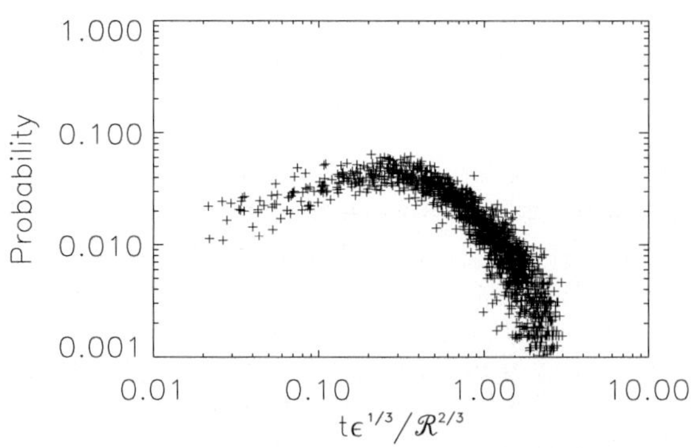

Fig. 10. Lagrangian transit time distribution, shown as a function of normalized temporal variable $t\epsilon^{1/3}/\mathcal{R}^{2/3}$. The figure contains 6 experimental conditions with ϵ = 62, 65, 135, 160, 225, and 279 mm^2/s^3, and each of these analysed for radii $\mathcal{R}$ = 5, 10, 15, and 20 mm

The present analysis, when applied to the predator-prey problem implicitly assumes that prey is captured with certainty. This might be an acceptable assumption for slow motions in the flow, but evidently it might become questionable when the flow is strongly turbulent. Little seems to be known about the capture probability of prey, when the relative velocities are large, although *some* observations have been quantified [21]. These results refer seemingly only to relatively large predators, fish larvae for instance. We can not here add to that discussion, but might note that one possibly essential part of the information, relevant for a detailed discussion, might be the distribution of transit times for prey through the sphere of interception. Also this question can be analysed on the basis of the present experimental data [15]. We can thus obtain the distribution of transit times taken as the time difference from a particle entering a reference sphere until it leaves it again for the first time (i.e. first passage time distributions). It turns out that also this distribution follows a seemingly universal scaling, as long as the radius of the moving reference sphere is within the universal subrange.

The problem discussed in the present communication is evidently of general interest. It has implications also for coagulation processes in turbulent colloids, for instance. A detailed investigation of this latter problem can, however, not be made by experiments like ours because the volume of the particles change upon coagulation, with a consequent change in their response to the turbulent flow motions. We can not reproduce this effect, for evident reasons. In standard studies of this problem [22], restricted to diffusion by Brownian motion, this effect is in part also ignored. With the same assumption it is possible to perform the relevant studies in experiments like ours, with results having implications for the formation rate of coagulants larger than the Kolmogorov scale η in turbulent flows.

Acknowledgement

Valuable discussions with Maria Pécseli in the initial stages of the project are gratefully acknowledged. This work was in part supported by the Danish Technical Research Council under contracts STVF-9601244 and 26-01-0087. Two of the authors (HLP and JT) were in part supported by the "Effects of North Atlantic Climate Variability on the Barents Sea Ecosystem" (ECOBE) project. The present study was completed while two of the authors (HLP and JT) were affiliated with the Norwegian Center for Advanced Studies.

References

1. G. K. Batchelor: Proc. Cambridge Philos. Soc. **48**, 345 (1952)
2. P. H. Roberts: J. Fluid Mech. **11**, 257 (1961)
3. S. Sundby, P. Fossum: J. Plankton Res. **12**, 1153 (1990)

4. T. Kiørboe, E. Saiz: Mar. Ecol. Prog. Ser. **122**, 135 (1995)
5. J. H. Muelbert, M. R. Lewis, D. E. Kelley: J. Plankton Res. **16**, 927 (1994)
6. B. J. Rothschild, T. R. Osborn: J. Plankton Res. **10**, 465 (1988)
7. T. Osborn: J. Plankton Res. **18**, 185 (1996)
8. J. Mann, S. Ott, H. L. Pécseli, J. Trulsen: Phys. Rev. E **65**, 026304 (2002)
9. J. Mann, S. Ott, J. S. Andersen: Technical Report No. Risø-R-1036(EN), Risø National Laboratory, DK-4000 Roskilde, Denmark (unpublished), can be downloaded from http://www.risoe.dk/rispubl/VEA/ris-r-1036.htm
10. S. Ott, J. Mann: J. Fluid Mech. **422**, 207 (2000)
11. J. O. Hinze: *Turbulence*, 2nd edn (McGraw-Hill, New York 1975)
12. A. S. Monin, A. M. Yaglom: *Statistical Fluid Mechanics* (The MIT press, Cambridge, Massachusetts, 1975), vol. 2
13. A. Babiano, J. H. E. Cartwright, O. Piro, A. Provenzale: Phys. Rev. Lett. **84**, 5764 (2000)
14. M. R. Maxey, J. J. Riley: Phys. Fluids **26**, 883 (1983)
15. J. Mann, S. Ott, H. L. Pécseli, J. Trulsen: Phys. Rev. E **67**, 056307 (2003)
16. L. F. Richardson: Proc. Roy. Soc. London, Ser. A **6**, 709 (1926)
17. G. Boffetta, A. Celani, A. Crisanti, A. Vulpiani: Phys. Rev. E **60**, 6734 (1999)
18. M. Virant, T. Dracos: Meas. Sci. Technol. **8**, 1529 (1997)
19. P. S. Hill, A. R. M. Nowell, P. A. Jumars: J. Mar. Res. **50**, 643 (1992)
20. S. Chandrasekhar: J. Madras Univ. B **27**, 251 (1957)
21. B. R. Mackenzie, T. Kiørboe, Limnol. Oceanogr. **45**, 1 (2000)
22. S. Chandrasekhar. In: *Selected Papers on Noise and Stochastic Processes*, ed by N. Wax (Dover, New York 1954), pp 3–91

Extinction Dynamics in Lotka–Volterra Ecosystems on Evolving Networks

Adam Lipowski[1] and Michel Droz[2]

[1] Department of Physics, Adam Mickiewicz University, 61-614 Poznań, Poland
lipowski@amu.edu.pl
[2] Department of Physics, University of Geneva, CH-1211 Geneva, Switzerland
Michel.Droz@physics.unige.ch

Summary. Lotka–Volterra models of interacting populations are typically used on a time scale that is shorter than the lifetime of species. Such long-term behaviour of ecosystems is very often studied but using models that operate at the level of species rather than individuals. Such an approach usually refers to the notion of fitness that is not commonly accepted. In the present paper we show that a Lotka–Volterra model can be used to describe long-term behaviour of multi-species ecosystems. Interactions among species in our model form a network whose evolution is determined by the dynamics of the model. Numerical simulations show a power-law for the distribution of intervals between extinctions, but only for ecosystems with a sufficient variability of species and with networks of connectivity above a certain threshold that is very close to the percolation threshold of the network. Effect of slow environmental changes on extinction dynamics, degree distribution of the network of interspecies interactions, and some emergent properties of our model are also examined

1 Introduction

Population dynamics, now an important part of biology and sociology, certainly has its roots in mathematics. This is because growth laws, that describe time evolution of the size of a given population, are very often formulated in terms of differential equations. In the simplest form one assumes that the growth of a population is proportional to the population itself. Such a natural at first sight assumption has an immediate and unpleasant consequence: exponentially diverging size of this population. It was a fundamental contribution of Verhulst to provide a remedy for such a drawback, by limiting the growth rate with a term that decreases with the size of the population [1]. Stimulated by such an important discovery, population dynamics became a rapidly developing field, aiming to describe also more complicated systems as e.g. interacting populations.

The earliest approach to study interacting populations was initiated by Lotka and Volterra [2], and their model quickly became a cornerstone of population dynamics [3]. Being inspired by a certain surprising, at that time, behaviour in some prey-predator systems, this model is typically used to describe populations on time scales shorter than the lifetime of species. It means that long-term properties of ecosystems (macro-evolution) are usually not captured within such an approach. On the other hand, models used to describe macro-evolution very often use the dynamics that operates at the level of species rather than individuals. Such coarse-grained models usually refer to the notion of fitness of a species that is however not commonly accepted [4].

Recently, there has been some attempts to study macro-evolution using models equipped with dynamics that operates at the level of individuals [5–7]. Taking into account that Lotka–Volterra models are relatively successful in describing many aspects of population dynamics it would be desirable to apply such an approach also to macro-evolution of large ecosystems.

Actually, Lotka–Volterra models were already used by May to study some aspects of the stability of ecosystems [8]. His results, based on the mathematical analysis of an equilibrium state, caused some concern among ecologists, since counter-intuitively, May predicted that complex ecosystems (i.e., those with a large number of species and a large number of interactions between these species) are more likely to be unstable. Ensuing debate tried to reconcile May's results with our intuition that is based on observations of e.g., tropical forests (that seem to be complex and stable), and agricultural ecosystems (simple and unstable). It was suggested that some of his assumptions (as e.g., the very existence of an equilibrium state of an ecosystem) might not be valid and that ecologically relevant definition of stability is different from the one used in the study of (typical) dynamical systems [9].

In addition to that, Abramson introduced a discrete version of the Lotka–Volterra ecosystem [10] and studied certain characteristics of extinctions. His model is an example of a one-dimensional food chain with $M(\sim 100)$ trophic levels and a single species occupying a given trophic level. Since in realistic food webs $M \sim 4-6$ with typically many species belonging to a given trophic level [11, 12], these are highly nonrealistic assumptions. Nevertheless, extinction dynamics in Abramson's model shows some features that are characteristic to Earth biosystem.

In the present chapter we introduce a Lotka–Volterra model that describes a simplified ecosystem of N species of predators and one species of preys [13]. Our model can thus be considered as a simple food web model with only two trophic levels. The competition between predator species is described by a certain network [14] of interactions whose evolution is governed by the dynamics of the model. Namely, when a certain species becomes extinct (i.e., its density falls below a certain threshold) it is replaced by a new species with a newly created set of interactions with some of the existing species. Despite obvious simplifications, the model exhibits some properties that are typical of more complicated ecosystems, as for example power-law distributions of the intervals between extinctions. Within our model we can also examine how robust this power-law distribution is. We find that under certain conditions, as for example very sparse interactions between species, or too strong dominance of a small group of species, these power-law characteristics disappear and the model is driven into a regime where extinctions have exponential distributions or where there are no extinctions and the ecosystem enters a steady state. In our opinion, such regimes might be relevant when a restricted (either in space or time) evolution of an ecosystem or its part is studied. Interestingly, a threshold value of the connectivity that separates power-law extinctions and steady state is very close to the percolation threshold of the random network of inter-species interactions.

According to a large class of statistical physics models of biological evolution, avalanches of extinctions do not require external factors to trigger them, but might be a natural consequence of the dynamics of the ecosystem. As a result, these external factors, as e.g. climate changes, solar activity or impact of a big meteorite, are very often neglected in such studies [15]. But such factors certainly affect the ecosystem and there is a good evidence of it [16]. Let us emphasize that even the basic mechanism that triggers avalanches of extinctions is not known and is a subject of an intensive multidisciplinary debate.

One possibility to take external factor(s) into account in our model is to modify the growth rate of the preys. Since the dynamics of the model is nonlinear, such a change might have more dramatic consequences than merely a change of the densities of the species. And indeed we noticed that the dynamics of extinctions is strongly dependent on the growth rate. It turns out, that in our model abundance of preys leads to a larger frequency of extinctions, and in periods of hunger there are less extinctions. This is clearly due to the nonlinearity of the dynamics. A larger growth rate increases the density of preys that in turn increases the densities of predators. With increased densities, the dynamics becomes more competitive and extinctions become more frequent. Such a periodically modulated growth rate leaves also some traces in the probability distribution of the extinctions. It might be interesting to notice that paleontological data also show some traces of periodic events, but their proper understanding is still missing [16, 17].

During evolution some species are favoured and selected at the expense of less fortunate ones. Evolution constantly searches for the best solutions, which resembles an optimization process. For example a large size of organisms of a given species might be of advantage in some situations, but it might cause some problems in other ones. What might be the Nature solution to this problem? Will it be middle-size species or rather two groups of species sitting at the extremes of conflicting requirements? In our opinion, this aspect of evolution is also often omitted in models of macroevolution. Within our model we look at such emergent properties of species selected by evolution. It turns out that depending on some dynamical details, our model can reproduce both types of solutions of such an optimization problem.

2 Model and Numerical Calculations

We study a Lotka–Volterra ecosystem that consists of N species of predators with densities ρ_i ($i = 1, 2, \ldots, N$) which are all feeding on one species of preys with density ρ_0. We assume that each predator species i is characterized by a parameter k_i ($0 < k_i < 1$) that enters the evolution equations of the model through death and growth terms

$$\dot{\rho_0} = g(t)\rho_0(1-\rho_0) - \frac{\rho_0}{N}\sum_{i=1}^{N} f(k_i)\rho_i \tag{1}$$

$$\dot{\rho_i} = -d(k_i)\rho_i(1-\rho_0) + f(k_i)\rho_i\rho_0\left(1 - \frac{k_i\rho_i + \sum_j' k_j\rho_j}{k_i + \sum_j' k_j}\right), \tag{2}$$

where $i = 1, 2, \ldots, N$. In our model we assume that species interact mainly through environmental capacity terms (the last term in (2)). Namely, the growth rate of a given species i is reduced not only due to its density but also due to weighted (with the factor k) densities of a group of randomly selected neighbouring species. In (2) the summation over these neighbouring species is denoted by ($\sum'$). A more detailed description of our model can be found elsewhere [13].

The differential (1)–(2) are solved using a fourth-order Runge–Kutta method. Multi-species Lotka–Volterra ecosystems were subject to intensive studies since the pioneering work of May [8]. It is known that such systems might evolve toward a steady state with positive densities. However, in some cases, in a steady state, the density of some species might be zero. Each time the density of a certain species in model (1)–(2) drops below a threshold value which we fix as $\varepsilon = 10^{-7}$ we consider such a species as extinct. Such a species is then replaced by a new one with a randomly assigned density (from the interval (0,1)), a coefficient k ($0 < k < 1$) that is randomly drawn from the distribution $p(k)$, and a new set of neighbours (all links of the 'old' species are removed). With such rules the model rather describes N niches, and we assume that the time to create a species that will occupy a niche is relatively short as compared to the typical lifetime of species.

We assume that a newly created species makes z links with randomly selected neighbours. Links are not directional, so a newly created species will also enter the evolution equation of species it is neighbouring. If the extinct species were chosen randomly, the network of interactions would be a random graph. However, it is the dynamics (1)–(2) that determines which species are extinct. Since the extinct species are not selected randomly, the resulting network is in general not a random graph.

3 Results

In the following we describe numerical results obtained for some particular cases of model (1)–(2).

3.1 Intervals Between Extinctions

Various paleontological data suggest that the dynamics of extinctions has some power-law distributions of sizes or durations [16]. In our model we measured time intervals t between successive extinctions. In this calculation we used a constant growth term of preys $g(t) \equiv 1$. We examined two cases: (i) model I: $f(k_i) \equiv 1$, $d(k_i) \equiv 1$ and (ii) model II: $f(k_i) = k_i$, $d(k_i) \equiv 1$. Unless specified otherwise we select k_i randomly with a homogeneous distribution on the interval (0,1) ($p(k) = 1$). Our results are shown in Fig. 1. In the simplest case, model I with $z = 4$ and $k_i \equiv 1$ (i.e., all species during the evolution have identical $k_i (= 1)$) we obtain an exponentially decaying distribution of intervals between extinctions $P(t)$. Such a decay is also seen for model I ($z = 4$) with a linear distribution of k_i namely $p(k) = 2k$. We expect that such a behaviour appears when the distribution of k_i in the ecosystem is relatively narrow and shifted toward unity. Such an effect might be due to the small

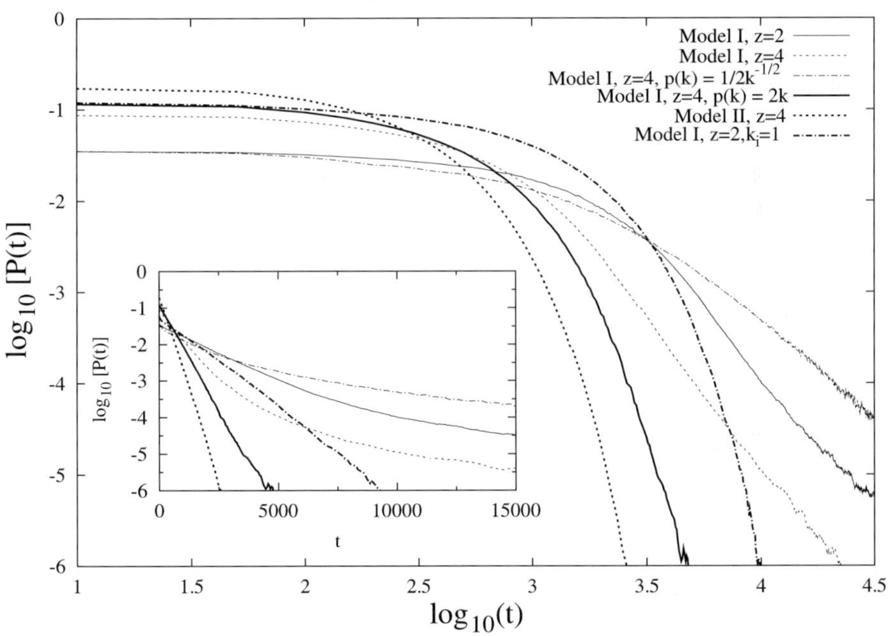

Fig. 1. Probability distribution of intervals between successive extinctions $P(t)$ calculated for some particular cases of model (1)–(2) for $N = 100$. Inset shows the same data but plotted on a lin-log scale

width of the distribution $p(k)$ (i.e., a distribution from which we draw k_i) or might be dynamically generated as in model II. In this case even though k_i are chosen from a homogeneous distribution, the dynamics favours large k_i species (due to their larger growth rate) and they dominate the ecosystem. When the distribution of k_i in the ecosystem is more uniform (model I with $p(k) = 1$) our simulations suggest that $P(t)$ decays as a power law. Let us notice, however, that a power-law behaviour is seen only on approximately one decade and we cannot exclude that on a larger time scale a different (perhaps exponential) behaviour appears as was already observed in some other macroevolutionary models [5]. Let us also notice that for model I with $p(k) = k^{-1/2}/2$ the power-law distribution $P(t)$ seems to decay as t^{-2}, i.e., with an exponent consistent with some paleontological data [16] as well as with the predictions of some other models [6]. However, one has to recognise that the error bars on experimental data are rather large and that a non-power law behaviour cannot be excluded.

However, a power-law decay of $P(t)$ is seen only for sufficiently large z. When z is too small, we observed that the ecosystem enters a steady state where all ρ_i are positive and there are no extinctions. This is probably due to the fact that the competition among predators is too weak (or rather too sparse). To examine the transition between these two regimes in more detail we measured the average time τ between extinctions and the results are seen in Fig. 2. One can see that τ diverges around $z \sim 1.8$. Such a value of the threshold parameter suggests that this transition might be related to

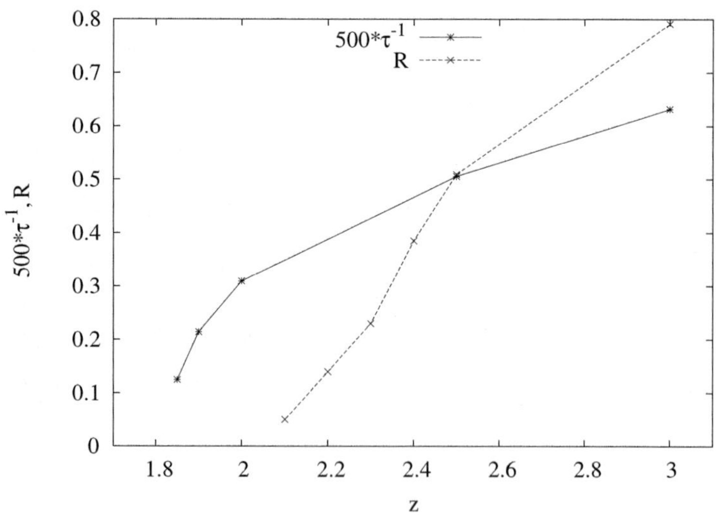

Fig. 2. The inverse average time between extinctions τ^{-1} and the percolation probability R as a function of z. The plotted results are based on calculations for $N = 100$, 200, 300 and 400 and extrapolation $N \to \infty$

the percolation transition in our network of interspecies interactions. To examine such a possibility we measured the average size of the largest cluster of connected links in the network R (normalized by the number of species N); the results are shown in Fig. 2. The vanishing of this quantity locates the percolation transition [18]. One can see that the percolation transition takes place at a larger value, namely around $z \sim 2.0$. Our results suggest that these two transitions take place at different values of z. However the analysis of finite size effects especially in the estimation of τ is rather difficult and we cannot exclude that these two transitions actually overlap, as might be suggested by their proximity. Such a result would show that the dynamical regime of an ecosystem (i.e., steady state or active with power-law distribution of extinctions) is determined by the geometrical structure of its interactions.

3.2 Effect of a Modulated Growth Rate

Now we examine the role of a modulated in time growth rate of preys. Such a modulation is supposed to mimic the influence of an external factor like a change of climate. One of the questions that one can ask in this context is how such a change affects the extinction dynamics. We studied model I with $p(k) = 1$ and $d(k_i) \equiv 1$. For the growth rate of preys we chose $g(t) = 1 + A\sin(2\pi t/T)$, where A and T are parameters. A typical behaviour in case of model I with such a growth rate is shown in Fig. 3. One can see that

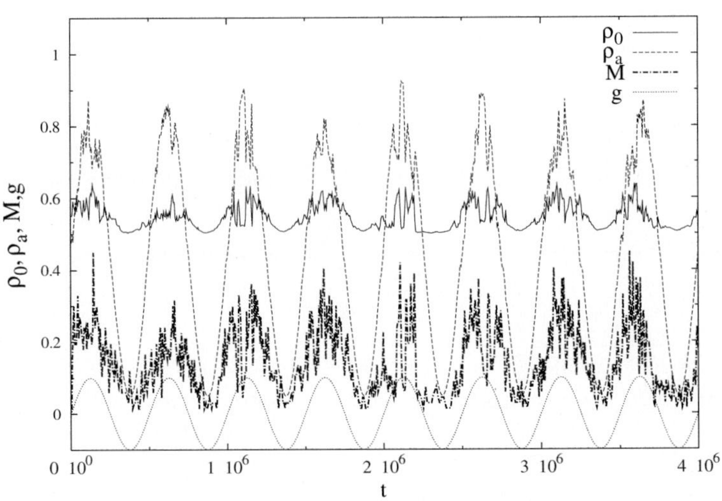

Fig. 3. A time evolution of the density of preys ρ_0, average density of predators $\rho_a = N^{-1}\sum_{i=1}^{N}$, and the number of extinctions M (divided by 20) in the time interval $\Delta t = 10^3$ for the model I with $N = 100$ and $z = 4$. A rescaled modulated growth rate $(g(t) - 1)/10 = 0.09\sin(2\pi t/T)$ $(T = 10^5)$ is also shown

an increased growth rate increases the density of preys ρ_0 that increases the density of predators. However, it increases also the frequency of extinctions. Such a behaviour, namely an increased extinction rate during abundance of food, might at first sight look as counterintuitive. This effect is related to the form of environmental capacity terms in the growth rate in (2), namely $1-(k_i\rho_i+\sum_j' k_j\rho_j)/(k_i+\sum_j' k_j)$. Such a term certainly has a larger variability for increased density of predators ρ_i, and for some species (depending on the distribution of links, coefficients k_i and densities) it causes faster extinction. Let us also notice that since the period of modulation T is quite large, there is no retardation effect between the density of preys and predators. We observed such a retardation for smaller values of $T(\sim 1000)$.

The modulated growth rate of preys also affects the probability distribution $P(t)$ of the intervals between extinctions as shown in Fig. 4. One can see that a period of modulation T is imprinted in $P(t)$. Let us notice that certain paleontological data do show some signs of periodicity but its origin still remains unclear [16, 17].

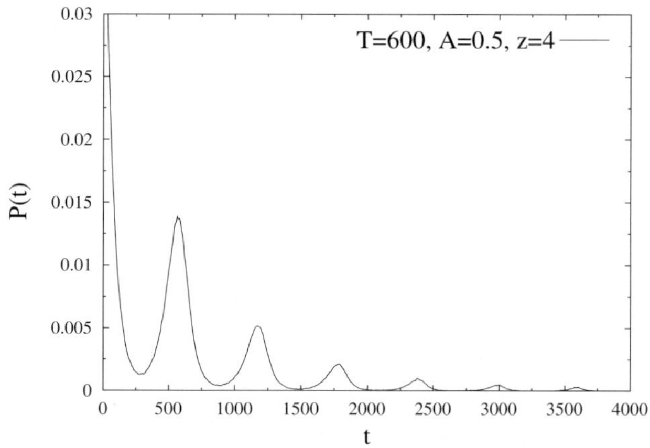

Fig. 4. Probability distribution $P(t)$ of the intervals between successive extinctions calculated for model I with a modulated growth rate ($N = 100$)

It is known that slowly changing ecosystems sometimes undergo catastrophic shifts [19]. As a result, the ecosystem switches to a contrasting alternative stable state. It would be interesting to examine whether multi-species ecosystems, as described by our model (1)–(2), might also exist in such alternative states. If so, one can ask whether for example the structure of the network of interspecies interactions or extinction dynamics are the same in such states.

3.3 Emergent Properties of Species

It might be interesting to ask what are the characteristics of species that are preferred by the evolution in our ecosystem. Since species are characterized only by the number k_i it is equivalent to calculate the distribution of k_i in the steady state of the model (1)–(2). Of course, due to selection, this distribution will in general be different from the distribution $p(k)$, i.e., the distribution from which we draw k_i for a newly created species. Some of our results are shown in Fig. 5 (all results for $g(t) \equiv 1$, $z = 4$, $N = 100$).

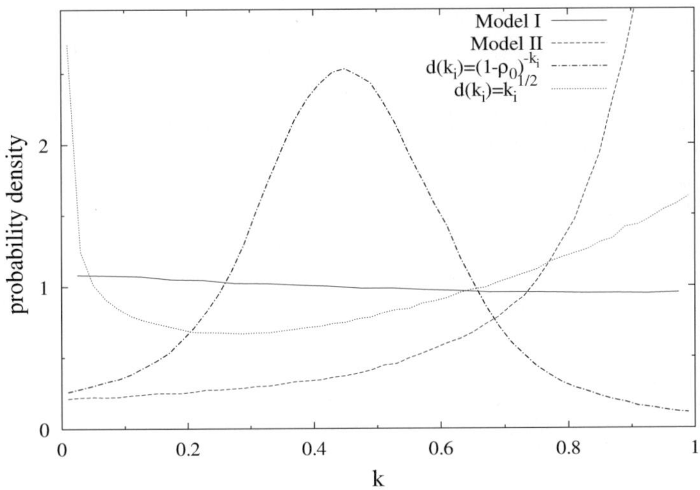

Fig. 5. Distribution of k_i in the steady state of some particular cases of model (1)–(2) (see text)

In the case of model I ($f(k_i) \equiv 1$, $d(k_i) \equiv 1$) with a homogeneous initial distribution of k_i ($p(k) = 1$) one can see that the steady state distribution is also approximately homogeneous (with a slight bias favoring small-k species). We checked that model I shows this behaviour also for other distributions $p(k)$ (what you put is what you get). A different behaviour appears for model II ($f(k_i) = k_i$, $d(k_i) \equiv 1$). In this case the growth rate factor $f(k_i)$ of the i-th species is proportional to k_i that certainly prefers species with large k_i. Numerical results for a homogeneous distribution $p(k) = 1$ confirm such a behaviour (Fig. 5). We observed similarly strong preferences of large k_i species also for Model II with other distributions $p(k)$.

We also examined the selection pattern in the presence of some competing effects. To compensate a strong preference toward large-k species we made simulations for our model with $f(k_i) = k_i$, $d(k_i) = \sqrt{k_i}$ and $p(k) = 1$. Such a term reduces the death rate of small-k species. Our results show (Fig. 5) that in this case the distribution of k_i has two maxima, one at each extremity of the interval (0,1). On the other hand with the same model but for $d(k_i) = (1-\rho_0)^{-k_i}$ (that also reduces the death rate of small-k species) we obtain a distribution with a single maximum around $k = 0.45$. It would be desirable to understand the origin of the qualitative difference between these two cases.

Actually, there is yet another property of our species that is subjected to evolutionary selection, namely the number of links l_i (degree) of a given species. Although at the beginning each species acquires z links this number changes during the evolution because some links of a given species might be created or removed due to creation or extinction of another species. And since it is the dynamics of our model and not the random process that determines which species are removed, one can expect that the degree distribution might be different from the Poissonian distribution that is characteristic for random graphs (see [20] for a precise definition of random graphs).

To check the statistical properties of the network of interactions in our model we calculated the degree distribution. Our results for model I with $z = 4$ and $N = 100$ are shown in Fig. 6. Let us notice that although each species has z links at the beginning it does not mean that the average number $\langle l_i \rangle$ of links connected to a given site equals z, since the dynamics of the model might preferentially remove sites having a certain connectivity. And indeed, numerical calculations show that in this case $\langle l_i \rangle = 2.98 < z = 4$, i.e., dynamics preferentially removes sites with a large connectivity. For comparison with the random graph we also plot the Poissonian distribution $r(l) = \mathrm{e}^{-\langle l_i \rangle} \langle l_i \rangle^l / l!$, where $\langle l_i \rangle = 2.98$. It should be emphasized that the

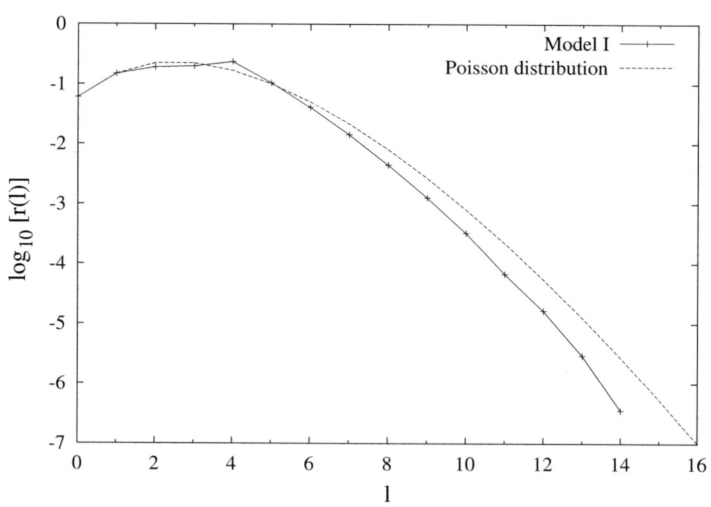

Fig. 6. Probability distribution $r(l)$ of sites with a given connectivity l for model I with $z = 4$ and $N = 100$ compared with the corresponding Poissonian distribution ($\langle l_i \rangle = 2.98$)

distribution might be approximately fitted using a Poisson distribution, for example with $\langle l_i \rangle = 2.65$. However, it is not a physically relevant distribution since the average connectivity $\langle l_i \rangle = 2.65$ differs from the value $\langle l_i \rangle = 2.98$ obtained from the simulations. In this sense the distribution is not Poissonian. One can see that for large connectivity the degree distribution decays faster than the Poissonian distribution. This result confirms that the dynamics of the model preferentially removes highly connected species. Such sites are probably more susceptible to fluctuations of the system due to extinctions and creations of new species. On the other hand, poorly connected species are more likely to arrive at a relatively stable state. Similar results concerning the degree distribution were obtained for some other cases of our model.

4 Conclusions

In the present chapter we studied the extinction dynamics of a Lotka–Volterra model for a two-level food web. In our model N species of predators feed on a single species of preys. The competition between predators, that is specified by a certain network of interactions, leads to their extinction and replacement by new species. The distribution of the intervals between successive extinctions in some cases has power-law tails and thus resembles extinction patterns of real ecosystems. However, when the network of interactions between predators is too sparse the ecosystem enters a steady state. We have shown that such a change of behaviour might be related to a percolation transition of the network. We also examined the influence of external factors on the evolution of the ecosystem. More specifically, we studied the evolution of our model in the case when the growth rate of preys is changing periodically in time. It turns out that such a modulation substantially changes the frequency of extinctions. Counterintuitively, periods with abundance of preys have higher frequency of extinctions than periods with lesser amount of preys. Moreover, we examined some properties of species that are preferentially selected by the dynamics of our model. Under some conditions preferred species are a compromise to the conflicting dynamics. Under some other conditions, preferred species form two antagonistic (with respect to the conflicting rules) groups. We also examined the degree distribution of the network of interactions between species. It turns out that the dynamics of the model has a slight preference to remove species of high connectivity. As a result the degree distribution shows some deviation from the Poissonian distribution that is characteristic to random graphs.

It would be desirable to examine some extensions of our model. For example one can introduce additional trophic levels or other forms of interspecies interactions. One can also examine a variable number of species that would allow one to create new species using certain mutation mechanism rather than assuming that they appear as soon as a niche becomes empty. Another possibility that is outside the scope of most macroevolutionary models would be to make further studies of emergent properties of the species. For example, one can imagine that a group of species in the ecosystem is well adapted and essentially not subjected to evolutionary changes. On the other hand there is a group of 'newcomers' for which evolutionary changes are much more frequent. How are the evolution and the properties of the 'newcomers' influenced by the properties of well-adapted species? Such problems might be easily approached within our model. The selection of a certain group of species (with a given value of k for example) can be considered as a selection of a certain strategy. One can examine models of this kind where species have multi-component parameters $[k = (k^a, k^b, \ldots)]$. Consequently, one can study an evolutionary selection with more complicated traits, strategies, or behaviours. Such an approach would provide an interesting link with certain evolutionary aspects of game theory [21].

Acknowledgements

Some part of our work was done in collaboration with François Coppex (University of Geneva). One of us (A.L.) was partially supported by the KBN grant 1 P03B 014 27.

References

1. P.-F. Verhulst: Corresp. Math. Phys. **10**, 113 (1838)
2. A.J. Lotka: Proc. Natl. Acad. Sci. USA **6**, 410 (1920); V. Volterra: *Leçon sur la Théorie Mathématique de la Lutte pour la vie* (Gauthier-Villars, Paris 1931); V. Volterra: Mem. Accad. Nazionale Lincei **26**, 31 (1926)
3. J.D. Murray: *Mathematical Biology*, (Springer, 1989); J. Hofbauer, K. Sigmund: *The Theory of Evolution and Dynamical Systems* (Cambridge University Press, Cambridge 1988)
4. P. Bak, K. Sneppen: Phys. Rev. Lett. **71**, 4083 (1993); B. Drossel: Adv. Phys. **50**, 209 (2001)
5. D. Chowdhury, D. Stauffer, A. Kunwar: Phys. Rev. Lett. **90**, 068101 (2003)
6. P.A. Rikvold, R.K.P. Zia: Phys. Rev. E **68**, 031913 (2003)
7. M. Hall, K. Christensen, S. A. di Collobiano, H. J. Jensen: Phys. Rev. E **66**, 011904 (2002)
8. R.M. May: Nature **238**, 413 (1972)
9. K.S. McCann: Nature **405**, 228 (2000)
10. G. Abramson: Phys. Rev. E **55**, 785 (1997)
11. B. Drossel, A.J. McKane: *Handbook of Graphs and Networks: From the Genome to the Internet* S. Bornholdt and H.G. Schuster (Eds) (Wiley-VCM, Berlin 2002), e-print: nlin.AO/0202034 D. Chowdhury and D. Stauffer, e-print: q-bio.PE/0311002
12. C. Quince, P.G. Higgs, A.J. McKane. In: *Biological Evolution and Statistical Physics*, ed by M. Lässig, A. Vallerian (Springer Verlag, Berlin Heidelberg 2002)

13. More detailed exposition of our results can be found in: F. Coppex, M. Droz, A. Lipowski: Phys. Rev. E **68**, 061901 (2004)
14. R. Albert, A.-L. Barabási: Rev. Mod. Phys. **74**, 47 (2002)
15. B.W. Roberts, M.E.J. Newman: J. Theor. Biol. **180**, 39 (1996) However, the model studied in this paper is a variant of the Bak–Sneppen model with dynamics operating at the level of species
16. M.E.J. Newman, R.G.O. Palmer: e-print: adap-org/9908002
17. D.M. Raup, J.J. Sepkoski: Proc. Natl. Acad. Sci. **81**, 801 (1984)
18. D. Stauffer, A. Aharony: *Introduction to Percolation Theory* (Taylor & Francis, London 1982)
19. M. Scheffer, S. Carpenter, J.A. Foley, C. Folks, B. Walker: Nature **413**, 591 (2001)
20. B. Bollobás: *Random graphs*, 2nd edn (Cambridge University Press, Cambridge 2001)
21. R. Axelrod: *The evolution of cooperation* (Basic Books, New York 1984); J. Maynard Smith: *Evolution and the theory of games* (Cambridge University Press, Cambridge 1982)

Exact Law of Live Nature

Mark Ya. Azbel'

School of Physics and Astronomy, Ramat Aviv, 69978 Tel Aviv, Israel;
Max-Planck-Institute für Festkorperforschung – CNRS, F–38042 Grenoble Cedex 9, France

Summary. Exact law of mortality dynamics in changing populations and environment is derived. The law is universal for all species, from single cell yeast to humans. It includes no characteristics of animal-environment interactions (metabolism etc.) which are a must for life. Such law is unique for live systems with their homeostatic self-adjustment to environment. Its universal dynamics for all animals, with their drastically different biology, evolutionary history, and complexity, is also unique for live systems – cf. different thermodynamics of liquids and glasses. The law which is valid for all live, and only live, systems is a life specific law of nature.

Mortality is an instrument of natural selection and biological diversity. Thus, the law which is preserved in evolution of all species is a conservation law of mortality, selection, evolution, biology. The law implies new kind of mortality and adaptation which dominate in evolutionary unprecedented protected populations and, in contrast to species specific natural selection, proceed via universal stepwise rungs. The law demonstrates that intrinsic mortality and at least certain aspects of aging are disposable evolutionary byproducts, and directed genetic and/or biological changes may yield healthy and vital Methuselah lifespan. This is consistent with experiments. Universality implies that yeast may provide a master key to the cellular mechanism of universal mortality, aging, selection, evolution, and its regulation in all animals. This suggests that one may look for its manifestations in animal cells, e.g., in their replicative senescence.

Arguably, universal biology emerged in response to major mass extinctions which posed universal threat to different species, and is related to disposable genes, which were beneficial for longevity in the wild, but became detrimental in evolutionary unprecedented conditions.

Further theoretical and experimental studies of the universal law and its implications are suggested.

1 Motivation and Approach

Life evolved via selection of the fittest. Selection posed different challenges for different species, thus yielded enormous biological diversity and complexity of survivors. In contrast, five major mass extinctions [1] were universal "rapidly adjust or die" threat to the very existence of large proportion of species (96% perished in the most drastic extinction about 248 to 238 million years ago). Universal threat could yield certain universality in selection. Indeed, presented physical approach unravels universality which underlies enormous diversity of evolutionary branches. Evolutionary data are sparse and largely qualitative. So, study universality of diverse living species. selection proceeded via death of the frail. Thus, quantify selection with mortality. To amplify universality, consider different human [2] and protected laboratory populations of med- and fruit flies [3,4], nematodes (including mutants and biologically amended) [5–8], yeast [5,9,10] in changing conditions. Their protection from elements of nature, predators, shortage of resources, diseases, etc. nearly eliminates extrinsic mortality, and thus selection, which dominate in the wild. Their predominantly intrinsic mortality is well quantified. It is heterogeneous and non-stationary (e.g., within human lifespan infant mortality increased 30-fold and life expectancy by 50%). Animal populations (especially genetically homogeneous) are relatively small, and their mortality significantly fluctuates[1]. So, consider life expectancy e and probability l for a live newborn to survive to a given age x. These quantities are robust to heterogeneity, non-stationarity and fluctuations. Indeed, suppose the population consists of the groups with the number $N^G(x)$ of survivors to age x. If $C_G = N^G(0)/N(0)$ and $l^G = N^G(x)/N^G(0)$ are correspondingly the ratio of the population and the survivability to x in the group G, then the population survivability l self-averages over population heterogeneity:

$$l = \frac{\sum N^G(x)}{\sum N^G(0)} = \sum C_G l^G = \langle l^G \rangle \tag{1}$$

($\langle \cdots \rangle$ denotes averaging). Since $l = p(0)p(1)\ldots p(x-1)$, where $p(x)$ is the probability to survive from x to $(x+1)$, so $\ln l(x)$ averages $\ln p$ over, and fluctuations with, time x. Similarly, life expectancy $e = \langle \mathrm{e}^G \rangle$ averages over population heterogeneity and entire lifespan. Life expectancy changes 20,000 times from yeast to humans. To account for this change, scale age x and life expectancy e with a single species specific parameter F. Choose $F = 1$ year for humans, $F = 0.5$ day for flies and nematodes, $F = 0.25$ generation for yeast (the choice of F see later). Then Fig. 1 for all animals manifests predominantly universal[2] dependence of survivability l on the scaled life expectancy e/F and age x/F.

[1] The lifespan of four populations of 623, 662, 248 and 5751 inbred 3×3 male fruit flies in 4-dram shell vials with weekly transfer to fresh medium [3] varied from 18.6 to 34.3 days. In the populations with close life spans (18.6 and 22 days) the probabilities to die on the 38-th day were 18 times different. In the largest population the mortality rate of 15 days olds was 17 times lower than of those 4 days younger and 3 days older. Such giant fluctuations may be related to vial difference and their weekly change. Nematode populations [5–8] include only 50–100 worms.

[2] Actuary Gompertz [11] in 1825 presented the first universal law of mortality for human advanced age. Thereafter the search for such law for all animals went on – see [12, 13] and references there. Accurate knowledge of human mortality is important for economics, taxation, insurance, gerontology, etc. So, demographic life tables present millions of mortality data in different countries over their history. To

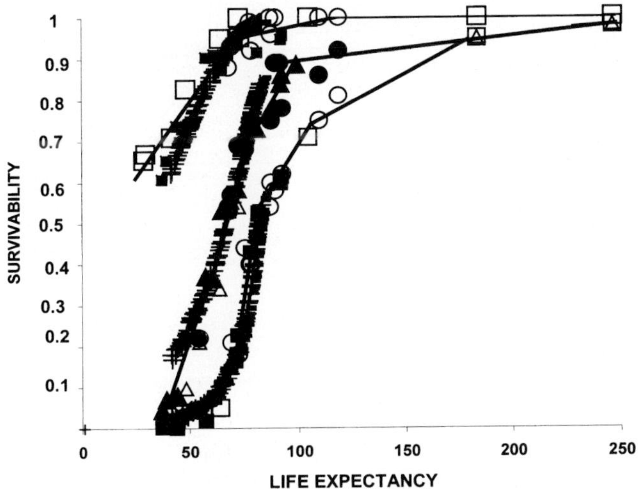

Fig. 1. Universal dependence of survivability (vertical axis) on scaled life expectancy e/F (horizontal axis) for given scaled ages x/F of 125 Swiss (1876–2001 years, crosses) and 50 Japanese (1950–1999, dashes) female [2]; 17 fly [3,4] (black) and 15 nematode [5–8] (white); 14 yeast [5,9,10] (circles) populations. Their scaling parameters F and ages x are correspondingly $F = 1$ year; 0.5 days; 0.25 generations and $x = 30$, 85, 73 (upper, lower and middle curves) years; 15, 45 (squares), 35 (triangles) days; 7, 21 (white), 16 (black) generations. Each sign presents raw experimental data for a population. Some of them overlap and are indistinguishable for humans and flies, nematodes and yeast. Few accidental deviations are omitted. Solid lines demonstrate the universal law. The difference between presented and other human data (e.g. those for, e.g., 252 Swedish female and 159 English male populations) is on the scale of difference between nematodes and yeast

In all cases total survivability $l = l^* + l'$, where l^* is universal, i.e. depends on e/F and x/F only, while non-universal l' which depends on all multiple factors affecting mortality is $\ll l^*$. (From now on, unless specified otherwise, only universal variables are considered, and the superscript star is omitted).

2 Universality Law: Derivation

Universality for different heterogeneous and homogeneous populations implies that the relation between universal l and e is the same as the relation between their values in any of the groups in the population, i.e. if $l = l(e/F, x/F)$, then $l^G = l(\mathrm{e}^G/F, x/F)$. So, by (1), $l = \langle l^G \rangle = \langle l(\mathrm{e}^G/F, x/F) \rangle$ and $l = l(e/F, x/F) = l(\langle \mathrm{e}^G/F \rangle, x/F)$, i.e.

---

better estimate and forecast mortality, demographers dropped the universal law and developed over 15 mortality approximations [14]. Yet, 180 years after Gompertz, the existence of the universal mortality law remains controversial.

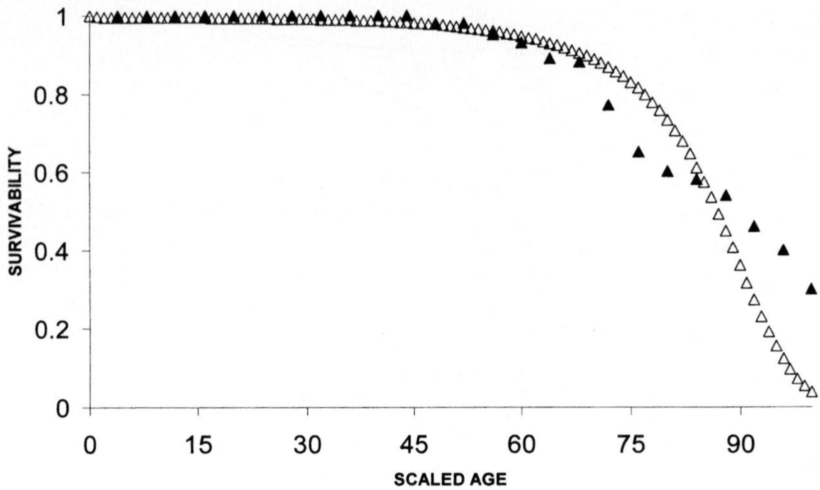

Fig. 2. Survivability vs scaled age for Japanese females who died in 1999 [2] (black triangles) and yeast [5] (white triangles). Their scaled life expectancies are correspondingly 84 and 89

$$\left\langle l\left(\frac{\mathrm{e}^G}{F},\frac{x}{F}\right)\right\rangle = l\left(\left\langle\frac{\mathrm{e}^G}{F}\right\rangle,\frac{x}{F}\right) \tag{2}$$

Such equation implies [15] that l is a universal piecewise linear function of e/F with simultaneous for all ages x/F intersections (denote such dependence as the universal law) and that at any age population heterogeneity, i.e. e^G/F in all groups, is restricted to a single interval ("echelon") of e/F between universal intersections (denote this as a restricted heterogeneity). The universal law agrees with Fig. 1, and restricted heterogeneity implies that dominant fraction of all its populations reduces to a single echelon.

The knowledge of exact analytical dependence on e allows one to establish species specific scales F which provide its minimal relative mean square deviation from experimental[3]. These scales demonstrate (see Fig. 2) proximity of human ($e/F = 84$) and yeast ($e/F = 89$) survival curves vs x/F despite enormous ($\sim 20,000$ times) difference in their life expectancies. Proximity is not always as good as in Fig. 1. Empirical study [16] demonstrates very different age dependence in different echelons (especially in young and old age), thus high sensitivity to contributions of few echelons. Ages and conditions with low mortality may be more universal (like high e/F in Fig. 2).

[3] The number of human data (whose statistics is by far the best) included in the approximations was chosen equal to the average number of data per each class; human e/F were chosen equidistant. Some scaled ages x/F in Fig. 1 are slightly different for different classes (e.g. 73 human years are compared to 35 rather than to $73/2 = 36.5$ fly days) to amplify universality. At certain ages some intersections in Fig. 1 are weakly pronounced and unobservable.

Poor animal statistics and unknown analytical formula of age dependence (thus insufficiently accurate scaling) do not allow to account for more than a single echelon. Large size and by far the best statistics of human populations allows it.

Consider the period probability $d(x) = [l(x) - l(x+1)]$ for a live newborn to die between x and $(x+1)$ years (note that human $F = 1$ and $x/F = x$). Similar to $l(x)$, the value of $d(x)$ self-averages over heterogeneity, but it is more time specific than l. The most time specific parameter is "infant mortality" $d(0) = q(0)$ which depends on the time from conception to $x = 1$ only. Similar to Fig. 1, the dependence of $d(x)$ on $q(0)$ for each human curve, is approximately piecewise linear, also with 5 (as in Fig. 1) intersections (see, e.g. Fig. 3) which are nearly simultaneous at all ages, but somewhat different in different countries. Since both $d(x)$ and $q(0)$ are self-averaging variables, previous analysis yields the universal law. Suppose the universal j-th echelon boundaries are

$$q_j < q^G(0) < q_{j+1}. \tag{3}$$

Since mortality is never negative, its ultimate minimum is $q_1 = 0$. An arbitrarily heterogeneous population may be distributed at several intervals, and piecewise linear law reduces $d(x)$ to the sum over all intersection crossover values $d_j(x)$:

$$d(x) = \sum c_j d_j(x), \text{ where } \sum c_j = 1. \tag{4}$$

3 Results

The number of population specific concentrations c_j of intersection $d_j(x)$ depends on the heterogeneity of the population. If it reduces to a single echelon, thus to two crossovers, then $d(0) = q(0)$ and $d_j(0) = q_j$ yield the universal law (which maps on coexistence of two phases):

$$d(x) = c_j d_j(x) + (1 - c_j) d_{j+1}(x); \qquad c_j = \frac{q_{j+1} - q(0)}{q_{j+1} - q_j}. \tag{5}$$

If a population reduces to two echelons, thus to three crossovers, then, by (4), $d(x)$ reduces to $q(0)$ and one population specific concentration. Simple algebra proves that then intersections in all such populations are situated at universal segments of the universal law or their extensions, and allow one to determine the universal law. This is the case in most developed countries (e. g., in 1948–1999 Austria, 1921–1996 Canada, 1921–2000 Denmark, 1841–1998 England, 1941–2000 Finland, 1899–1997 France, 1956–1999 West Germany, 1906–1998 Italy, 1950–1999 Japan, 1950–1999 Netherlands, 1896–2000 Norway, 1751–2000 Sweden, 1876–2001 Switzerland). The resulting universal law in Fig. 4 is verified with few percent accuracy with ~ 3000 human curves [2] (18 countries, two sexes, ages from 1 to 95) – see Fig. 3, where straight lines

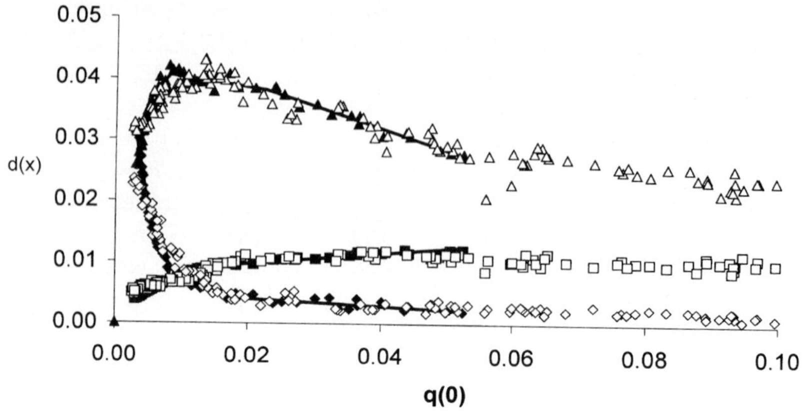

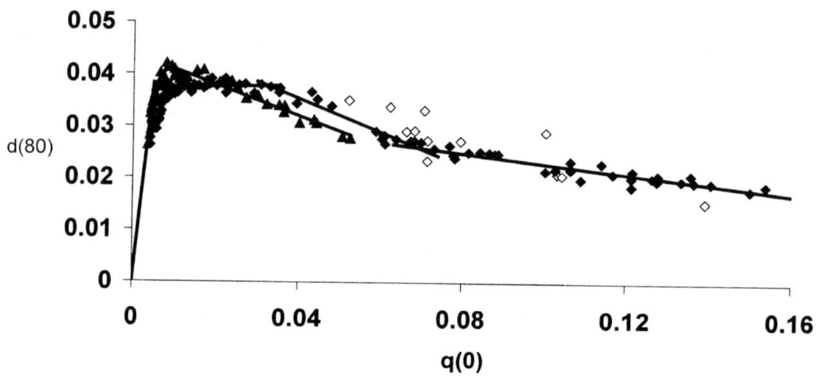

Fig. 3. (*Upper plot*). Period probabilities for live newborn Japanese (*black*) and Swedish (*white*) females to die (year of death from 1950 to 1999 and 1751 to 2002) between 60 and 61 (*squares*), 80 and 81 (*triangles*), 95 and 96 (*diamonds*) years of age vs. infant mortality $q(0)$. Japanese relative mean squared deviations from the universal law with two echelons (*straight lines*) are correspondingly 2.4%, 2.3% and 10%. Significant Swedish deviations are related to 1918 flu pandemic in Europe. (*Lower plot*). Same for French (*diamonds*) and Japanese (*triangles*) females (year of death from 1898 and 1950 to 2001 and 1999) between 80 and 81 years of age. Empty diamonds correspond to 1918 flu pandemic and World Wars. They are disregarded in the universal law with two echelons (*straight lines*), which yield relative mean square deviations from black signs to generic 5%. When Japanese $q(0) = 0$, its extrapolation yields $d(80) = 0$

demonstrate the accuracy of approximations with two echelons. (Deviations are more pronounced when infant mortality significantly changes from one calendar year to another due to wars, epidemics, crop failures etc., and/or is relatively large, i.e. when conditions are insufficiently protected. The accuracy may be improved if all echelons in Fig. 4 are accounted for). General

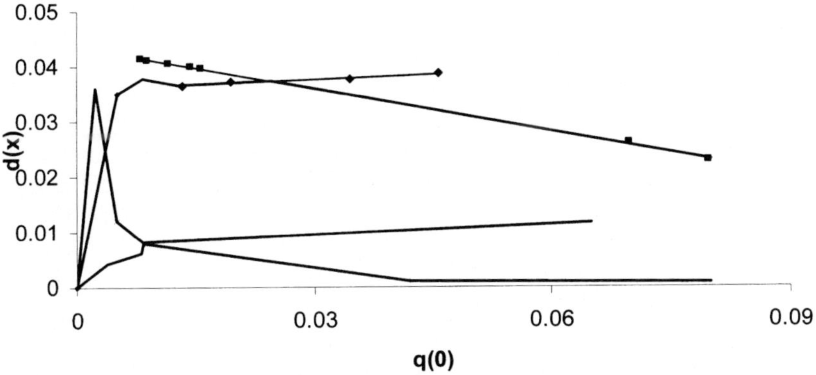

Fig. 4. Universal law (*thick lines*) of human mortalities $d(60)$, $d(80)$ and $d(95)$ vs $q(0)$-middle, lower and upper curves. At $q(0) < 0.003$ it is extrapolated. *Thin lines* extend its linear segments. Intersections (similar to those in Fig. 3) are exemplified by diamonds and squares for England (two successive intersections), France, Italy and Japan, Finland, Netherlands, Norway, Denmark, France, England. Note that all intersections are close to universal straight lines

universality in Fig. 1 suggests that (properly scaled) law in Fig. 4 is universal for all protected animal populations. Consider its predictions and implications. (Earlier these results were predicted empirically [13,15,16] and derived analytically [17]).

4 Discussion and Conclusions

Derived law is universal for all species, from single cell yeast to humans. At a given age x it depends in Fig. 1 on a single population specific parameter – life expectancy e – and a single species specific parameter F. The law includes no characteristics of non-stationary and heterogeneous animal- environment interactions (e.g. via metabolism) which are necessary conditions of life. Such law is specific for live systems with their homeostatic self-adjustment to environment. Its dynamics which is universal for all animals, with their drastically different biology, evolutionary history, and complexity, is also unique for live systems – cf. different thermodynamics of liquids and glasses. The law which is valid for all live, and only live, systems is a life specific law of nature.

Mortality is an instrument of natural selection and biological diversity. The law which is preserved in evolution of species from humans to yeast is a conservation law of selection, evolution, and biology. It suggests their universal mechanism which dominates in evolutionary unprecedented protected populations with predominantly intrinsic mortality. Then the contribution of all other mechanisms is either relatively small or indirect, via the universal mechanism. Its universality in all animals implies that single cell yeast may be a master key to it and its regulation. Universal law demonstrates that species specific natural selection is replaced in protected populations by predominantly universal adaptation of intrinsic survivability to genotypes, phenotypes, life history, environment, etc., via properly scaled life expectancy. Universal adaptation is stepwise and proceeds via universal "ladder" of "rungs" with simultaneous for all ages crossovers. The number of its crossovers equals the number of major extinctions. (Note that each live species in the course of its entire history survived all extinctions). Less universal extinctions may yield mini-rungs, and possibly punctuated evolution [18]).

Universal Fig. 1 establishes universal scale of ages for different species, and suggests that life expectancy in all existing species is restricted to around 100 human years, while minor directed genetic and/or biological changes increase it to the Methuselah 250 (healthy and vital [8]) human years.

When infant mortality vanishes, the universal law yields, according to Fig. 4, zero universal mortality till certain age ($\sim$ 80 years for humans), thus correspondingly low total mortality and high life expectancy. Mortality on the scale of stochastic fluctuations, i.e. consistent with zero universal mortality, was indeed observed in humans, flies, nematodes, yeast. In 2001 Switzerland only 1 (out of 60,000) girl died at 5, 9, and 10 years; 5 girls died in each age group from 4 to 7 and from 9 to 13 years; 10 or less from 2 till 17 years; no more than 16 from 2 till 26. Statistics are similar in all 1999–2002 Western developed countries [2]. Similarly, mortality of dietary restricted flies at 8 days was $\sim$ 0.0004 [19]. Yeast mortality [9] was zero during half of its mean life span (Jazwinski et al. [9] presented the first model which stated that a sufficient augmentation of aging process resulted in a lack of aging). The probability to survive from 80 to 100 years increased in Western Europe 20-fold in the last 50 years [20]. Mean life expectancy increased almost three times in the last 250 years with improving (medical included) human conditions [2]; 2.4-fold with genotype change in Drosophila [3]. None of nematodes with changes in small number of their genes and tissues [8] died till 27 days, i.e. during 54 human years on Fig. 1 scale; from 58 till 90, from 126 till 162 "human years". 25% of amended nematodes survived till 296 and thereafter did not die till 318 "human years". All human data demonstrate remarkably low non-universal (in particular, accidental) mortality, whose fraction does not increase with the decrease in total mortality. Zero mortality till certain age implies zero universal mortality at any age (unless it has a singularity at certain age [21]), thus very low total mortality, and the Methuselah life expectancy. Indeed, mean life span of mutant nematodes increased to 90 days [6, 7] and to the Methuselah 124 days [8] (248 years on "human" scale), with no apparent loss in health and vitality.

An important implication of the universal law is its plasticity. Universal mortality at any age is related to infant mortality[4] (see Fig. 4). Thus, it rapidly adjusts to, and is determined by, current living conditions if they do not significantly change in 2 years, from conception till 1 year, for humans; few percent of the life span for any species. So, universal mortality is independent of the previous life history ("short mortality memory" of it) and, together with infant mortality, it may be rapidly reduced and reversed to its value at a much younger age. Indeed, following unification of East and West Germany, within few years mortality in the East declined toward its levels in the West, especially among elderly with $\sim$ 45 years of their different life histories. Mortality of the female cohort, born in 1900 in neutral Norway, at 59 years restored its value at 17 years, i.e. 42 years younger [2]. Note that such mortality decrease, similar to the one in East Germany after its unification, is not dominated by death of the frail. The latter alters composition of the cohort, and the resulting change in mortality depends on life history rather than on current conditions only. Thus, it contributes to the deviations from the universal laws (which are relatively small) rather than to the universal mortality. Mortality plasticity is also very explicit in experiments where dietary restriction in rats [22] and flies [19] is switched on. However, when dietary restriction is switched off and changes to full feeding, their longevity remains higher than in the control group of animals fully fed throughout life. Also, when fly temperature was lowered from 27 to 18 degrees or vice versa, the change in mortality, driven by life at previous temperature, persisted in these flies compared to the control ones. Such long memory of life history may be related to rapid changes in temperature or feeding, since universal law is valid when infant mortality little changes within a day for flies, a month for rats, a year for humans. This calls for comprehensive tests of mortality adaptation to such conditions. Similar tests may verify a possibility to reverse and reset mortality of a homogeneous cohort to a much younger age.

Restricted heterogeneity implies that at the intersections population homogenizes. This agrees with experimental data [15].

[4]Thus, eliminating all deaths before age 50 would not yield just about a 4-year rise in current life expectancy at birth, as it would if mortality at higher ages were little correlated with lower age mortality. Demo- and biodemographers consider the most specific mortality variable – the probability to die between ages x and $(x+1)$. It equals

$$q(x) = \frac{l(x) - l(x+1)}{l(x)} = \frac{d(x)}{1 - d(0) - d(1) - \cdots - d(x-1)},$$

thus its universality is not as explicit as that of $d(x)$ and was unnoticed.

5 Outstanding Problems

Vanishing and highly plastic universal mortality calls for evolutionary and biological explanation. In the wild competition for sparse resources is fierce, and only relatively few genetically fittest animals survive to their evolutionary "goal"-reproduction. Even human life expectancy at birth was around 40–45 years just over a century ago, and 17.2 years for males in 1773 (crop failure year) Sweden. There are no evolutionary benefits from genetically programmed death and/or aging of tiny number of survivors to old age. Since prior to and during reproductive age (when survival is evolutionary beneficial) mortality, and even aging (thus irreparable damage), may be negligible in protected populations (see above) and there are no evolutionary benefits in switching off repair mechanisms later, so intrinsic mortality and aging are presumably disposable evolutionary byproducts. Such byproducts may be related to genes, which are beneficial for non-universal longevity in the wild, but are detrimental in evolutionary unprecedented protected conditions where longevity is predominantly universal (new kind of Williams antagonistic pleiotropy). Such genes are relatively easy to alter or switch off. This is consistent with healthy and vital Methuselah age in nematodes. Universality suggests that its mechanism may be reduced to genetically regulated universal processes in cells, and related to a certain universal genome (cf. "longevity genes" [23–26]. Single cell yeast may provide a master key to the cellular mechanism of Methuselah age, adaptation, and their regulation, in all animals (see cartoon in Fig. 5). One may look for its manifestations in animal cells, e.g., in their replicative senescence (see review [27] and references therein), apoptosis, possibly even in certain aspects of cancer ([28] and references therein) and cancer gene therapy (e.g., inhibition of ontogenes and activation of tumor suppressor genes).

In protected populations non-universal mortality is relatively small, thus all other mechanisms are less important or correlated with universal mechanism. Conservation law of universal evolution allows for its quantitative study with current survivors, as well as for accurate definition of species, families etc. according to their intersection scales in Figs. 1 and 4. Remarkably simple species specific scales in Fig. 1 suggest the existence of their "quantization law".

Universal piecewise linear dependence on e/F is related to its invariance to restricted population heterogeneity. Invariance which yields analytical formula of the universal dependence on age, remains to be established, as well as accurate species specific scaling of mortality dependence on age and infant mortality.

Universal law presents universal demo- and biodemographic approximation, which may be important for economics, taxation, insurance, gerontology, etc.

Interconnection between universal evolution, selection, mortality, aging and its vitality, and presumably mass extinctions, suggests certain univer-

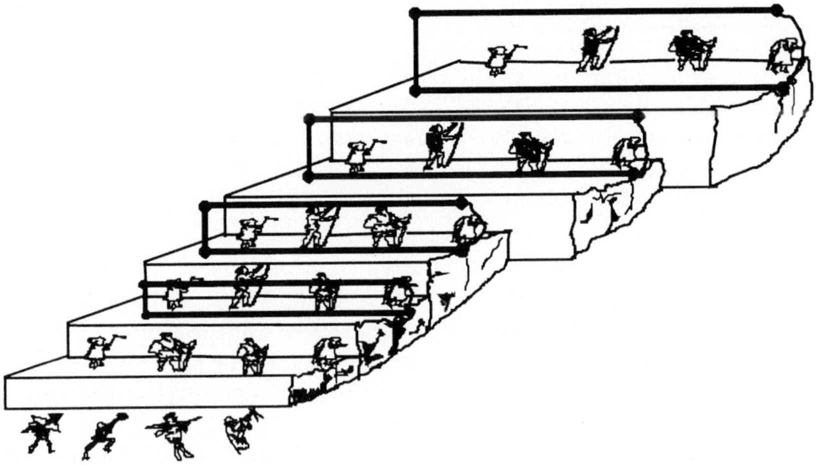

Fig. 5. The ladder of rungs in the human "bridge of death". Better social and medical protection at its successive rungs implies higher "protective walls" against, thus delay in, death and aging, but does not shift the precipice of the maximal life span at the bridge end. Biological amendments increase the maximal life span and shift the "bridge of death" end

sality in biology at large and calls for multidisciplinary (evolutionary, biological, demographic, physical and mathematical) study. Universal law, its implications, and predictions may be comprehensively verified and refined theoretically (with available mortality data) and experimentally. Other outstanding problems include partitioning of universal intrinsic and non universal extrinsic mortality and their relative contribution; "quantization law" of evolutionary scales in Figs. 1, 4; establishing genes, cellular mechanism and the origin of plasticity of universal mortality; physical and biological nature of intersections and echelons.

Acknowledgement

I am very grateful to I. Kolodnaya for assistance and cartoon. Financial support from A. von Humboldt award and R. & J. Meyerhoff chair is highly appreciated.

Appendix

Demographic life tables present mortality data in different countries over their history. For males and females, who died in a given country in a given calendar year, the data list, in particular, "period" probabilities $q(x)$ (for survivors to x) and $d(x)$ (for live newborns) to die between the ages x and $(x+1)$

[note that $d(0) = q(0)$]; the probability $l(x)$ to survive to x for live newborns; the life expectancy $e(x)$ at the age x. The tables also present the data and procedures which allow one to calculate the values of $q(x)$, $d(x)$, $l(x)$, $e(x)$ for human cohorts, which were born in a given calendar year. Populations, their conditions and heterogeneity are different, yet demographic approximations reduce period mortality of any given population to few parameters. Assumption that under certain conditions a dominant fraction of period mortality in all heterogeneous populations is universal is sufficient to derive the universal Fig. 1, as well as (5) and its conditions (3). According to Fig. 4, until $\simeq 65$ years, $d(x)$ decreases when $q(0)$ increases. Beyond $\simeq 85$ years, $d(x)$ increases together with $q(0)$. In between, $d(x)$ exhibits a well pronounced maximum (naturally, smeared by generic fluctuations). Consider the origin of such dependence on age. The value of $d(x)$ is proportional to the probability for a newborn to survive to x and to die before the age $(x + 1)$. When living conditions improve, the former probability increases, while the latter one decreases. In young age the probability to survive to x is close to 1, so $d(x)$ is dominated by the mortality rate, and thus monotonically decreases together with $q(0)$. For sufficiently old age, low probability to reach x dominates. It increases with improving living conditions, i.e. with decreasing $q(0)$, thus $d(x)$ increases with decreasing $q(0)$. At an intermediate age, when improving living conditions sufficiently increase survival probability, $d(x)$ increase is replaced with its decrease. Then $d(x)$ has a maximum at a certain value of $q(0)$. Thus, minor genetic and/or biological changes should yield the $d(x)$ maximum at 95 years and beyond. To quantify the accuracy of the results, consider the number $D(x)$ of deaths at a given age x in each calendar year. According to statistics, the corresponding stochastic (i.e. minimal) error is $\sim 2/[D(x)]^{1/2}$. At 10 years of age it increases from $\sim 20\%$ in 1976 to $\sim 200\%$ in 2001 Switzerland and leads to large fluctuations in $q(10)$. At 40 years it is $\sim 20\%$; at 80 years it is $\sim 6\%$ in Switzerland and $\sim 2\%$ in Japan. Universal Fig. 4 and accuracy of the universal $d(x)$ vs $d(0)$ with two echelons may be refined with larger number of echelons in populations. The total number of (4) is $2XT$, where 2 is the number of sexes, $(X-1)$ is the maximal considered age, $T = \sum T_g$, where T_g is the number of calendar years in the period life tables of the country g. The total number of (4) variables with 5 intersections is $10T+5X$. Since $T \sim 2000$, $X \sim 100$, the number of variables is ~ 20 times less than the number of equations. So, consider non-universal mortality with the crossover concentrations which change with age (e.g., every five years) to provide the same number of equations and variables. The resulting change in the concentrations, calculated according to life tables and (4), determines the relative non-universal mortality.

References

1. See, e.g., L. Villier and D. Korn, Science **306**, 264 (2004) and Refs. therein

2. Human Mortality Database, Univ. of California, Berkeley (USA) and MPI of Demographic Research (Germany), http://www.mortality.org (2003)
3. J.W. Curtsinger, H.H. Fukui, D.R. Townsend, J.W. Vaupel: Science **258**, 461 (1992)
4. J.R. Carey: *Applied demography for biologists* (Oxford university press, New York 1993)
5. J.C. Jiang, E. Jaruga, M.V. Repnevskaya, S.M. Jazwinski: FASEB J.14, 2135 (2000)
6. K. Houthoofd et al: Exp. Geront. **37**, 1371 (2002)
7. K. Houthoofd, B.P. Braeckman, T.E. Johnson, J.R. Vanfleteren: Exp. Geront. **38**, 947 (2003)
8. A. Arantes-Oliveira, J.R. Berman, C. Kenyon: Science **302**, 611 (2003)
9. S.M. Jazwinski, S. Kim, C.-Y. Lai, A. Benguria: Exp. Geront. **33**, 571 (1998)
10. S. Shama, C.-Y. Lai, J.M. Antoniazzi, J.C. Jiang, S.M. Jazwinski: J. Exp. Cell Res. **245**, 379 (1998)
11. B. Gompertz: Phil. Tranc. R. Soc. A **115**, 513 (1825)
12. S.J. Olshansky, B.A. Carnes, Demography **34**, 1 (1997)
13. M.Ya Azbel': Exp. Geront. **37**, 859 (2002)
14. A.J. Coale, P. Demeny, B. Vaughan: *Regional Model Life Tables and Stable Populations*, 2nd edn (Academic Press, New York 1993)
15. M.Ya. Azbel', Phys. Rev E **66**, 01607-1 (2002)
16. M.Ya. Azbel', PNAS USA 95, 9037 (1997); 96, 3303, 15368 (1999)
17. M.Ya. Azbel', Physica A **329**, 436 (2003)
18. S.J. Gould, N. Eldridge: Nature (London) **366**, 223 (1993), see also A. Cossing, Nature (London) **369**, 309 (1994).
19. W. Mair, P. Goymer, S.D. Pletcher, L. Partridge: Science, **301**, 1731 (2003)
20. J.W. Vaupel, J.R. Carey, K. Christiansen: Science **301**, 1679 (2003)
21. M.Ya. Azbel': Physica A **341**, 629 (2004).
22. B.P. Yu, E.J. Masoro, C.A. McMahon: J. Gerontology **40**, 657 (1985)
23. S.M. Jazwinski: Science **263**, 54 (1996)
24. A.A. Puca et al: Proc. Natl. Acad. Sci. **98**, 10505-10508 (2001)
25. G. Atzmon, P. Rabizadeh, R. Gottesman, A.R. Shuldiner, N. Barzilai: Am. J. Hum. Genetics **73**, 1183B (2003)
26. T.S. Nawrot, J.A. Staessen, J.P. Gardner, A. Aviv: Lancet **363**, 507 (2004)
27. V.J. Cristofalo et al, Mechanisms of Aging and Development **125**, 827 (2004)
28. M.M. Gottesman: Cancer Gene Therapy **10**, 501 (2002)

Manifestation of Chaos in Real Complex Systems: Case of Parkinson's Disease

Renat M. Yulmetyev[1], Sergey A. Demin[1], and Peter Hänggi[2]

[1] Department of Physics, Kazan State Pedagogical University, 420021 Kazan, Mezhlauk Street, 1 Russia
rmy@dtp.ksu.ras.ru
rmy@theory.kazan-spu.ru

[2] Department of Physics, University of Augsburg, Universitätsstrasse 1, D-86135 Augsburg, Germany

In this chapter we present a new approach to the study of manifestations of chaos in real complex systems. Recently we have achieved the following result. In real complex systems the informational measure of the chaotic character (IMC) can serve as a reliable quantitative estimation of the state of a complex system and helps to estimate the deviation of this state from its normal condition. As the IMC we suggest the statistical spectrum of the non-Markovity parameter (NMP) and its frequency behaviour. Our preliminary studies of real complex systems in cardiology, neurophysiology and seismology have shown that the NMP has diverse frequency dependence. It testifies to the competition between Markovian and non-Markovian, random and regular processes and makes a crossover from one relaxation scenario to the other possible. On this basis we can formulate the new concept in the study of the manifestation of chaoticity. We suggest the statistical theory of discrete non-Markov stochastic processes to calculate the NMP and the quantitative evaluation of the IMC in real complex systems. With the help of the IMC we have found the evident manifestation of chaoticity in a normal (healthy) state of the studied system, its sharp reduction in the period of crises, catastrophes and various human diseases. It means that one can appreciably improve the state of a patient (of any system) by increasing the IMC of the studied live system. The given observation creates a reliable basis for predicting crises and catastrophes, as well as for diagnosing and treating various human diseases, Parkinson's disease in particular.

1 Introduction

Today the study of the manifestations of chaos in real complex systems of diverse nature has acquired great importance. The analysis of some properties and characteristics of real complex systems is impossible without a quantitative estimate of various manifestations of chaos. The dynamics or evolution of the system can be predicted by the change of its chaoticity or regularity. The discovery of the phenomenon of chaos in dynamic systems has changed the attitude with regard to the functioning of complex systems, a human organism in particular. The chaos is the absence of regularity. It characterizes the randomness and the unpredictability of the changes of the behaviour of a system. At the same time, the presence of chaos in dynamic systems does not mean it cannot be taken under control. Instability of dynamic systems in the state of chaos creates special sensitivity to both external and internal influences and perturbations. The series of weak perturbations of the parameters of the system allows to change its characteristics in the required direction. "Chaos" is frequently understood as a determined dynamic chaos, that is, the dynamics depend on the initial conditions, parameters.

Lasers, liquids near the threshold of turbulence, devices of nonlinear optics, chemical reactions, accelerators of particles, classical multipartite systems, some biological dynamic models are examples of nonlinear systems with manifestations of determined chaos. Now manifestations of chaos are being studied in different spheres of human activity.

The control of the behaviour of chaotic systems is one of the most important problems. Most of the authors see two basic approaches to solve the problems [1,2]. Both directions envisage a preliminary choice of a certain perturbation. The selected perturbation is used to exert influence on the chaotic system. The first direction relies on an internal perturbation, the choice of which is based on the state of the system. The perturbation changes the parameter or the set of parameters of the system, which results in the ordered behaviour of the chaotic systems. The methods focusing on the choice of such parameters (perturbations) are referred to as "methods with a feedback" [1]-[6]. They do not depend on the studied chaotic system (model) as these parameters can be selected by observing the system for some period of time. One also considers that the methods with a delayed feedback [3, 7] belong to the first direction. The second approach presupposes that the choice of the external perturbation does not depend on the state of the chaotic system under consideration. By affecting the studied system with the similar perturbation, it is possible to change its behaviour. The present group of methods is an alternative to the first one. These methods can be used in cases when internal parameters depend on the environment [1, 8, 9].

Generally, when choosing internal (external) perturbations it is possible to determine three basic stages: the estimation of the initial information, the choice of the perturbation and the bringing of the chosen strategy of control into action (its practical realization). At the first step the information on the state of the studied system is collected. At the second stage the received information is processed according to the plan or strategy of the control. On the basis of the achieved results the decision on the choice of the internal (of the external) perturbation is accepted. After that the chosen strategy of chaos control is put into practice [2].

The initial idea of the present concept was to separate Markov (with short-range time memory) and non-Markov (with long-range time memory) stochastic processes. However, the study of real complex systems has revealed additional possibilities of the given parameter. Actually, the non-Markovity parameter represents a quantitative measure of the chaoticity or regularity of various states of the studied system. An increase of the given parameter ($\varepsilon_1(0) \gg 1$) corresponds to an increase of the chaoticity of the state of the system. A decrease of the non-Markovity parameter means a greater ordering (regularity) of the state of the system. The given observation allows one to define a new strategy for estimating the chaoticity in real systems. This new approach in chaos theory can be presented as an alternative to the existing methods. Further analysis of the non-Markovity parameter allows one to define the degree of chaoticity or regularity of a state of the system.

In this work the new strategy for the study of manifestations of chaoticity is applied to real complex systems. The possibilities of the new approach are revealed at the analysis of the experimental data on various states of a human organism with Parkinson's disease. Parkinson's disease is a chronic progressing disease of the brain observed in 1–2% of elderly people. The given disease was described in 1817 by James Parkinson in the book "An essay on the shaking palsy". In the 19th century the French neurologist Pierre Marie Charcot called this disease "Parkinson's disease". The steady progress of the symptoms and yearly impairment of motor function is typical of Parkinson's disease. Complex biochemical processes characteristic of Parkinson's disease result in a lack of dopamine, a chemical substance which carries signals from one nerve cell to another. The basic symptoms typical of Parkinson's disease form the so-called classical triad: tremor, rigidity of muscles (disorder of speech, amimia), and depression (anxiety, irritability, apathy). The disease steadily progresses and eventually the patient becomes a helpless invalid. The existing therapy comprises a set of three basic treatments: medical treatment, surgical treatment and electromagnetic stimulation of the affected area of the brain with the help of an electromagnetic stimulator. Today this disease is considered practically incurable. The treatment of patients with Parkinson's disease requires an exact estimate of the current state of the person. The offered concept of research of manifestations of chaoticity allows one to track down the least changes in the patient with the help of an exact quantitative level of description.

Earlier we found out an opportunity for defining the predisposition of a person to the frustration of the central nervous system due to Parkinson's disease [13]. Our work is an extension and development of the informational possibilities of the statistical theory of discrete non-Markov random processes and the search for parameters affecting the health of a subject.

2 The Statistical Theory of Discrete Non-Markov Random Processes. Non-Markovity Parameter and its Frequency Spectrum

The statistical theory of discrete non-Markov random processes [10–12] forms a mathematical basis for our study of complex live systems. The theory allows one to calculate the wide quantitative set of dynamic variables, correlation functions and memory functions, power spectra, statistical non-Markovity parameter, kinetic and relaxation parameters. The full interconnected set of these variables, functions and parameters creates a quantitative measure of chaoticity used for the description of processes, connected with the functioning of alive organisms.

We use the non-Markovity parameter ε as a quantitative estimate of the non-Markov properties of the statistical system. The non-Markovity parameter allows to statistical processes into Markov processes ($\varepsilon \to \infty$), quasi-Markov processes ($\varepsilon > 1$) and non-Markov processes ($\varepsilon \sim 1$). Besides the non-Markovity parameter we also use the spectrum of the non-Markovity parameter. We define the spectrum as the set of all values of the physical parameter used for describing the state of a system or a process. Let's consider the first and the nth kinetic equations of the chain of connected non-Markov finite-difference kinetic equations [10, 11]:

$$\frac{\Delta a(t)}{\Delta t} = \lambda_1 a(t) - \tau \Lambda_1 \sum_{j=0}^{m-1} M_1(j\tau) a(t - j\tau) \, , \tag{1}$$

$$\cdots$$

$$\frac{\Delta M_n(t)}{\Delta t} = \lambda_{n+1} M_n(t) - \tau \Lambda_{n+1} \sum_{j=0}^{m-1} M_{n+1}(j\tau) M_n(t - j\tau) .$$

The first equation is based on the Zwanzig–Mori kinetic equation of nonequilibrium statistical physics:

$$\frac{\mathrm{d}a(t)}{\mathrm{d}t} = -\Omega_1^2 \int_0^t \mathrm{d}\tau M_1(j\tau) a(t - j\tau) \, .$$

Here $a(t)$ is a normalized time correlation function (TCF):

$$\lim_{t \to 0} a(t) = 1, \qquad \lim_{t \to \infty} a(t) = 0 \, .$$

The zero memory function $a(t)$ and the first-order memory function $M_1(t)$ in (1),

$$M_0(t) = a(t) = \frac{\langle \boldsymbol{A}_k^0(0)\boldsymbol{A}_{m+k}^m(t)\rangle}{\langle |\boldsymbol{A}_k^0(0)|^2\rangle}, \quad t = m\tau,$$

$$M_1(j\tau) = \frac{\langle \boldsymbol{A}_k^0(0)\hat{L}_{12}\{1+i\tau\hat{L}_{22}\}^j\hat{L}_{21}\boldsymbol{A}_k^0(0)\rangle}{\langle \boldsymbol{A}_k^0(0)\hat{L}_{12}\hat{L}_{21}\boldsymbol{A}_k^0(0)\rangle}, \quad M_1(0) = 1.$$

$$\boldsymbol{A}_k^0(0) = (\delta x_0, \delta x_1, \delta x_2, \cdots, \delta x_{k-1}),$$

$$\mathbf{A}_{m+k}^m(t) = (\delta x_m, \delta x_{m+1}, \delta x_{m+2}, \cdots, \delta x_{m+k-1}),$$

describe statistical memory in complex systems with a discrete time ($\boldsymbol{A}_k^0(0)$ and $\boldsymbol{A}_{m+k}^m(t)$ are the vectors of the initial and final states of the studied system). The operator $\hat{L}$ is a finite-difference operator:

$$i\hat{L} = \frac{\Delta}{\Delta t}, \qquad \Delta t = \tau ,$$

where τ is a discretization time step, $\hat{L}_{ij} = \Pi_i \hat{L} \Pi_j$ $(i,j = 1,2)$ are matrix elements of the splittable Liouville quasi-operator, $\Pi_1 = \Pi, \Pi_2 = P = 1 - \Pi$ and Π are projection operators.

Let us define the relaxation times of the initial TCF and of the first-order memory function $M_1(t)$ as follows:

$$\tau_a = Re \int_0^\infty a(t)\mathrm{d}t, \tau_{M_1} = Re \int_0^\infty M_1(t)\mathrm{d}t, \ldots, \tau_{M_n} = Re \int_0^\infty M_n(t)\mathrm{d}t \,.$$

Then the spectrum of the non-Markovity parameter $\{\varepsilon\}$ is defined as an infinite set of dimensionless numbers:

$$\{\varepsilon_i\} = \{\varepsilon_1, \varepsilon_2, \ldots, \varepsilon_n, \ldots\} ,$$

$$\varepsilon_1 = \frac{\tau_a}{\tau_{M_1}}, \varepsilon_2 = \frac{\tau_{M_1}}{\tau_{M_2}}, \ldots, \varepsilon_n = \frac{\tau_{M_n}}{\tau_{M_{n+1}}} ,$$

$$\varepsilon = \frac{\tau_{\mathrm{rel}}}{\tau_{\mathrm{mem}}} \,. \tag{2}$$

Note that $a(t) = M_0(t)$. The number ε_n characterizes the ratio of the relaxation times of the memory functions M_n and M_{n+1}. If for some n the value of the parameter $\varepsilon_n \to \infty$, then this relaxation level is Markovian. If ε_n changes in limits from zero to a unit value, then the relaxation level is defined as non-Markovian. The times τ_{rel} (relaxation time) and τ_{mem} (memory life time) appear when the effects of the statistical memory in the complex discrete system are taken into account by means of the Zwanzig–Mori method of kinetic equations. Thus, the non-Markovity parameter spectrum is defined by the stochastic properties of the TCF.

In [10] the concept of generalized non-Markovity parameter for a frequency-dependent case was introduced:

$$\varepsilon_i(\omega) = \left(\frac{\mu_{i-1}(\omega)}{\mu_i(\omega)}\right)^{\frac{1}{2}} \,. \tag{3}$$

Here as $\mu_i(\omega)$ we have the frequency power spectrum of the ith memory function:

$$\mu_1(\omega) = |Re \int_0^\infty M_1(t) e^{i\omega t} dt|^2, \ldots, \mu_i(\omega) = |Re \int_0^\infty M_i(t) e^{i\omega t} dt|^2 .$$

The use of $\varepsilon_i(\omega)$ allows one to find the details of the frequency behaviour of the power spectra of the time correlations and memory functions.

3 The Universal Property of Informational Manifestation of Chaoticity in Complex Systems

In our work the discussion of the manifestation of chaoticity is carried out on the basis of a statistical invariant which includes a quantitative informational measure of chaoticity and pathology in a covariant form. The existence of this invariant is very important for taking decisions in the problems related to medicine as well as for analysing a wide area of physical problems related to complex systems of various nature.

In each live organism there is a universal informational property of the following form:

$$\mathrm{IMC} + \mathrm{IMP} = \mathrm{Invariant} . \tag{4}$$

Here IMC is an informational (quantitative) measure of chaoticity for the concrete live system, IMP is an informational measure of a pathological state of a live organism. As an informational (quantitative) measure of the degree of chaoticity (regularity) we propose to use the first point of the non-Markovity parameter at zero frequency: $\varepsilon_1(0) = [\mu_0(0)/\mu_1(0)]^{1/2}$. The physical sense of the parameter consists in comparing the relaxation scales of the time correlation function $(a(\omega))$ to the memory functions of the first order $(\mu(\omega))$. Depending on the values of this parameter one can discriminate Markov processes (with short-range memory) and non-Markov processes (with long-range memory effects). Thus, the phenomena distinguished by the greatest chaoticity correspond to Markov processes. Non-Markov processes are connected with greater regularity. The informational measure of a pathological state (IMP) defines the qualitative state of a real live system.

The quantitative estimate IMC of the degree of chaoticity of a system contains the information on a pathological state of the system. It testifies to the close interrelation of the given quantities. A high degree of chaoticity is characteristic of a normal physiological state. In a pathological state the degree of chaoticity decreases. A high degree of regularity is typical of this condition. Thus, the quantitative estimate of the chaoticity in live systems allows one to define their physiological or pathological state with a high degree of accuracy. In the right-hand side of (4) we have a statistical invariant, which reveals the independence of the physical (as well as biophysical, biochemical and biological) laws in the given live organism from the concrete situations as well as the methods of description of these situations. The invariance, submitted in (4), is formulated as the generalization of the experimental data. Among other physical laws the properties of invariance reflect the most general and profound properties of the studied systems and characterize a wide sphere of phenomena. Equation (4) reflects an informational observation. It consists of two informational measures: the measure of chaoticity and the measure of pathology (disease).

Let us use the operator of transformation $T(S', S)$ in both parts of (4). It realizes the transition of the system from one state S to another one S'. By taking into account the statistical invariance $I(S') = T(S', S)I(S) = I(S)$ in the right-hand side of (4) we get:

$$\Delta P = P(S') - P(S) = -\Delta C = -\{C(S') - C(S)\}, \Delta P + \Delta C = 0\,. \quad (5)$$

Here the following designations are introduced: $I(S)$ = Invariant, $P(S)$ = IMP(S) is an informational measure of pathology (disease) for the state S, $C(S)$ = IMC(S) is an informational measure of the chaoticity for the state of patient S. Besides in (5) we take into account the rules of transformation:

$$C(S') = T(S', S)C(S), \qquad P(S') = T(S', S)P(S)\,. \quad (6)$$

Equations (4)–(6) are rather simple but they make the quantitative description of the state of a patient possible, both during the disease and under the medical treatment. Equations (4)–(6) have a general character. They are true for many complex natural and social systems. It is possible to develop the algorithms of prediction of various demonstrations of chaos in complex systems of diverse nature on the basis of these equations.

4 The Quantitative Factor of Quality of a Treatment

One of the major problems of the medical physics consists in the development of a reliable criterion to estimate the quality of a medical treatment, a diagnosis and a forecasting of the behaviour of real live complex systems. As one can see from the previous section, such a criterion should include the parameter of the degree of randomness in a live organism. The creation of a quantitative factor for the quality of a treatment Q_T is based on the behaviour law of the non-Markovity parameter $\varepsilon_1(0)$ in the stochastic dynamics of complex systems. The greater values of the parameter $\varepsilon_1(0)$ are characteristic of stable physiological states of systems; the smaller ones are peculiar for pathological states of live systems. Thus, by the increase or reduction of the non-Markovity parameter one can judge the physiological state of a live organism with a high degree of accuracy. Therefore the non-Markovity parameter allows one to define the deviation of the physiological state of a system from a normal state.

The factor Q_T defines the efficacy or the quality of the treatment and is directly connected with the changes in the quantitative measure of chaoticity in a live organism. We shall calculate it in a concrete situation. Let us consider **1** as the patient's state before therapy, and **2** as the state of the patient after certain medical intervention. Then $\varepsilon_1(\mathbf{1})$ and $\varepsilon_1(\mathbf{2})$ represent quantitative measures of the chaoticity for the physiological states **1** and **2**. The ratio δ of these values ($\delta = \varepsilon_1(\mathbf{2})/\varepsilon_1(\mathbf{1})$) will define the efficacy of the therapy. Various jth processes occur simultaneously in the therapy. Therefore the total value of δ can be defined by the following way:

$$\delta = \prod_{j=1}^{n} \frac{\varepsilon_1^j(\mathbf{2})}{\varepsilon_1^j(\mathbf{1})} , \tag{7}$$

where $j = 1, 2 \ldots n$ is the number of factors affecting the behaviour of the non-Markovity parameter. However, the natural logarithm $\ln \delta$ is more convenient for use.

Then we have:

$$\delta > 1, \ln \delta > 0 ;$$
$$\delta = 1, \ln \delta = 0 ;$$
$$\delta < 1, \ln \delta < 0 .$$

The three values of δ mentioned above correspond to the three different qualities of treatment: effective, inefficient and destructive treatment. They reflect an increase, preservation and reduction of the measure of the chaoticity in the therapy. Thus, one can define $Q_T(\varepsilon) = \ln \delta$ according to (7) as follows:

$$Q_T(\varepsilon) = \ln \prod_{j=1}^{n} \frac{\varepsilon_1^j(\mathbf{2})}{\varepsilon_1^j(\mathbf{1})} . \tag{8}$$

However, the total factor Q_T is defined both by the quantitative measures of the chaoticity and by other physiological and biochemical data. Now we shall consider the transition of the patient from state **1** into state **2**. Then by analogy, one can introduce the physiological parameter $k(\mathbf{1})$, determined for state **1**, and $k(\mathbf{2})$ for state **2**. In the case of Parkinson's disease one can introduce the amplitude or the dispersion of the tremor velocity of some extremities (hand or leg) of the patient as this parameter. In other cases any medical data, which is considered for diagnostic purposes, can be used. For greater reliability it is necessary to use the combination of various parameters $k^j(\mathbf{1})$ and $k^j(\mathbf{2})$.

The value:

$$Q_T = \ln \prod_{j=1}^{n} \frac{\varepsilon_1^j(\mathbf{2})}{\varepsilon_1^j(\mathbf{1})} \frac{k^j(\mathbf{2})}{k^j(\mathbf{1})} \tag{9}$$

will be considered as a generalized quantitative factor of quality of the therapy.

However in real conditions it is necessary to increase or weaken the magnitude of chaotic, or physiological contributions to (9). For this purpose we shall take the simple ratio:

$$\ln \prod (a^n b^m \ldots) = n \ln a + m \ln b + \cdots$$

By analogy, we can reinforce or weaken various contributions depending on the concrete situation:

$$Q_T = \ln \prod_{j=1}^{n} \left(\frac{\varepsilon_1^j(\mathbf{2})}{\varepsilon_1^j(\mathbf{1})} \right)^{m_j} \left(\frac{k^j(\mathbf{2})}{k^j(\mathbf{1})} \right)^{p_j} . \tag{10}$$

If incomplete experimental data are available in some situations, one can assume $p_j = 1$ (attenuation of the physiological contribution). A value $m_j > 1$ can mean an amplification of the chaotic contribution. Otherwise, if we want to weaken the chaotic contribution, we should take $m_j = 1$ and if we want to reinforce the physiological contribution we come towards $p_j > 1$. We have presented the results of the calculation of the quantitative factor Q_T below in Sect. 6

5 Experimental Data

We have taken the experimental data from [15]. They represent the time records of the tremor velocity of the index finger of a patient with Parkinson's disease (see, also http://physionet.org/physiobank/database/). The effect of chronic high frequency deep brain stimulation (DBS) on the rest tremor was investigated [15] in a group of subjects with Parkinson's disease (PD) (16 subjects). Eight PD subjects with high amplitude tremor and eight PD subjects with low amplitude tremor were examined by a clinical neurologist and tested with a velocity laser to quantify time and frequency domain characteristics of tremor. The participants received DBS of the internal globus pallidus (GPi), the subthalamic nucleus (STN) or the ventrointermediate nucleus of the thalamus (Vim). Tremor was recorded with a velocity laser under two conditions of DBS (on–off) and two conditions of medication (L-Dopa on–off).

All the subjects gave informed consent and institutional ethics procedures were followed. The selected subjects were asked to refrain from taking their medication at least 12 h before the beginning of the tests and were not allowed to have more than one coffee at breakfast on the two testing days. Rest tremor was recorded on the most affected side with a velocity-transducing laser [16, 17]. This laser is a safe helium–neon laser. The laser was placed at about 30 cm from the index finger tip and the laser beam was directed perpendicular to a piece of reflective tape placed on the finger tip. Positive velocity was recorded when the subjects extended the finger and negative velocity when the subjects flexed the finger.

The conditions, counterbalanced across subjects, included the following:

1. The L-Dopa condition (no stimulation).
2. The DBS condition (stimulation only).
3. The "off" condition (no medication and no stimulation).
4. The "on" condition (medication on and stimulation on).
5. The effect of stopping DBS on tremor (time record of the tremor 15, 30, 45 and 60 min after having switched the stimulator off).

In Fig. 1 the time records of the velocity of changing tremor of the index finger of the second patient's hand (man, 52 years old) under various conditions of influence on the organism are submitted as an example. High tremor velocity is observed: 1) in a natural condition of the patient (a), 2) 15 (45) minutes after the stimulator was switched off. Lower tremor speed occurs: 1) when both methods (stimulation, medication) are used, 2) when each of

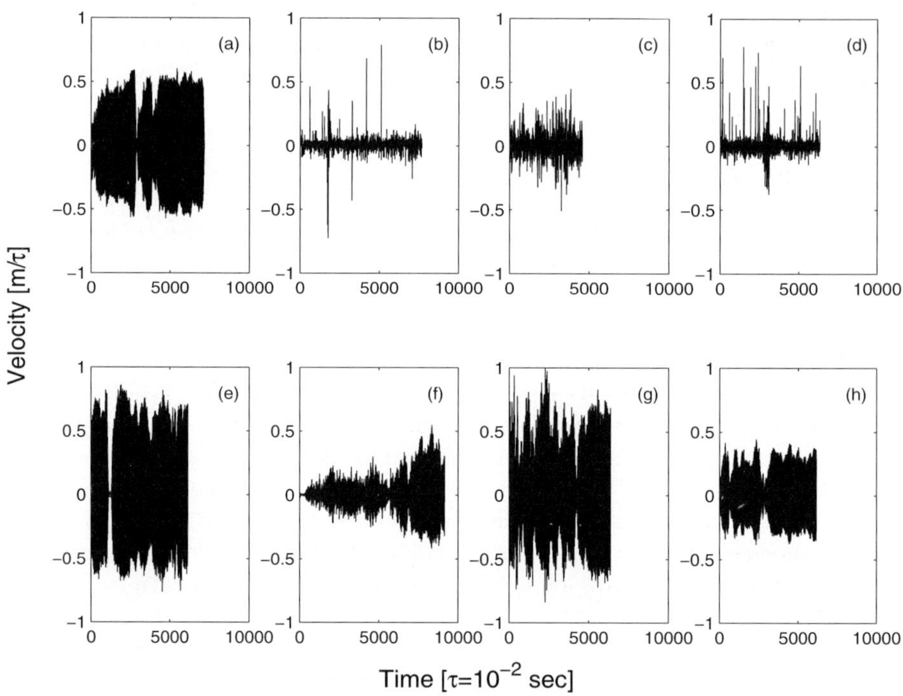

Fig. 1. The velocity of the change of tremor of the index finger of the patient's right hand (the second subject) with Parkinson's disease under various conditions of the experiment. **(a)** deep brain stimulation off, medication off; **(b)** the subject was receiving stimulation of the GPi, medication on; **(c)** deep brain stimulation off, medication on; **(d)** the subject was receiving stimulation of the GPi, medication off; **(e)**–**(h)** the recording of rest tremor in the right index finger of the subject 15 (30, 45, 60) minutes after the stimulator was switched off, this subject was off medication for at least 12 hours

these methods is used separately, 3) 30 (60) minutes after the stimulator was switched off. Similar results are presented in [15].

6 Results

In this section the results obtained by processing the experimental data for one of the patients (subject number 2) are shown. Similar or related pictures are observed in the experimental data of other subjects.

6.1 The Non-Markovity Parameter as a Quantitative Measure of Defining Chaoticity

In this subsection the technique to calculate quantitative and qualitative criteria under various conditions influencing the state of a patient is given. The basic idea of the approach consists in defining the quantitative ratio between chaoticity and regularity of the observed process. It allows one to judge the physiological (pathological) state of a live system by the degree of chaoticity or of regularity. The highest degree of chaoticity in the behaviour of a live system corresponds to a normal physiological state. Higher degree of regularity or specific ordering is characteristic of various pathological states of a live system. In the given work we use the non-Markovity parameter $\varepsilon_1(0)$ as a special quantitative measure defining chaoticity or regularity of the studied process. The examples [10–14], [18] which have been investigated by us earlier serve as a basis for such a reasoning. As one of the examples we shall consider the tremor velocity of the changing of the subject's index fingers in the case of Parkinson's disease.

The comparative analysis of the initial time record and the non-Markovity parameter for all the submitted experimental data allows one to discover the following regularity. The value of the non-Markovity parameter $\varepsilon_1(0)$ decreases with the increase of the tremor velocity of the patient's fingers (deterioration of the physiological state) and grows with the decrease of the tremor velocity (improvement of the state of the patient). We shall also consider the power spectrum $\mu_0(\omega)$ of the initial TCF under various conditions influencing an organism, the window-time behaviour of the power spectrum $\mu_0(\omega)$ and the non-Markovity parameter $\varepsilon_1(\omega)$, the time dependence local averaging relaxation parameter $\lambda_1(t)$ as additional sources of information.

Figure 2 represents the power spectrum of the initial TCF for various experimental conditions. One can observe a powerful peak in all the figures at the characteristic frequency $\omega = 0.07$ f.u.($\omega = 2\pi\nu, 1$ f.u. $= 2\pi/\tau, \tau = 10^{-2}$ s). The amplitude values of this peak for $\mu_i(\omega)$ $(i = 1, 2, 3)$ are given in Table 1. The given peak testifies to a pathological state of the studied system. A similar picture is observed in patients with myocardial infarction [11]. The comparison of these values reflects the amplitude of the tremor velocity at the initial record of time.

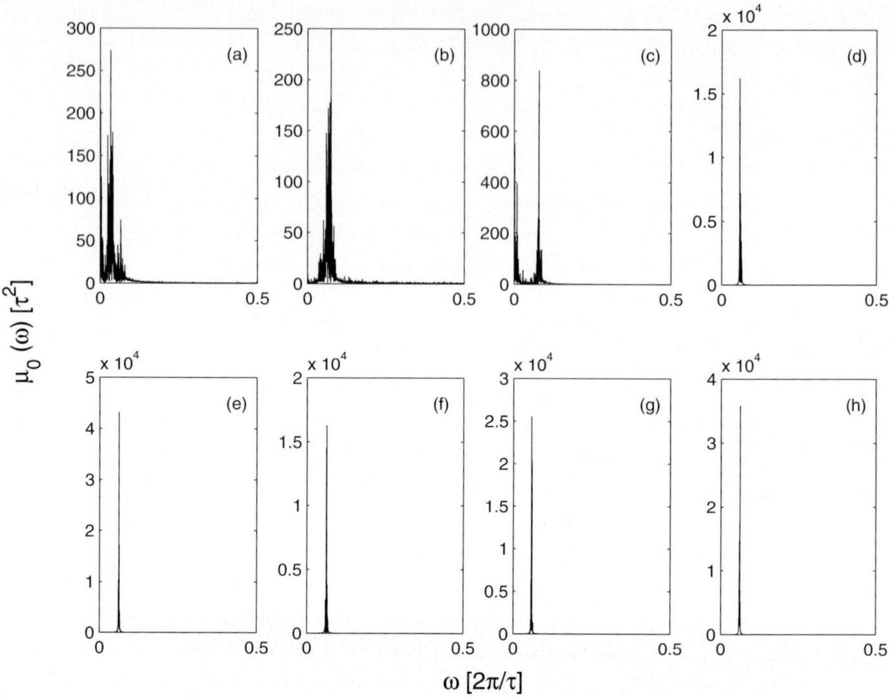

Fig. 2. The power spectrum $\mu_0(\omega)$ of the initial TCF for the velocity of changing of tremor of subject number 2 under various conditions that influence an organism. **(a)** deep brain stimulation on, medication on; **(b)** deep brain stimulation on, medication off; **(c)** deep brain stimulation off, medication on; **(d)** deep brain stimulation off, medication off; **(e)–(h)** the power spectrum $\mu_0(\omega)$ of the initial TCF for the recording of rest tremor in the right index finger of the subject 15 (30, 45, 60) minutes after the stimulator was switched off, medication off. At the frequency $\omega = 0.07$ f.u., with 1 f.u. = 100 Hz (the characteristic frequency), a peak is found. The presence and amplitude of this peak are determined by the state of the patient

Table 1. The value $\mu_0(\omega)$ for the initial TCF and $\mu_i(\omega)$ $(i = 1, 2, 3)$ for the memory functions of junior orders at the frequency $\omega = 0.07$ f.u. 1: Deep brain stimulation; 2: Medication (subject number 2). For example, OFF OFF – no DBS and no medication

	ON ON	ON OFF	OFF ON	OFF OFF	15 OFF	30 OFF	45 OFF	60 OFF
μ_0	75	250	812	1.71×10^4	4.34×10^4	1.53×10^4	2.51×10^4	3.68×10^4
μ_1	19	52	1.28×10^3	1.17×10^4	3.21×10^4	1.32×10^4	1.82×10^4	2.8×10^4
μ_2	42	60	113	71	300	62	137	224
μ_3	37	54	141	73	147	74	152	186

In Table 1 the second patient's amplitude values $\mu_0(\omega)$ for the initial TCF and the memory functions of junior orders $\mu_i(\omega)$ $(i = 1, 2, 3)$ at the frequency $\omega = 0.07$ f.u. are submitted. The terms of the first row define the conditions under which the experiment is carried out. Under all conditions a power peak at the frequency $\omega = 0.07$ f.u. can be observed. The amplitude values of the given peak (in particular in the power spectrum $\mu_0(\omega)$) reflect the amplitude of the tremor velocity. For example, the least amplitude 75 τ^2 corresponds to the condition (ON, ON; or: deep brain stimulation on, medication on). The highest amplitude $4.34 \times 10^4 \tau^2$ corresponds to the greatest tremor speed (see Figs. 1e, 2e). Thus, the given parameter can be used to estimate the physiological state of a patient. A similar picture is observed in all other patients.

In Fig. 3 the initial time record (the normal state of the subject; OFF, OFF) and the window-time behaviour of the power spectrum of the TCF (the technique of the analysis of the given behaviour is considered in Ref. [18]) are submitted. In these figures regions 1, 2, 3, which correspond to the least

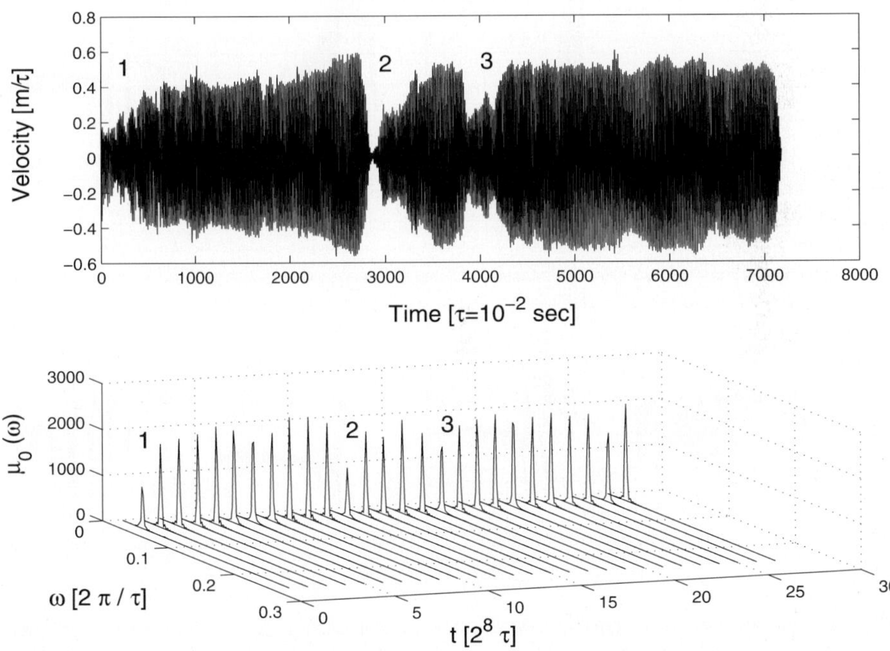

Fig. 3. The initial time series and the window-time behaviour of the power spectrum $\mu_0(\omega)$ of the TCF. Two figures are submitted to illustrate the case of subject number 2: no stimulation of the brain and no medication. The change of regimes in the initial time series is reflected in the decrease of the tremor velocity (regions 1, 2 and 3) and becomes visible as a sharp reduction of the power spectrum $\mu_0(\omega)$ (see, the 1st, 12th, 17th windows for more details)

values of the tremor velocity are shown. The minimal amplitude of the peaks of the power spectrum $\mu_0(\omega)$ corresponds to the regions with the least tremor velocity.

In Fig. 4 the frequency dependence of the first point of the non-Markovity parameter $\varepsilon_1(\omega)$ is submitted for the second subject under various experimental conditions. The value of the parameter $\varepsilon_1(0)$ at zero frequency is of special importance for our study of manifestations of chaoticity. It is possible to judge the change of the state of a subject by the increase (or by the decrease) of this value. The comparative analysis of the initial time records allows one to come to similar conclusions. In Figs. 4d–h a well-defined frequency structure of the non-Markovity parameter can be seen. This structure is completely suppressed and disappears only when during the treatment. The characteristic frequency of the fluctuations corresponds approximately to $\omega = 0.06$ f.u.

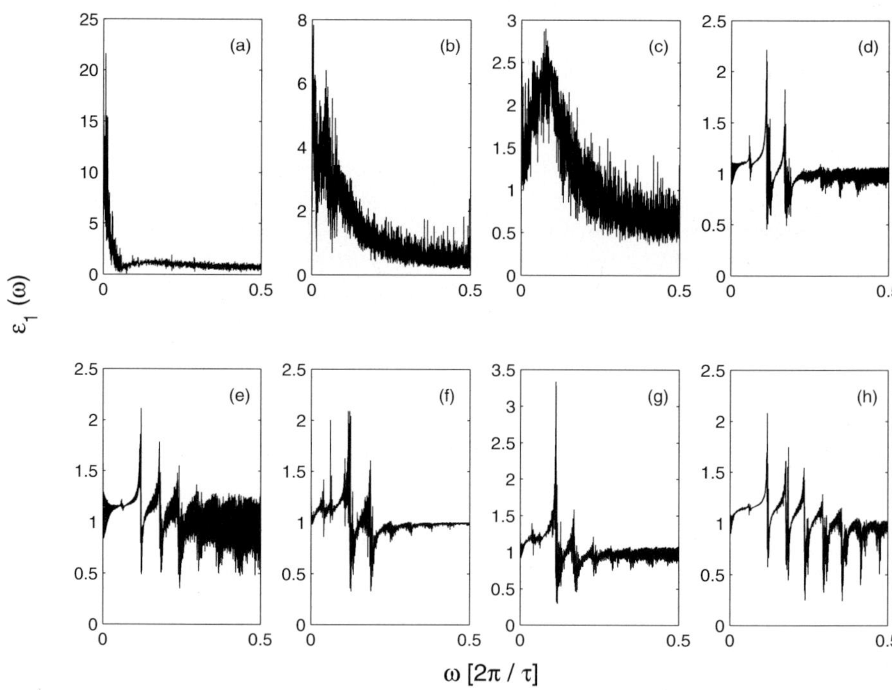

Fig. 4. The first point of the non-Markovity parameter $\varepsilon_1(\omega)$ for the second subject under various experimental conditions: **(a)** deep brain stimulation off, medication on; **(b)** deep brain stimulation on, medication on; **(c)** deep brain stimulation on, medication off; **(d)** deep brain stimulation off, medication off; **(e)**–**(h)** the recording of rest tremor in the right index finger of the subject 15 (30, 45, 60) minutes after the stimulator was switched off, medication off. The non-Markovity parameter at zero frequency $\varepsilon_1(0)$ plays a special role. These values (6.02 in the first case and 1.0043 in the last one) define the chaoticity or the regularity of the studied states. The amplitudes of these values also characterize the state of the subject

These multiple peaks are the most appreciable at low frequencies. At higher frequencies these fluctuations are smoothed out. As can be seen in these figures, the 2nd subject has a strong peak which remains stable over time. As our data show, the comb-like structure with multiple frequencies can be observed in all patients with high tremor velocity. In a group of patients with low tremor velocity it disappears, and a wider spectrum that presents some fluctuations over time is observed. The present structure testifies to the presence of characteristic frequency of fluctuations of tremor of human extremities.

In Table 2 the dispersion interval of the values and the average value $\varepsilon_1(0)$ for the whole group of subjects (16 subjects) are submitted. Let us consider 2 conditions: OFF, OFF and OFF, ON. In the first case the dispersion interval and the average value $\varepsilon_1(0)$ are minimal. It means the presence of a high degree of regularity of the physiological state of the patient. The degree of regularity is appreciably reduced when applying any method of treatment. Here the degree of chaoticity grows. The maximal degree of chaoticity corresponds to the condition OFF, ON (medication is used only). The difference in $\varepsilon_1(0)_{\mathrm{av.val}}$ with medication and without it (OFF, OFF) is 3.8 times (!). On the basis of the comparative analysis of the given parameters the best method of treatment for each individual case can be found. It is necessary to note, that the given reasoning is true only for the study of the chaotic component of the quantitative factor of the quality of treatment Q_T. The most trustworthy information about the quality of treatment can be given by the full quantitative factor Q_T which takes account of other diagnostic factors.

Table 2. The dispersion interval $\varepsilon_1(0)_{\mathrm{int}}$ of the values and the average value $\varepsilon_1(0)_{\mathrm{av.val}}$ of the first point of the non-Markovity parameter under various experimental conditions for the group of 16 subjects. 1: Deep brain stimulation; 2: Medication

	OFF OFF	ON OFF	OFF ON	ON ON	15 OFF	30 OFF	45 OFF	60 OFF
$\varepsilon_1(0)_{\mathrm{int}}$	1–1.8	2–18	2–22	1.5–8	1.5–3	1.8–5	1.7–4.5	2–6
$\varepsilon_1(0)_{\mathrm{av.val}}$	1.41	4.14	5.31	3.17	2.43	2.92	2.76	2.93

The results of the calculation of the quantitative factor Q_T are shown in Table 3. The data are submitted for a single patient and for the whole group. Here $Q_T(\varepsilon)$ is the chaotic contribution to the quantitative factor (see (8)). Q_T is the total quantitative factor (see (10)), where $\varepsilon^{(1)}(\mathbf{1})$ and $\varepsilon^{(1)}(\mathbf{2})$ are the chaotic contributions for the tremor amplitudes $k^{(1)}(\mathbf{1}), k^{(1)}(\mathbf{2})$; $\varepsilon^{(2)}(\mathbf{1})$ and $\varepsilon^{(2)}(\mathbf{2})$ are the dispersions of the tremor amplitudes $k^{(2)}(\mathbf{1}), k^{(2)}(\mathbf{2})$ (physiological contributions). The full factor Q_T provides detailed information about the quality of the treatment. The present factor includes both the chaotic

Table 3. The quantitative factor $Q_T(\varepsilon)$ and the total quantitative factor Q_T for the second patient and for the whole group (16 subjects). 1: Deep brain stimulation; 2: Medication. $m_j = 1, p_j = 1$

	OFF OFF	ON OFF	OFF ON	2nd patient ON ON	15 OFF	30 OFF	45 OFF	60 OFF
$Q_T(\varepsilon)$		0.758	2.556	1.756	0.291	0.438	0.041	0.017
Q_T		1.763	2.013	2.654	–0.013	0.883	–0.004	0.856
				Whole group				
$Q_T(\varepsilon)$		1.077	1.326	0.810	0.544	0.728	0.671	0.731
Q_T		3.661	2.883	4.071	1.47	1.734	1.624	1.742

component $Q_T(\varepsilon)$, and the physiological contribution $Q_T(k)$. The calculation of $Q_T(k)$ is described in Sect. 4. One can define the quality of a treatment by means of Q_T. The positive value of the given factor defines an effective treatment. For a separate patient and for the whole group, Q_T reaches its maximal value under the conditions ON, ON. The total quantitative factor is supplemented by a diagnostic (physiological) component. It allows one to take into account those features of the system which the chaotic component does not contain. For the second patient under condition 15 OFF (see Table 3) the factor Q_T has a negative value. It testifies to the negative influence of the given treatment on the organism of the patient. The best treatment is thus the combination of the two medical methods: electromagnetic stimulation and medication.

Figure 5 reflects the behaviour of the parameter $\varepsilon_1(0)$ for four different subjects. The points lying above the horizontal line testify to an improvement of the state of the subject and the efficacy of the treatment. The points, lying below the horizontal line testify to a deterioration of the state of the subject and the inefficiency of the applied treatment. For example, Fig. 5b corresponds to the sevenfold change of the quantitative measure of chaoticity for the 9th patient. In the case of the 8th patient (see Fig. 5c) no influence could change the measure of the chaoticity. Therefore there was practically no change in the state of the subject either. In some cases (see Figs. 5b, 5d) the DBS or the medication reduces the measure of chaoticity which testifies to a deterioration of the state of the subject. This approach allows one to define the most effective (or inefficient) treatment in each individual case.

6.2 The Definition of a Predictor of Sudden Changes of the Tremor Velocity

In this subsection the window-time behaviour of the non-Markovity parameter $\varepsilon_1(\omega)$ for a certain case (the second patient, two methods of medical

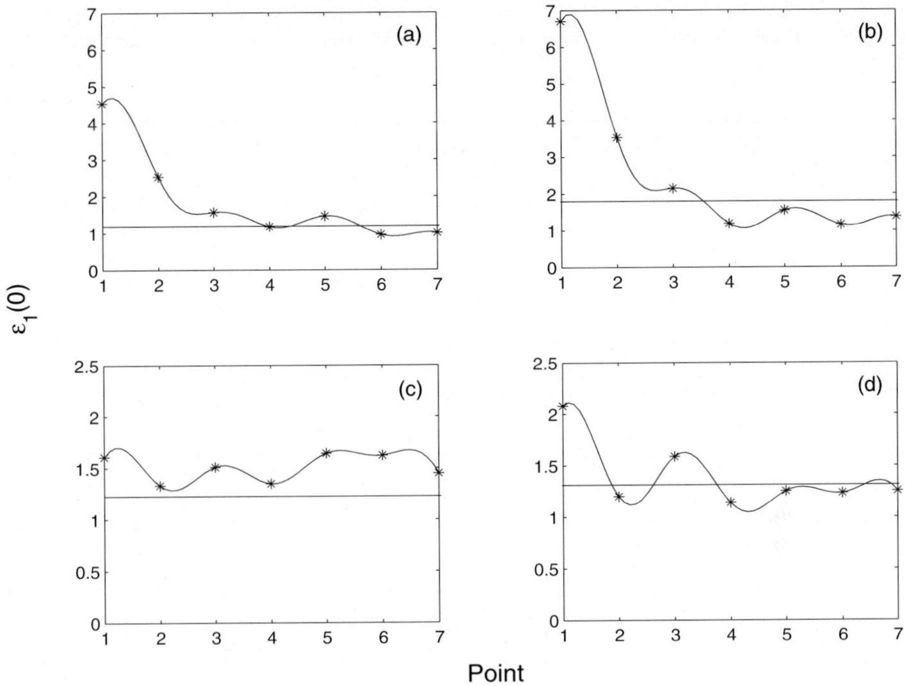

Fig. 5. The behaviour of the parameter $\varepsilon_1(0)$ for four various patients: **(a)** the second subject, high amplitude tremor, **(b)** the 9th subject, low amplitude tremor (the stimulation of the GPi); **(c)** the 8th subject, high amplitude tremor, **(d)** the 15th subject, low amplitude tremor (the stimulation of the STN). The value of $\varepsilon_1(0)$ for seven consecutive conditions of the experiment: 1 - both methods are used; 2 - treatment by medication is applied only; 3 - the DBS is used only; 4 (5, 6, 7) - value of the parameter 15 (30, 45, 60) minutes after the stimulator was switched off; the horizontal line corresponds to the value of the parameter when no method is used. This representation allows one to define the most effective treatment for each patient

treatment were used) and the procedure of local averaging of the relaxation parameters are considered. These procedures allow one to determine specific predictors of the change of regimes in the initial time records.

The idea of the first procedure is, that the optimum length of the time window ($2^8 = 256$ points) is found first. In the studied dependence (in our case the frequency dependence of the first point of the non-Markovity parameter) the first window is cut out. Then the second window is cut out (from point 257 to point 512), etc. This construction allows one to find the local time behaviour of the non-Markovity parameter. At the critical moments when the tremor velocity increases the value of the non-Markovity parameter comes nearer to a unit value. One can observe that the value of

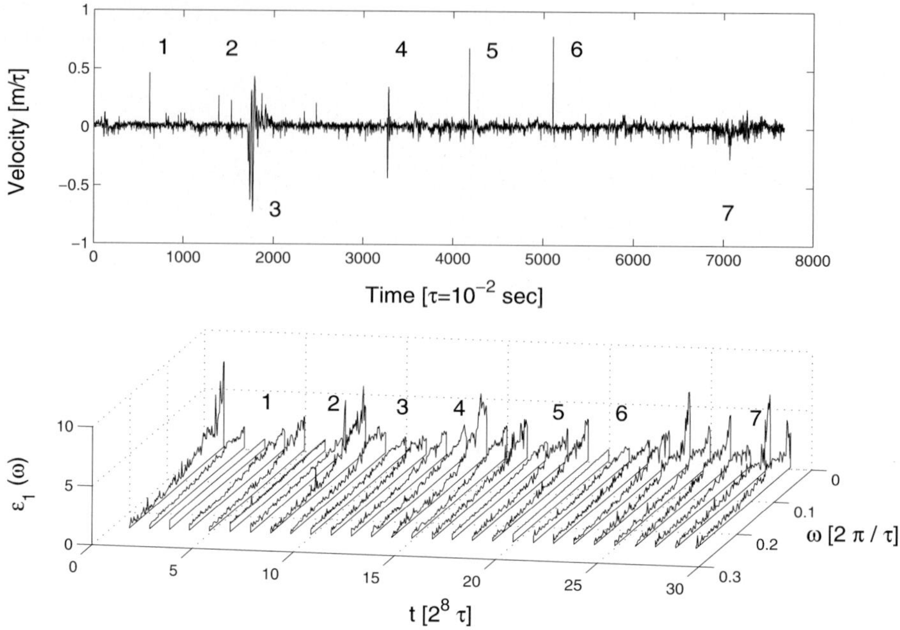

Fig. 6. The initial signal – the change of tremor velocity when the second patient is treated by two methods – and the window-time behaviour of the first point of the non-Markovity parameter $\varepsilon_1(\omega)$. At the time of a sharp change of the mode (sharp increase of the tremor velocity) in the behaviour of the initial time series (regions 1–7) a gradual decrease of the non-Markovity parameter down to a unit value (the 3rd, 6th, 10th, 14th, 17th, 20th, 27th windows) is observed. The decrease of the non-Markovity parameter begins 2–2,5 s earlier than the tremor acceleration on an initial series

the non-Markovity parameter starts to decrease 2–2.5 s before the increase of the tremor velocity (see Fig. 6).

The idea behind the second procedure is the following: one can consider the initial data set and take an N-long sample. We can calculate kinetic and relaxation parameters for the given sample. Then we can carry out a "step-by-step shift to the right". Then we calculate kinetic and relaxation parameters. After that we execute one more "step-by-step shift to the right" and continue the procedure up to the end of the time series. Thus, the local averaged parameters have a high sensitivity to the effects of intermittency and non-stationarity. Any non-regularity in the initial time series is instantly reflected in the behaviour of the local average parameters. The optimal length of the sample is 120 points. In Figs. 7, 8 the initial time record and the time dependence of the local relaxation parameter $\lambda_1(t)$ are shown in two cases. The change in the time behaviour of the parameter $\lambda_1(t)$ begins 2–3 s prior to the change of the regimes of the time record of the tremor velocity.

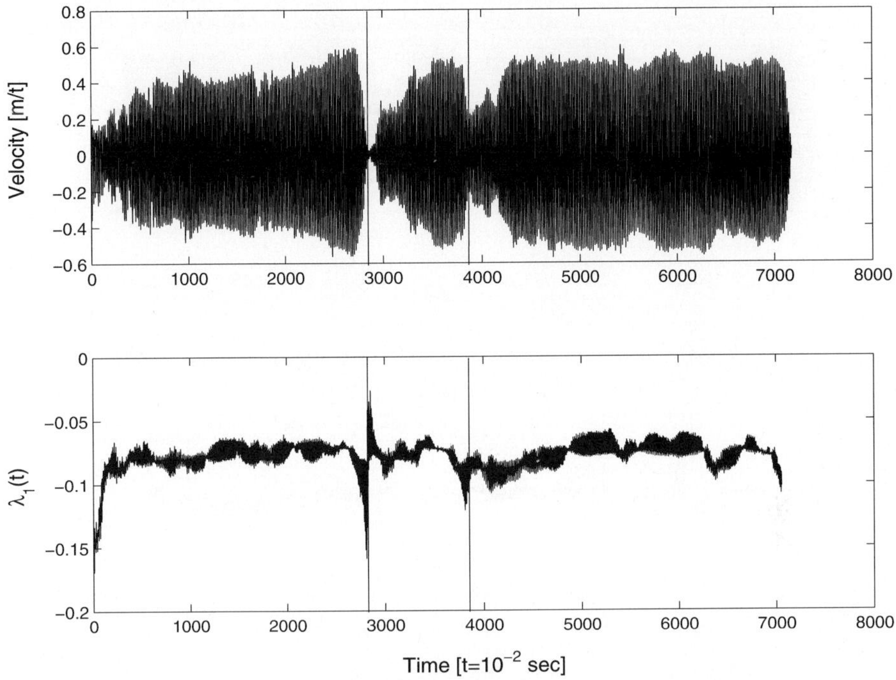

Fig. 7. The change of the tremor velocity for the second patient (stimulation of the brain and medication are not used) and the time dependence of the local relaxation parameter $\lambda_1(t)$. The localization procedure allows one to find sudden changes of relaxation regimes of the system under consideration. The largest amplitude values of the local relaxation parameter are in the region of the lowest tremor velocity. The change of the time behaviour of the parameter $\lambda_1(t)$ begins 2–3 s earlier than the sharp change of the regimes in the initial time series appears

The increase of speed of the local relaxation parameter ($\lambda_1(t)$) testifies to a decrease of tremor velocity.

7 Conclusions

In this chapter we have proposed a new concept for the study of manifestations of chaoticity. It is based on the application of the statistical non-Markovity parameter and its spectrum as an informational measure of chaoticity. This approach allows one to define the difference between a healthy person and a patient by means of the numerical value of the non-Markovity parameter. This observation gives a reliable tool for the strict quantitative estimates that are necessary for the diagnosis and the quantification of the treatment of patients. As an example we have considered the changes of

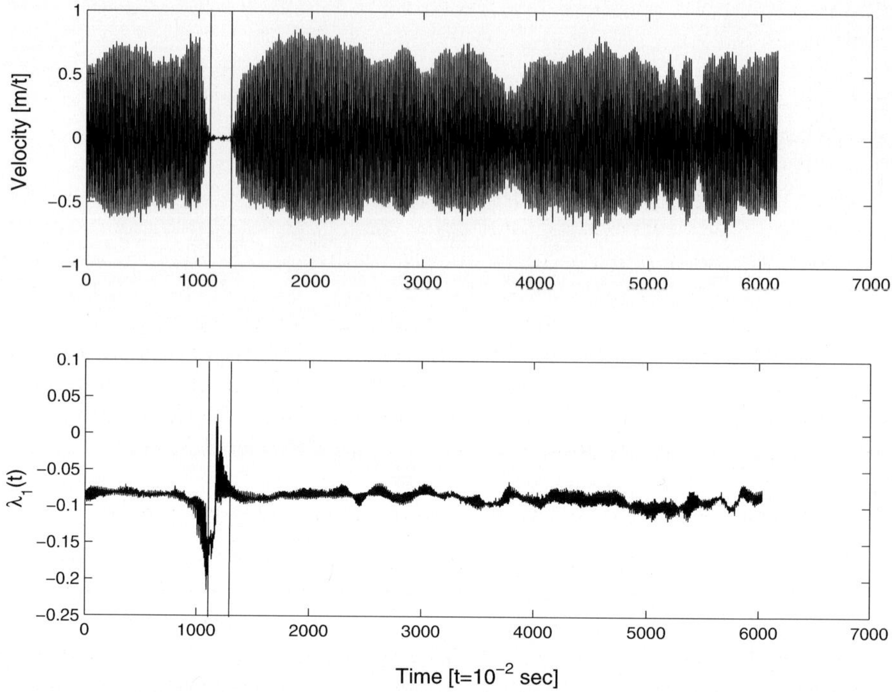

Fig. 8. The change of the tremor velocity for the second patient (15 minutes after the stimulator was switched off, medication off) and the time dependence of the local relaxation parameter $\lambda_1(t)$. The site characterizing the minimal tremor velocity is allocated. The increase and decrease of the local relaxation parameter occur 2.5 s before the decrease or increase of the tremor velocity. The similar behaviour of the parameter $\lambda_1(t)$ can be explained by its high sensitivity to the presence of nonstationarity of the initial signal

various dynamic conditions of patients with Parkinson's disease. The quantitative and qualitative criteria used by us for the definition of chaoticity and regularity of the investigated processes in live systems reveal new informational opportunities of the statistical theory of discrete non-Markov random processes. The new concept allows one to estimate quantitatively the efficacy and the quality of the treatment of different patients with Parkinson's disease. It allows one to investigate various dynamic states of complex systems in real time.

The statistical non-Markovity parameter $\varepsilon_1(0)$ can serve as a reliable quantitative informational measure of chaoticity. It allows one to use $\varepsilon_1(0)$ for the study of the behaviour of different chaotic systems. In the case of Parkinson's disease the change of the parameter defines the change of a quantitative measure of chaoticity or regularity of a physiological system. The increase of the chaoticity reflects the decrease of the quantitative measure of pathology and the improvement of the state of the patient. The increase of the regularity defines high degree of manifestation of pathological states of live systems. The combined power spectra of the initial TCF $\mu_0(\omega)$, the three memory functions of junior orders and the frequency dependence of the non-Markovity parameter compose an informational measure which defines the degree of pathological changes in a human organism.

The new procedures (the window-time procedure and the local averaging procedure) give evident predictors of the change of the initial time signal. The window-time behaviour of the non-Markovity parameter $\varepsilon_1(\omega)$ reflects the increase of the tremor velocity 2–2.5 s earlier. It happens when the non-Markovity parameter approaches a unit value. The procedure of local averaging of the relaxation parameter $\lambda_1(t)$ reflects the relaxation changes of physiological processes in a live system. The behaviour of the local parameter$\lambda_1(t)$ reacts to a sudden change of relaxation regimes in the initial time record 2–3 s earlier. These predictors allow to lower the probability of ineffective use of different methods of treatment.

In the course of the study we have come to the following conclusions:

- The application of medication for the given group of patients proved to be the most efficient way to treat patients with Parkinson's disease. Used separately stimulation is less effective than the use of medication.
- The combination of different methods (medication plus electromagnetic stimulator) is less effective than the application of medication or of stimulation. In some cases the combination of medication and stimulation exerts a negative influence on the state of the subject.
- After the stimulator is switched off its aftereffect has an oscillatory character with a low characteristic frequency corresponding to a period of about 30 min.
- The efficacy of various medical procedures and the quality of a treatment can be estimated quantitatively for each subject separately with utmost precision.

However, if we take both chaotic and physiological components into account, the general estimation of the quality of treatment will be more universal. The combination of two methods (DBS and medication, $Q_T = 4.071$) produces the most effective result in comparison with the effect of DBS (3.661) or of medication (2.883) given separately. This is connected with additional aspects of the estimation of the quality of treatment due to the study of both chaotic and diagnostic components of a live system. This conclusion corresponds to the results of [15].

In conclusion we would like to state that our study gives a unique opportunity for the exact quantitative description of the states of patients with Parkinson's disease at various stages of the disease as well as of the treatment and the recovery of the patient. On the whole, the proposed concept of manifestations of chaoticity opens up great opportunities for the alternative analysis, diagnosis and forecasting of the chaotic behaviour of real complex systems of live nature.

Acknowledgements

This work supported by the RHSF (Grant No. 03-06-00218a), RFBR (Grant No. 02-02-16146) and CCBR of Ministry of Education RF (Grant No. E 02-3.1-538). The authors acknowledge Prof. Anne Beuter for stimulating criticism and valuable discussion and Dr. L.O. Svirina for technical assistance.

References

1. S. Boccaletti, C. Grebogi, Y.-C. Lai et al.: Phys. Reports **329** ,103 (2000)
2. H. Touchette, S. Lloyd: Physica A **331**, 140 (2004)
3. K. Pyragas: Phys. Lett. A **170**, 421 (1992)
4. E.R. Hunt: Phys. Rev. Lett. **67**, 1953 (1991)
5. V. Petrov, V. Gaspar, J. Masere et al.: Nature **361**, 240 (1993)
6. B.B. Plapp, A.W. Huebler: Phys. Rev. Lett. **65**, 2302 (1990)
7. W. Just, H. Benner, E. Reibold: Chaos **13**, 259 (2003
8. R. Lima, M. Pettini: Phys. Rev. A **41**, 726 (1990)
9. Y. Braiman, J. Goldhirsch: Phys. Rev. Lett. **66**, 2545 (1991)
10. R.M. Yulmetyev, P. Hänggi, F.M. Gafarov: Phys. Rev. E **62**, 6178 (2000)
11. R.M. Yulmetyev, P. Hänggi, F. Gafarov, Phys. Rev. E **65**, 046107 (2002)
12. R.M. Yulmetyev, F.M. Gafarov, P. Hänggi et al.: Phys. Rev. E **64**, 066132 (2001)
13. R.M. Yulmetyev, S.A. Demin, N.A. Emelyanova et al.: Physica A **319**, 432 (2003)
14. R.M. Yulmetyev, N.A. Emelyanova, S.A. Demin et al.: Physica A **331**, 300 (2003)
15. A. Beuter, M. Titcombe, F. Richer et al.: Thalamus & Related Systems **1**, 203 (2001); M. Titcombe, L. Glass, D. Guehl et al.: Chaos **11**, 766 (2001)
16. A. Beuter, A. de Geoffroy, P. Cordo: J. Neurosci. Meth. **53**, 47 (1994)
17. K.E. Norman, R. Edwards, A. Beuter: J. Neurosci. Meth. **92**, 41 (1999)
18. R.M. Yulmetyev, P. Hänggi, F.M. Gafarov: JETP, **123**, 643 (2003)

Monte Carlo Simulations of Ageing and Speciation

Suzana Moss de Oliveira[1,2] and Dietrich Stauffer[1,3]

[1] Laboratoire de Physique et Mécanique des Milieux Hétérogènes, Ecole Supérieure de Physique et de Chimie Industrielles de la Ville de Paris, 10, rue Vauquelin, 74231 Paris Cedex 05, France.
[2] Visiting from Instituto de Física, Universidade Federal Fluminense, Av. Litorânea s/n, Boa Viagem, Niterói 24210-340, RJ, Brazil
`suzana@if.uff.br`
[3] Visiting from Institute for Theoretical Physics, Cologne University, D-50923 Köln, Euroland
`stauffer@thp.uni-koeln.de`

Summary. We present results from computer simulations of ageing and speciation, using mostly the Penna bit-string model which is based on the mutation-accumulation hypothesis.

1 Introduction

Verhulst would be 200 years old now, and the combined age of the present authors is more than half that age. Therefore, it is appropriate if we apply Verhulst death probabilities to simulations of biological ageing. But there is more to life than eating, drinking, giving conference talks, and dying. We also want to be better than others, like the Brazilians from São Paulo would like when compared to those from Rio de Janeiro, or Homo Neanderthalis (living in Düsseldorf) would like to be better than Homo Sapiens from Cologne. This differentiation is called speciation if members from one group can no longer have viable children with members of the other group. (The opposite is not true: Both authors count themselves among homo sapiens and are male and female but have no children together [1]). We concentrate here on widespread simulations of ageing and on the few speciation theories which take into account the ageing of the individuals. In particular we want to know which ageing simulations explain the exponential growth of adult mortality with increasing age, i.e. the Gompertz law from about the same time as the Verhulst probability, and which theories explain the branching into two species without the help of any river or mountain.

The Chowdhury model assumes ageing and then puts it into a foodweb model. This model is reviewed elsewhere [2] and thus will not be given in this chapter.

Sections 2.1 and 2.2 deal with the asexual Penna model of biological ageing and some of its applications and in Sects. 2.3 and 2.4 we present its sexual variant including some important results. Recent reviews are published in [3–5], and an earlier exhaustive review including complete Fortran programs and more biological literature is given in [6]. In Sect. 2.5 we present alternative computer simulations performed in order to explore some other possible reasons for biological ageing, different from mutation accumulation. Finally, Sect. 3 is dedicated to more recent versions of the sexual Penna model adapted to the study of sympatric speciation.

2 The Penna Model

2.1 Asexual Version

According to the mutation accumulation-theory, Darwinian selection pressure tries to keep our genomes as clean as possible until reproduction starts. For this reason we age: bad mutations that appear early in life are not transmitted and disappear from the population, while those that become active late in life can accumulate, decreasing our survival probability but without risking the extinction of the species. The Alzheimer disease is a good example of such a mechanism: although its corresponding defective gene is present in the genetic code since birth, its effects generally appear at old ages (around 80). As a consequence, this disease is a very common one. The Penna model's dynamics can be regarded as a computational realization of the mutation accumulation theory.

In the original asexual version of the model [7] the chronological genome of each individual is represented by a bit-string of 32 bits. Each bit corresponds to one "year" in the individual lifetime, and consequently each individual can live at most for 32 "years". A bit set to one means that the individual will suffer from the effects of a deleterious inherited mutation (genetic disease) in that and all following years. As an example, an individual with a genome 10010... would start to become sick during its first year of life and would become worse during its fourth year when a new disease appears. One step of the simulation corresponds to reading one bit of all chronological genomes. Whenever a new bit of a given genome is read, we increase by one the individual's age.

Selection pressure is modelled by the introduction of a threshold T for the number of deleterious mutations that can be simultaneously active in a living individual's genome. Usually, the probability of death because of genetic causes is assumed to be given by a step function $\Theta(N_{\text{mut}} - T)$, with $\Theta(x) = 1$ if $x \geq 0$ and $= 0$ otherwise; N_{mut} corresponds to the number of accumulated diseases up to the current individual's age. If $T = 2$, for instance, the individual of the example above would die at age 4. Smoother functional forms have also been used [8] but, except at very old ages, the fundamental results of the Penna model do not appear to be sensitive to this choice.

Death for non-genetic causes, representing the outcome of intra-specific competition for the limited resources of the environment, is modelled by a density- and time-dependent quantity, the Verhulst factor. This is a mean-field death probability, given by

$$V(t) = \frac{N(t)}{N_{\max}} ,$$

where $N(t)$ is the total population at the beginning of time step t and $N_{\max}$ is the carrying capacity, that is, the maximum population size the environment could support if no other causes of death would exist. We usually consider $N_{\max}$ ten times larger than the initial population $N(0)$. At each time step and for each individual a random number between zero and one is generated and compared with V: if it is greater than V, the individual dies independently of its age or genome. The larger the population size is, the higher is the probability of an individual to die due to this death roulette.

The introduction of the Verhulst factor, or some equivalent form of limiting factor for the total population, is a necessity in simulational models to avoid population overflow. The one just presented has been criticized by some authors, since really random deaths in nature are hardly observed. In [9], for instance, the authors adopted the alternative strategy of killing randomly only the newborns. In [10], simulations of the Penna model on a lattice present an equivalent strategy.

If the individual survives until a minimum reproduction age R, it generates b offspring in that and all following years. The offspring genome is a copy of the parent's one, except for M mutations at randomly chosen genome positions introduced at birth. Although the model allows good and bad mutations, generally we consider only the bad ones. The reason is that harmful mutations are around 100 times more frequent than the backward ones – reverse mutations deleting harmful ones [11]. In this case, if a bit position holding a bit 1 is randomly tossed in the parent's genome, it remains 1 in the offspring genome; however, if a bit zero is randomly tossed, it is set to 1 in the mutated offspring genome. In this way, for the asexual reproduction, the offspring is always as good as or worse than the parent. Even so, a stable population is obtained, provided the birth rate b is greater than a minimum value [12].

2.2 Applications of the Asexual Version

A crucial test of ageing theories is the Pacific Salmon, which dies shortly after producing offspring. Alternative theories of ageing, based on wear and tear, on damage due to oxygen radicals, or on telomere depletion leading to the Hayflick limit, can hardly explain this sudden death of previously healthy animals. Within the mutation accumulation hypothesis as implemented in the Penna model, it is explained very simply by introducing a maximum reproduction age: Random mutations acting for ages above that age are not weeded out by Darwinian selection pressure, and thus the bits for ages above the cessation of reproduction are all mutated (if back mutations are prohibited). Thus the animals die necessarily shortly after producing offspring. [13, 14].

In the opposite direction, a good ageing model should also explain the Gompertz law of the 19th century, which is well obeyed by modern humans of industrialized countries:

$$q \simeq \mu \propto \exp(ba) ,$$

where q is the mortality at age a, i.e. the fraction of people dying between ages a and $a+1$, but alive at age a. By definition this q cannot become bigger than one and, for $q > 1/2$, it is better to use the mortality function $\mu \simeq -\ln(1-q)$, also called the hazard factor [15]. This mortality function can increase above unity and is defined as $-\mathrm{d}\ln S(a)/\mathrm{d}a \simeq \ln[S(a)/S(a+1)]$, where $S(a)$ is the probability to survive from birth to age a. Figure 1 shows that the mortality function for Germany around the year 2000 agrees, for adult men, very well with the Gompertz law, and Fig. 2 shows that the Penna model reproduces this law. For children, the standard Penna model, the Gompertz law, and reality all differ from each other.

For old age, the last dozen years has raised the hope that the mortality of the oldest old shows a maximum, a plateau, or at least a mortality deceleration (i.e. a negative second derivative $\mathrm{d}^2\mu(a)/\mathrm{d}a^2$ with respect to age.) For flies and some other animals this deviation is widely believed to hold [16]; in the case of humans, for which demographic statistics are the most complete,

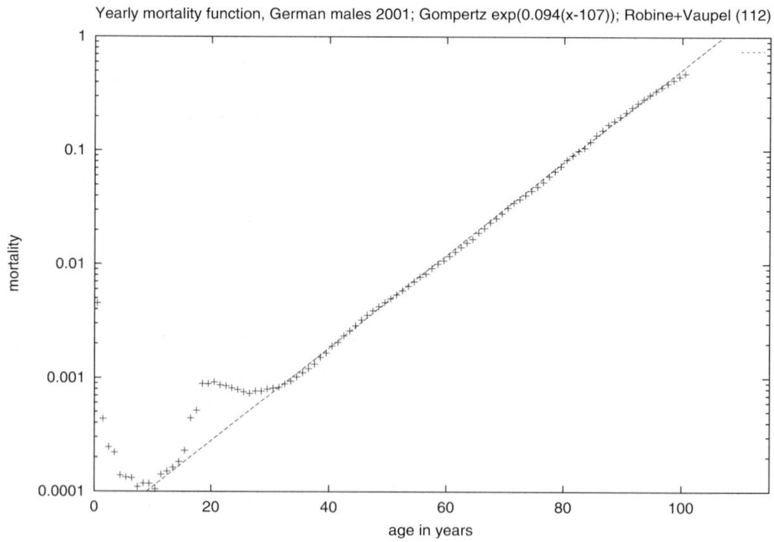

Fig. 1. Mortality function μ for German men (from www.destatis.de) compared with a Gompertz fit in this semilogarithmic plot

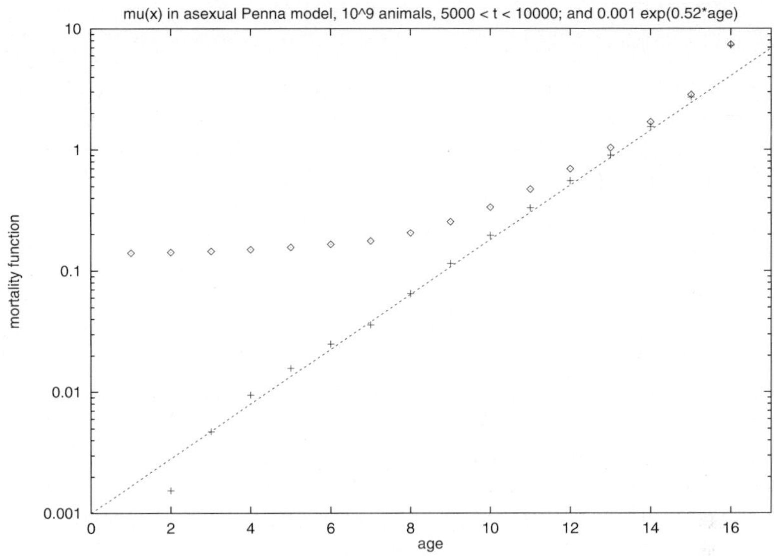

Fig. 2. Mortality function μ for the standard asexual Penna model, counting (upper data) or not (lower data) the deaths from the Verhulst factor. From [5]

lots of exaggerations are published in [14] which contradict existing reliable data for Swedish or German men, Fig. 1. (Women usually obey less strictly the Gompertz law.) Only above the age of 110 years does a plateau possibly exist [15, 17] but the statistics there are sparse, and systematic errors due to age exaggeration seem still to play an important role in many industrialized countries [17]. Perhaps with better statistics and better estimates of systematic errors the deviations of human mortalities from the Gompertz law will have vanished in years to come. Nevertheless, several modifications of the Penna model were published to give downward deviations from the Gompertz law, first in [18] and perhaps best in [19]. For ecosystems with prey and predator, He et al. [20] found strong oscillations with time on a lattice, similar to those found in the solutions of the Lotka–Volterra equations.

The reason for ageing in the mutation accumulation theory is the high selection pressure before the minimum age of reproduction is reached, and the lack of such selection after the offspring is produced. This does not mean, however, that only the minimum age of reproduction fixes the maximum lifespan. Figure 3 shows that it also depends slightly on the length of the genomic bit-string. Only if besides the minimum age of reproduction also the mutation rate per bit is kept constant, do the results for long bit-strings become independent of the bit-string length. There seem to be no simple scaling laws how the mortality curves depend on this bit-string length [21]. Their simulations with sex are complemented here by our asexual simulations in Fig. 4.

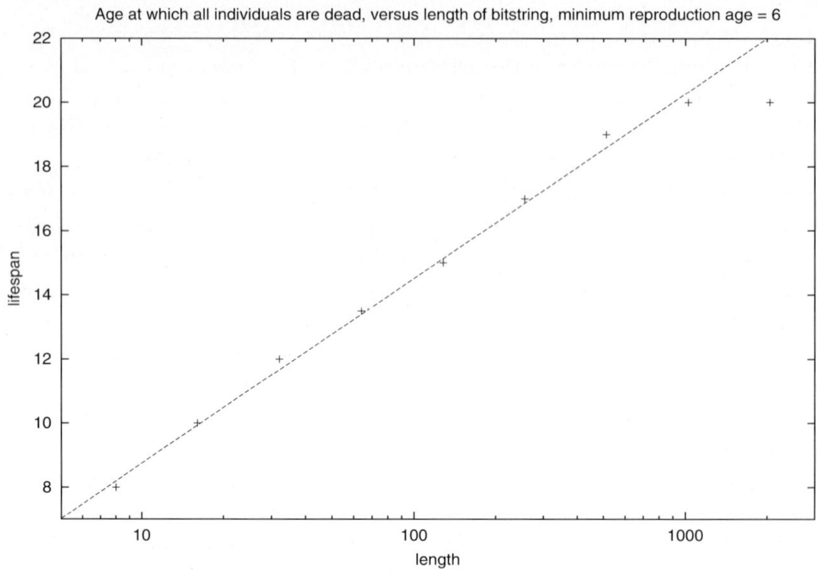

Fig. 3. Lifespan versus genome length at fixed minimum age of reproduction

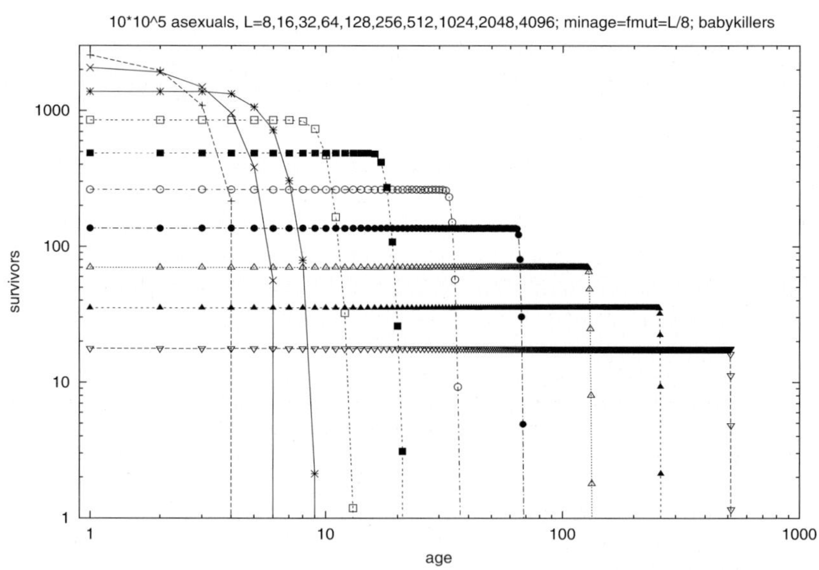

Fig. 4. Number of asexual individuals as a function of age, for widely varying lengths of the genomic bit-string. 10 samples with 10^4 individuals were simulated up to mostly 20,000 iterations (longer for the longest genome). Four births were given at each iteration by each adult

This figure shows that for the longer genomes, the survival curve becomes more rectangular than for the shorter genomes. This compression of mortality was actually observed during the medical progress in most of the last two centuries, though perhaps no longer in the last few decades in Europe. In the extreme case it means that we remain healthy until our sudden death, similar to Pacific Salmon [13]. But the genetic reasons for this rectangular survival curve are different: Now nearly all bits become mutated to one, independent of their position (age), for long genomes. Thus the individuals die if their age reaches the threshold of tolerated diseases at which they die. This threshold, as well as the mutation rate per genome and the minimum age of reproduction, were taken as 1/8 of the genome length in this scaling test. Following [9] we applied the Verhulst deaths only to the babies.

These simulations used the program given below, based on the one published and explained in [6]. The bit-string consists of w computer words of n_{byte} bytes each such that the overall genome lengths is $8wn_{\text{byte}}$ bits. This parameter n_{byte} is defined in line 6 and the same number must be put into line 7 after the `integer*`, except that `integer*1` does not exist in usual Fortran and needs to be replaced by `byte`. Thus the word length varies from 1 to 8 bytes. (The random integers `ibm` always use 8 bytes.)

```
c      asexual version (females only); variation of genome size c
many words w for each genome; Verhulst\index{Verhulst} variable or
constant c      Verhulst at birth only
       implicit none
       integer popdim,nbyte,nbit,nshift,w,nbiw
       parameter(w=2,nbyte=8,popdim=100000,nbit=8*nbyte,nbiw=nbit*w)
                    integer*8 gen1f(popdim,w), gene1, p, bit(0:nbit)
c                    byte      gen1f(popdim,w), gene1, p, bit(0:nbit)
       integer popmax,inipop,maxstep,medstep,minage,fage,lim,fmut,
      1 n6,t,i,seed,fa,n,imut,age,nmut,fpop,j,k,irun,nrun,agm,birth,
      2 fnumber(0:nbiw),dataf(popdim)
       parameter(popmax=popdim/10, inipop=popmax/10, maxstep=20000,
      1 medstep=maxstep/2, minage= nbiw/8, fage=nbiw, lim=nbiw/8,
      2 fmut=nbiw/8, birth=4, seed=2, nrun=1  )
       real factor
       integer*8 ibm,verhu,mult,fymed(0:nbiw),count(0:nbiw),ict
       data n6/'00007FFF'X/
       if(nbyte.eq.8) nshift=58
       if(nbyte.eq.4) nshift=59
       if(nbyte.eq.2) nshift=60
       if(nbyte.eq.1) nshift=61
c
       ict=0
       ibm=2*seed-1
       factor=(0.5/2147483647)/2147483647
       mult=13**7
```

```
      mult=mult*13**6
      print *, popmax,inipop,maxstep,medstep,minage,
     1          fage,lim,fmut,birth,seed,nbit,w,nrun
      bit(0)=1
      do 2 i=0,nbiw
        if(i.gt.0.and.i.lt.nbit) bit(i)=ishft(bit(i-1),1)
        fymed(i)=0
        fnumber(i)=0
        count(i)=0
 2      ibm=ibm*16807
      do 29 irun=1,nrun
      print *, irun
      call flush(6)
      do 28 i=0,nbiw
 28     fnumber(i)=0
      fpop=inipop
      fnumber(0)=fpop
      do 6 i=1,fpop
        dataf(i)=0
        do 6 k=1,w
 6        gen1f(i,k)=0
c     dataf: age at bits 0 to 14, mutations at 15 to 29
c
      do 7 t=1,maxstep
      if((t.le.100.and.t.eq.(t/10)*10).or.t.eq.(t/1000)*1000.or.
     1    (t.le.1000.and.t.eq.(t/100*100)).or.t.le.10) then
        if(irun.eq.1) print *, irun, t,fpop
        call flush(6)
      end if
c     verhu=2147483648.0d0*(fpop*4.0/popmax-2.0)*2147483648.0d0
      i=1
      fa=fpop
 9    age =iand(n6,dataf(i))
      verhu=2147483648.0d0*(fpop*4.0/popmax-2.0)*2147483648.0d0
      k=1+age/nbit
      agm=mod(age,nbit)
      nmut=iand(n6,ishft(dataf(i),-15))
      n=iand(1,ishft(gen1f(i,k),-agm))
      if(n.ge.1) nmut=nmut+1
      fnumber(age)=fnumber(age)-1
      age=age+1
      if(nmut.ge.lim.or.age.eq.nbiw) then
c     death
        if(fpop.le.1) goto 1
        do 21 k=1,w
 21       gen1f(i,k)=gen1f(fpop,k)
        dataf(i)=dataf(fpop)
        fpop=fpop-1
```

```
          if(fpop.ge.fa) then
            i=i+1
          else
            fa=fa-1
          endif
        else
c       survival
          fnumber(age)=fnumber(age)+1
          dataf(i)=ior(age,ishft(nmut,15))
          if(age.ge.minage.and.age.le.fage) then
            do 12 n=1,birth
              ibm=ibm*16807
              if(ibm.lt.verhu) goto 12
c           birth
              fnumber(0)=fnumber(0)+1
              fpop=fpop+1
              if(fpop.gt.popdim) goto 1
              dataf(fpop)=0
              do 23 j=1,w
 23             gen1f(fpop,j)=gen1f(i,j)
              do 13 imut=1,fmut
                ibm=ibm*16807
                k=iabs(ibm)*factor*w+1
                gene1=gen1f(i,k)
                ibm=ibm*16807
                p=bit(ishft(ibm,-nshift))
c               mutations in mother
c               gen1f(fpop,k)=ior(gene1,p)
                gen1f(fpop,k)=ior(gen1f(fpop,k),ior(gene1,p))
 13           continue
 12         continue
c           if(female suitable) then
          endif
          i=i+1
        endif
c       if(death) .. else (survival, birth) ..
        if(i.le.fa) goto 9
c       end of selection and birth, now start averages
        if(t.lt.maxstep-medstep .or. t.ne.(t/10)*10) goto 7
        do 10 i=0,nbiw
 10       fymed(i)=fymed(i)+fnumber(i)
        do 25 i=0,nbit
          do 26 j=1,w
          do 26 k=1,fpop
            if(i+j.eq.1) ict=ict+1
            gene1=gen1f(k,j)
            age=i+nbit*(j-1)
 26         count(age)=count(age)+iand(1,ishft(gene1,-i))
```

```
25    continue
7     continue
29    continue
      print100,(i,fymed(i)*1.0/medstep,count(i-1)*1.0/ict,i=1,nbiw)
100   format(1x,i5,2f13.5)
      stop
1     print *, 'error' ,t, fpop
      stop
      end
```

2.3 Sexual Version

In the sexual version of the Penna model [22, 23] the population is divided into males and females. Individuals are diploids, with their genomes represented by two bit-strings that are read in parallel. Each bit-string contains the genetic information inherited from one of the parents. In order to count the accumulated number of mutations and compare it with the threshold T, it is necessary to distinguish between recessive and dominant mutations. A mutation is counted if two bits set to 1 appear at the same position in both bit-strings (homozygous) or if it appears in only one of the bit-strings (heterozygous) but at a dominant position (locus). The dominant positions are randomly chosen at the beginning of the simulation and are the same for all individuals.

After reaching the minimum reproduction age R, a female randomly chooses a male with age $\geq R$ to breed. To construct one offspring genome first the two bit-strings of the mother are cut in a random position (crossing), producing four bit-string pieces. Two complementary pieces are chosen to form the female gamete (recombination). Finally, M_f deleterious mutations are randomly introduced. The same process occurs with the male's genome, producing the male gamete with M_m deleterious mutations (see Fig. 5). These two resulting bit-strings form the offspring genome. The sex of the baby is randomly chosen, with a probability of 50% for each one. This whole strategy is repeated b times to produce the b offspring. The Verhulst killing factor works in the same way as in the asexual reproduction.

2.4 Applications of Sexual Version

From the example of Pacific Salmon one could first guess that women should not survive menopause, or more generally, females should not survive the end of their reproductive phase (which is observed in many animals: [24], Austad in [14]). Simulations [6] show that this is not so: Men are assumed to be able to reproduce until their death, and then they also help women to survive after menopause since the sex of the child is selected randomly; thus, within this model, mutations relevant for old age accumulate equally for both men and women, prohibiting to selectively kill women but not men.

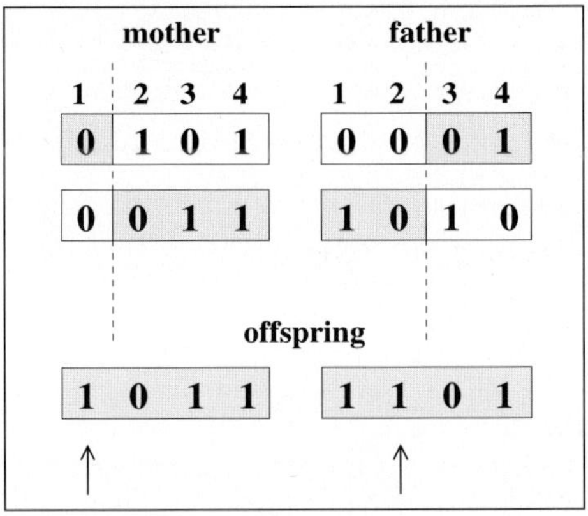

Fig. 5. Schematic representation of sexual reproduction in the Penna model using bit-strings of only four bits (individuals would live at most for four "years"). Vertical lines indicate crossover positions. Arrows indicate where random mutations were introduced in each gamete. The union of the two gametes forms the offspring genome

These simulations only explain why menopause is possible and not why it arises. For humans, cultural effects may be important [25]: Grandmothers help in raising their grandchildren instead of giving birth. For animals, in general, this explanation does not hold. Fortunately, simulations of the sexual Penna model, with a lethal risk of giving birth increasing with age, and with a certain age interval where offspring can survive only if their mother takes care of them, have shown that cessation of female reproductive ability emerges by itself, without being put in at the beginning [26], Fig. 6. This self-organisation explains menopause or its analogues without recourse to any traits special to humans; these computer simulations [26] were, to our knowledge, the first quantitative explanations of menopause analogues for animals. Correlations like those in Fig. 1 of [25] are also found in the Penna model (to be published).

Why do women live longer than men, in nearly all countries at present? One of the authors claims that men drink too much wine and eat too many steaks, as opposed to her. The other claims to be oppressed by women in general and Mother Nature in particular, since he got only one X chromosome while women got two. Only the second claim thus far is confirmed by computer simulations [27]: A bad mutation in one X chromosome does not reduce the health if the other X chromosome is not mutated and if the bit-position is recessive; for men with only one X chromosome this helpful effect is impossible. However, the actual difference between male and female life expectancies at birth in Sweden (no war since nearly two centuries) shows drastic variations in time, Fig. 7, though always favouring women. Thus presumably both genetic and life-style reasons are important and may even be coupled [28].

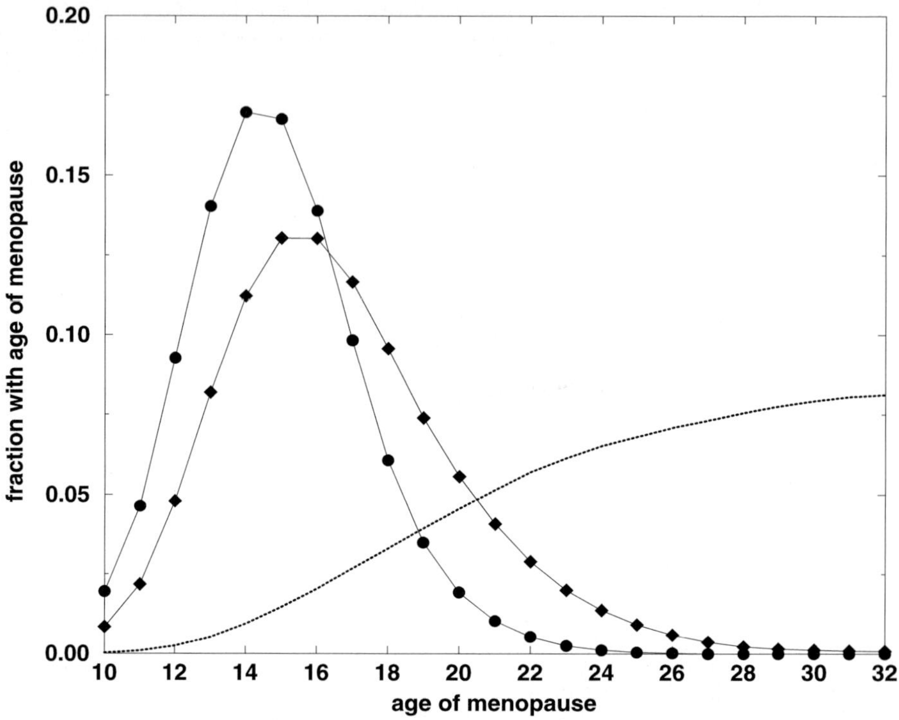

Fig. 6. Fraction of females with a given age of menopause as a function of the age of menopause. Circles: parental care for 5 years; Diamonds: parental care for 4 years; Line: no parental care

If men are slightly defective versions of women, why has Mother Nature invented them? Why not follow bdelloid rotifers and some fungi [29], who have lived without males since $\sim 10^8$ years. Redfield [30] already questioned the usefulness of males and was largely ignored by fellow biologists [31], while her paper triggered many simulations of physicists reviewed in [3]. By assuming that mutations only kill the individual when enough of them have accumulated, but that already before they reduce the survival probabilities in addition to the Verhust deaths, an advantage of sexual versus asexual reproduction was found [32], in spite of the loss of births by a factor two (since men don't become pregnant). Much clearer is the situation for bacteria where, without a separation of males and females, a clear advantage was found if sometimes they exchange parts of their genome [33]. Also changes in the environment favours sex [34]. However, no explanation was found yet explaining why hermaphroditic reproduction is not clearly dominating over other forms of life [3]. A good review of the biological questions is given in [35].

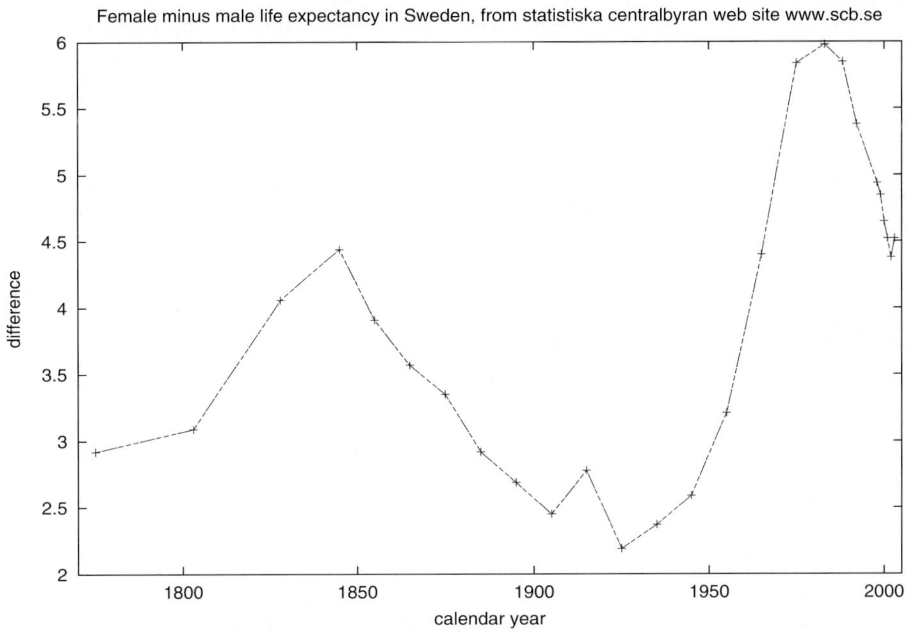

Fig. 7. Difference between the female and male life expectancies at birth for Sweden, over more than two centuries

2.5 Other Models

The Dasgupta model [36] uses no bit-string as a genome but instead works with survival probabilities. Originally it was thought not to give a large number of different ageing steps, but years later minor modifications solved this problem. It even was successfully generalized to sexual reproduction.

Mueller and Rose [37] tried to find a mortality plateau for the oldest old; it seems however that their simulations were not run long enough to get into a good equilibrium. Charlesworth made simulations with the same aim, but not many follow-up simulations are known to us [38]. Promislow [39] has very recently published a senescence theory based on pleiotropic protein networks, using the scale-free networks investigated by physicists. More similar to the Penna model is the one of [40].

More follow-up was published for the Stauffer model [4] which revived the century-old Weismann idea that we die to make place for our children. This idea works if one assumes that the total number of offspring during the life of an individual remains constant [41]. Thus random mutations which increase the life span simultaneously reduce the birth rate per unit time. Normally this model gives a mortality increasing linearly (not exponentially) with age, as for mayflies (but not for humans). It was used for another confirmation [42] that menopause and its analogues are not necessarily connected with grandmothers and human civilization: Death risk increasing with age was sufficient to let menopause emerge by itself.

All these theories including the Penna model use the accumulation of mutations over many generations, sometimes combined with antagonistic pleiotropy, which has good effects at young age and bad effects at old age. There may be entirely different reasons for ageing, like the wear and tear which often ends the careers of athletes. Or oxygen radicals produced by normal metabolism may slowly damage the cells and their DNA; perhaps the polyphenol resverotrol in red wine helps to scavenge these dangerous molecules. But we are not aware of quantitative theories allowing a comparison with real mortality functions.

The most recent quantitative theory known to us uses telomeres [43]. These are end segments of the DNA string; at each cell division, the duplication of the DNA causes the loss of one of these segments. If none of them is left the cell no longer divides and has reached its Hayflick limit *in vitro*. The enzyme telomerase may restore lost telomeres *in vivo*. Aviv et al suggested, against widespread opinion, that telomere loss may be the reason for ageing [44]. Masa et al [43] combined telomeres with cancer and showed that genetic manipulation for longer telomeres may increase the risk of dying due to cancer, which can prevail over the positive effect of longer telomeres on the longevity. Figure 8 shows one of the resulting mortality functions, in striking similarity with reality, Fig. 1.

3 Sympatric Speciation

3.1 Model with a Single Phenotypic Trait

Speciation involves the division of an already well-adapted species, so that each part moves onto a new adaptive peak. This process is easily understood if a species becomes subdivided by a physical barrier, like a river: Each part experiences different mutations, population fluctuations and selective forces, in what is called allopatric speciation. In contrast, conceiving the division of a single population and radiation onto separate peaks without geographical isolation, in what is called sympatric speciation, is intuitively more difficult. Through which mechanism can a single population be converted into two reproductively isolated segments in the absence of spatial barriers to prevent gene exchange?

Many papers have appeared during the last 5 years, mainly produced by biologists (see for instance [45, 46] and also [47] for a review), proposing different mechanisms to explain the origins of sympatric speciation. We concentrate here on simulations including ageing. The first successful attempt to simulate this phenomenon through the Penna model [48] was performed using the same ingredients proposed in [45, 46], which are: competition for resources (related to some ecological phenotypic trait of the individuals) and

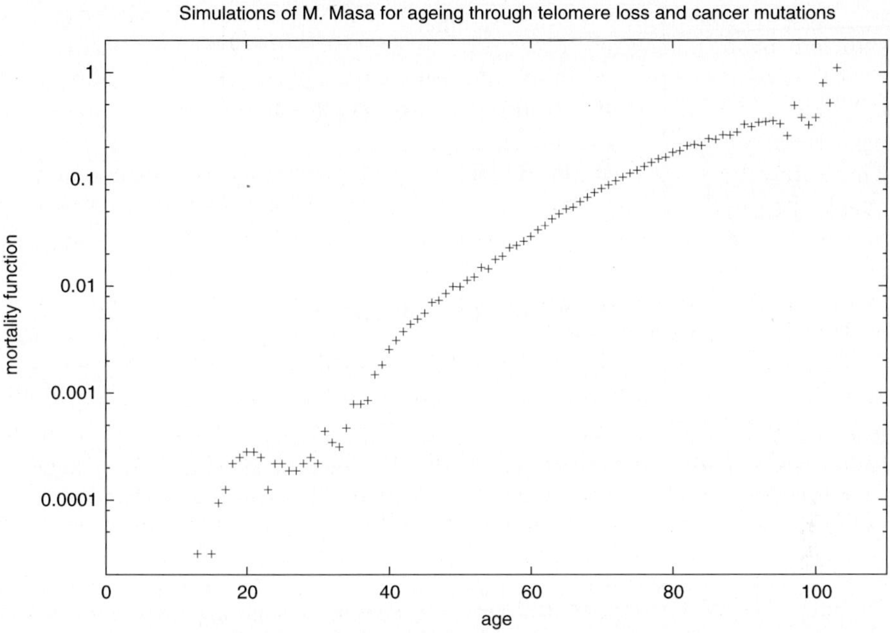

Fig. 8. Mortality function in a telomere model with cancer, from [43]. Note the similarity with the plot for German men, Fig. 1

sexual selection (related to some phenotypic trait that determines sexual preferences).

In the simulations presented in [48], the ecological and the sexual selection traits were chosen to be the same. A sudden change of the food resources provokes a change in the individuals' phenotypes, favouring two substantially different phenotypes (disruptive selection). Assortative mating (non random mating) then leads to reproductive isolation between these two phenotypes, followed by the elimination of all intermediate phenotypes (speciation). The biological motivation of these simulations was to mimic a real phenomenon observed in the Galapagos Islands [49–51], where the size of the available seeds is dictated by the amount of rain. Depending on the season, the distribution of these seeds changes from a broad one, centred at middle-sized seeds, to a double-peaked distribution, of only small and large seeds. It has a great impact on the morphology of beak sizes in the population of ground finches that feed on these seeds: the distribution of beak sizes follows that of the seeds, in a very fast process of adaptation.

The first step to implement the above ingredients was to introduce a phenotype into the Penna model. It was done by adding to the chronological genome an extra-pair of non-age structured bit-strings of 32 bits each, representing the beak. The dynamics of reproduction and mutations are the same

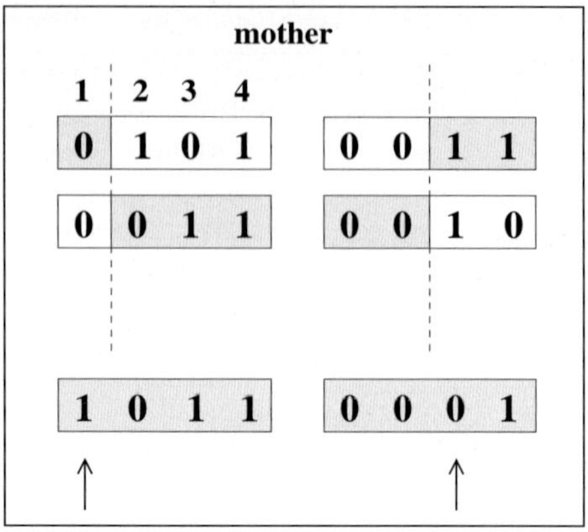

Fig. 9. Schematic representation of gamete formation in sexual reproduction, adding now a non age-structured part to the chronological genome. Arrows indicate where random (good and bad) mutations occurred. The same scheme applies to the father's genome (who cares about males?)

for both the age-structured and the new strings – for the latter, a mutation that changes a bit from 1 to 0 is also allowed (Fig. 9).

The beak size is determined by counting, in this non-structured pair of bit-strings, the number of recessive bit-positions (chosen as 16) where both bits are set to 1, plus the number of dominant positions with at least one of the two bits set. It will be a number n between 0, meaning a very small beak, and 32, for a very large one. The second step was to make use of the Verhulst factor to attribute a selective character to the beak size and to change the competition between individuals from a general one to an intra-specific competition, where animals with large (small) beaks dispute only among them.

The new Verhulst factor is given by:

$$V(t) = \frac{N_n(t)}{N_{\max}F(n)} ,$$

where $N_n(t)$ accounts for the population that competes for resources available to individuals of beak size n. The fitness function $F(n)$ is a measurement of the individual's ability to make use of food resources, depending on the current distribution of seeds and its beak size. $N_{\max}$ is the same carrying capacity of the original Verhulst factor and so the practical role of the function $F(n)$ is to increase or decrease this carrying capacity perceived by each individual, depending on its beak size.

The simulations were done with two different functional forms for the function $F(n)$. At the beginning of the simulations, $F(n)$ is a single-peaked function with a maximum at $n = 16$, representing large availability of medium-sized seeds. The whole population of finches will compete for the same resources and $N_n(t)$ is the total population. After some number N_{step} of iterations, the function $F(n)$ changes to a two-peaked shape, with maxima at $n = 0$ and $n = 32$, and now the food resources concentrate on either small or large seeds, with a vanishing number of medium-sized ones. Only small(large)-beaked individuals – those with $n < (>)16$ – can compete for the small(large) seeds. For that reason, the death probability $V(t)$ of an individual with $n < (>)16$ is computed by assigning to $N_n(t)$ the number of individuals with $n < (>)16$ plus half of the population that has $n = 16$. An individual with $n = 16$ competes either for small or large seeds, and this choice is random.

Finally, in order to simulate sexual selection a single locus was introduced into the genome that codes for this selectiveness, also obeying the general rules of the Penna model for genetic heritage and mutation. If it is set to 0, the individual will not be selective in mating (panmictic mating), and it will be selective (assortative mating) if this locus is set to 1. The mutation probability for this locus was set to 0.001 in all simulations. Females that are selective will choose mating partners that satisfy some criterion related to the beak size.

At the beginning of the simulations, when the fitness function $F(n)$ is single-peaked, there is no selective pressure for mating selectiveness, and the population mates randomly: all females are non-selective. After $F(n)$ becomes double-peaked, females that mutate into selectiveness will choose mating partners that have beak sizes similar to their own: if a female has $n < (>)16$ and is selective, she will only mate with a partner that also has $n < (>)16$.

In Fig. 10 we show the distribution of beak sizes of the populations at time step $N_{\text{step}} = 12\,000$, up to which the fitness function was single-peaked, and at time step 50 000, after it has been double-peaked for 30 000 time steps. A stable polymorphism has clearly been established as a result of the duplicity of food resources. The fraction of selective females in the population, which was 0 at the start of the run, also increases to nearly 1.0 after the establishment of a double-peaked $F(n)$. Now there are two distinct populations, each of which does not mate with a partner from the other.

3.2 Model with Two Phenotypic Traits

Not always the ecological trait, determining individual fitness to the environment, is the same used by females to choose a partner. In case of fish, for instance, sexual selection is related to the colour, while fitness depends on the size [52]. In order to make the simulations more realistic a second phenotypic trait – another pair of non age-structured bit-strings – was added to the Penna chronological genome [53].

Dynamics of death and birth follow the same rules as before, and the phenotype space for this new sexual trait is mapped again onto an integer between 0 and 32. For mating, a female chooses, among a random selection

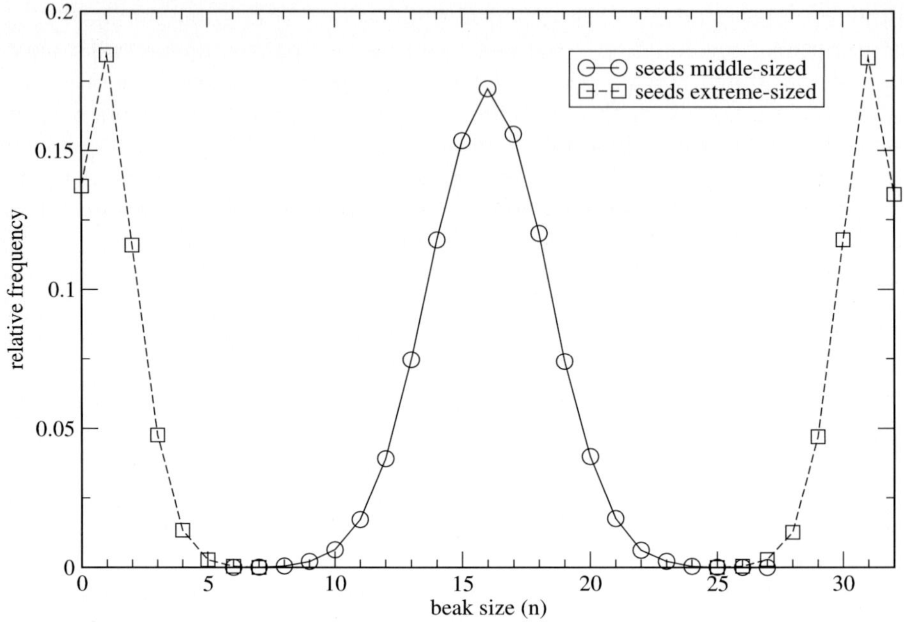

Fig. 10. Distribution of beak sizes when middle-sized seeds are abundant (circles – time step = 12,000) and when only small and large-sized seeds are available (squares – time step = 50,000). We show the fraction of the population with each beak size

of a fixed number of males, chosen to be 6 in the simulations, a suitable mating partner for whom the phenotype for this second trait (colour, say) matches her own. Let us call f the phenotype of the female and m the one for the male, for the sexual selection trait. Then, if the female has mutated into selective, it follows the rules:

If $f < 16$ then it selects the male with the smallest m;
If $f > 16$ then it selects the male with the largest m;
If $f = 16$ then the female chooses randomly to act as one of the above.

If we think about this mating trait as colour, for instance, and assign $f < (>) 16$ to a blue (red) character, a blue (red) female will choose the male that lies deepest into the blue (red) region.

Figure 11 shows that, like in the former case, the distribution of the fitness trait is single-peaked at $n = 16$ up to step N_{step}, as a consequence of the number of loci (16) where the 1 allele is dominant, and moves into a polymorphism after the ecology becomes bi-modal. The sexual selection trait also shows a single peak until step N_{step}, and splits the population in two groups afterwards. But now a strong correlation develops between these traits, and the individuals with sexual selection phenotype $< (>) 16$

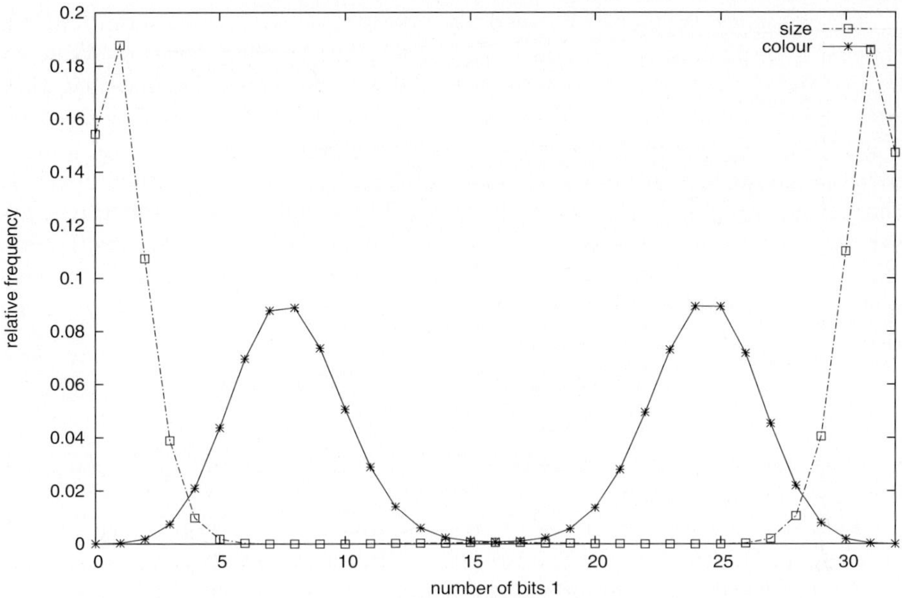

Fig. 11. Distribution of fitness (size = squares) and sexual (colour = *) traits at the end of the simulation. These traits are correlated, and the population with fitness trait to the left has its sexual trait also to the left of the plot

have their fitness trait $<$ ($>$)16. Sexual selectiveness also develops as a result of evolutionary dynamics, and at the end of the simulation all females are selective.

3.3 Model Without Phenotypic Traits

A much simpler strategy, without considering phenotypes but adopting the same ingredients of competition and sexual selection, has also succeeded in obtaining speciation through the Penna model. In this case a single bit position, which was taken as position 11 in the Penna chronological genome, determines the mating [53]. Each diploid individual has $k = 0, 1,$ or 2 bits set at this position. A female with k such bits at position 11 selects only males with the same number k of such "speciation" bits to mate. The simulation starts with a single species, that is, all individuals have $k = 0$. Due to the randomness of mutations and crossover, offspring do not necessarily have the same k speciation bits set to one as their parents, and this randomness allows the emergence of new species out of the original one. At every time step t three populations N_k, depending on the number $k = 0, 1, 2$ of speciation bits set to one, are now co-evolving, and each of these three sub-populations is half male and half female.

Co-existence is again obtained by adopting intra-specific competition, that is, by replacing the standard Verhulst factor by three separate Verhulst factors for the separate populations $k = 0, 1, 2$. Suppose, for instance, that the original population $k = 0$ is vegetarian, and that the second population $k = 2$ emerging out of it consists of carnivores. Both populations are limited by the amount of food, but their food sources are completely different; thus, there is no competition between the two different populations, but the meat-eating females will not select any herbivore males for mating, and vice versa. The small population with $k = 1$ can be regarded as one that feeds in both niches. It is added half to $k = 0$ and half to $k = 2$ for the evaluation of the two intra-specific Verhulst factors,

$$V_0 = \frac{(N_0 + N_1/2)}{N_{\rm max}} \text{ and } V_2 = \frac{(N_2 + N_1/2)}{N_{\rm max}} ,$$

and has the arithmetic average of these two Verhulst factors as its own food-limiting Verhulst factor.

Figure 12 shows how males of the new species N_2 emerges from the old species N_0. For the females the results are about the same.

Similar results were found using a square lattice without Verhulst factors; intra-specific competition is introduced through some rules that determine whether a place for a newborn exists [54].

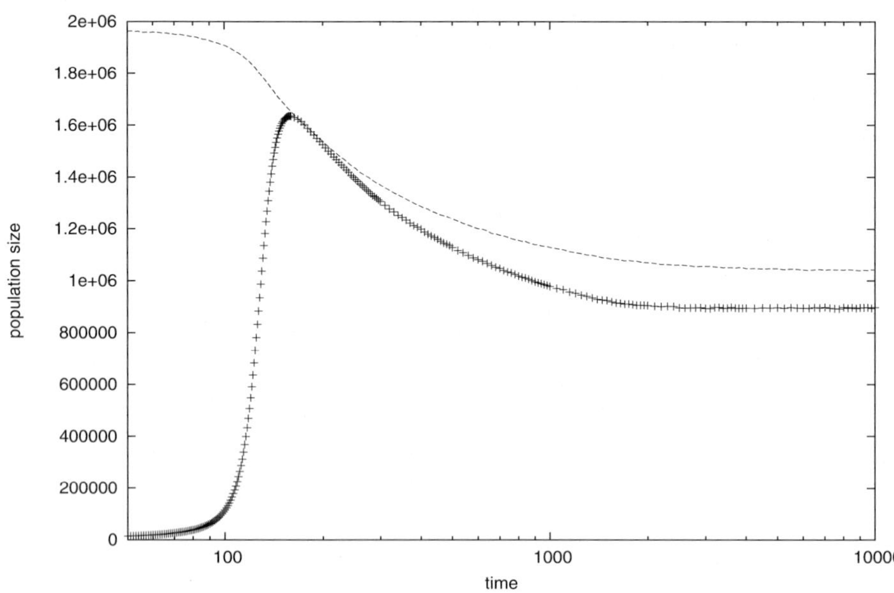

Fig. 12. Variation in time of N_0 (dashed line, original species) and N_2 (+, new species). Note linear, not logarithmic, vertical scale. The intermediate population N_1, that competes both with N_0 and N_2, is only about one percent of the total and is not shown

The influence of recombination rate in speciation has also been investigated making use of the Penna model [55].

Acknowledgments

We thank ESPCI/PMMH for the warm hospitality during the period when this manuscript was produced. We also thank many people, in particular P.M.C. de Oliveira, T.J.P. Penna, A.T. Bernardes, S. Cebrat, J.S. Sá Martins and A.O. Sousa for helpful discussions over many years. S.M.O. thanks J.S. Sá Martins and K. Luz-Burgoa for the collaboration in the results presented in Sects. 3.1 and 3.2 and the Brazilian agency CNPq for financial support.

References

1. P.M.C. de Oliveira: Physica A **306**, 351 (2002)
2. D. Chowdhury, D. Stauffer: J. Biosci. (India) **30**, 277 (2005) and Physica A **352**, 202 (2005); A. Kunwar: Int. J. Mod. Phys. C **15**, 1449 (2004)
3. D. Stauffer, P.M.C. de Oliveira, S. Moss de Oliveira, T.J.P. Penna, J.S. Sá Martins: An. Acad. Bras. Ci. **73**, 15 (2001) = cond-mat/0011524
4. D. Stauffer. In: *Biological Evolution and Statistical Physics*, ed by M. Lässig and A. Valleriani (Springer, Berlin Heidelberg 2002) p 258
5. D. Stauffer: The Complexity of Biological Ageing. In *Thinking in Patterns*, ed by M.M. Novak (World Scientific, Singapore 2004) p 131
6. S. Moss de Oliveira, P.M.C. de Oliveira, D. Stauffer: *Evolution, Money, War and Computers* (Teubner, Leipzig 1999)
7. T.J.P. Penna: J. Stat. Phys. **78**, 1629 (1995)
8. J. Thoms, P. Donahue, N. Jan: J. Physique I **5**, 1697 (1995)
9. J.S. Sá Martins, S. Cebrat: Theory Biosci. **119**, 156 (2000); A. Łaszkiewicz et al.: Int. J. Mod. Phys. C **13**, 967 (2002)
10. D. Makowiec: Physica A **289**, 208 (2001)
11. P. Pamilo P, M. Nei, W.H. Li: Genet. Res. Camb. **49**, 135 (1987)
12. T.J.P. Penna, S. Moss de Oliveira: J. Physique I **5**, 1697 (1995)
13. T.J.P. Penna, S. Moss de Oliveira, D. Stauffer: Phys. Rev. E **52**, 3309 (1995); see also remark of S. Tuljapurkar on page 20 in [14]
14. K.W. Wachter, C.E. Finch: *Between Zeus and the Salmon. The Biodemography of Longevity* (National Academy Press, Washington DC 1997)
15. A.R. Thatcher, V. Kannisto and J.W. Vaupel: *The Force of Mortality at Ages 80 to 120* (Odense University Press, Odense 1998)
16. J.W. Vaupel et al.: Science **280**, 855 (1998)
17. J.-M. Robine, J.W. Vaupel: Exp. Gerontology **36**, 915 (2001). See also K. Suematsu, M. Kohno: J. Theor. Biol. **201**, 231 (1999)
18. S. Moss de Oliveira, P.M.C. de Oliveira, D. Stauffer: Physica A **221**, 453 (1995)
19. D. Stauffer: Int. J. Mod. Phys. C **10**, 1363 (1999); J.B. Coe, Y. Mao, M.E. Cates: Phys. Rev. Lett. **89**, 288103 (2002)

20. M.F. He, Q.L. Pan and S. Wang: Int. J. Mod. Phys. C **16**, N_o 1 (2005) - to appear
21. E. Brigatti, J.S. Sá Martins, I. Roditi: Eur. Phys. J. B **42**, 431 (2004); for sex see A. Łaszkiewicz, S. Cebrat, D. Stauffer: Adv. Complex Syst. **8**, 7 (2005)
22. A.T. Bernardes: Ann. Physik **5**, 539 (1996)
23. D. Stauffer, P.M.C. de Oliveira, S. Moss de Oliveira, R.M. Zorzenon dos Santos: Physica A **231**, 504 (1996)
24. D.J. Holmes, M.A. Ottinger: Experimental Gerontology **38**, 1365 (2003)
25. M. Lahdenperä et al.: Nature **428**, 178 (2004)
26. S. Moss de Oliveira, A.T. Bernardes, J.S. Sá Martins: Eur. Phys. J. B **7**, 501 (1999)
27. J. Schneider, S. Cebrat, D. Stauffer: Int. J. Mod. Phys. C **9**, 721 (1998); see also D.W.E. Smith: Biol. Rev. **64**, 1 (1989)
28. T. Klotz, D. Stauffer: The Aging Male **4**, 95 (2001); D. Stauffer, Theory in Biosciences **120**, 87 (2001)
29. T.E. Pawlowska, J.W.Taylor: Nature **427**, 733 (2004)
30. R. Redfield: Nature **369**, 145 (1994)
31. S. Siller: Nature **411**, 689 (2001); S.P. Otto, S.L. Nuismer: Science **304**, 1018 (2004)
32. J.S. Sá Martins, D. Stauffer: Physica A **294**, 191 (2001)
33. S. Moss de Oliveira, P.M.C. de Oliveira, D. Stauffer: Physica A **322**, 521 (2003)
34. Mingfen He, Hongbo Ruan, Changliang Yu, Lei Yao: Int. J. Mod. Phys C **15**, 289 (2004)
35. *Science et Vie*, issue 1043, August 2004, pp 28-35
36. S. Dasgupta: J. Physique (France) I **4**, 1563 (1994); M. Heumann, M. Hötzel: J. Stat. Phys. **79**, 483 (1995); N.G.F. de Medeiros, R.N. Onody: Phys. Rev. E **64**, 041915 (2001); A.O. Sousa: Theory in Biosciences **122**, 303 (2003)
37. L.D. Mueller, M.R. Rose: Proc. Natl. Acad. Sci. USA **93**, 15249 (1996); S.D. Pletcher, J.W. Curtsinger: Evolution 52, **454** (1998)
38. B. Charlesworth: J. Theor. Biol. **210**, 47 (2001)
39. D.E.L. Promislow: Proc. Roy. Soc. B **271**, 1225 (2004)
40. M. Magdoń-Maksymowicz and A.Z. Maksymowicz: Proc. of 4th International Conference on Computational Science - ICCS 2004, Poland, ed by M. Bubak, G.D. van Albada, P.M.A. Sloot, J.J. Dongarra (Springer Lecture Notes in Computer Science, 3036 Part IV, Berlin Heidelberg 2004)
41. N.S. Gavrilova et al.: Ann. N.Y. Acad. Sci. **1019**, 513 (2004)
42. A.O. Sousa: Physica A **326**, 233 (2003)
43. M. Masa, S. Cebrat, D. Stauffer: preprint = q-bio.PE/0408026
44. A. Aviv, D. Levy, M. Mangel: Mech. Ageing Dev. **124**, 929 (2003)
45. A.S. Kondrashov, F.A. Kondrashov: Nature **400**, 351 (1999)
46. U. Dieckmann, M. Doebeli: Nature **400**, 354 (1999)
47. M. Turelli, N.H. Barton, J.A. Coyne: Trends in Ecology & Evolution **16**, 330 (2001)
48. J.S. Sá Martins, S. Moss de Oliveira, G.A. de Medeiros: Phys. Rev. E **64**, 021906 (2001)
49. P.T. Boag, P.R. Grant: Nature **274**, 793 (1978); P.T. Boag, P.R. Grant: Science **214**, 82 (1981)
50. P.R. Grant: *Ecology and evolution of Darwin's finches* (Princeton University Press, Princeton 1986)

51. D. Lack: *Darwin's finches* (Cambridge University Press, Cambridge 1983)
52. A.B. Wilson, K. Noack-Kunnmann, A. Meyer: Proc. R. Soc. Lond. B **267**, 2133 (2000)
53. K. Luz-Burgoa, S. Moss de Oliveira, J.S. Sá Martins, D. Stauffer, A.O. Sousa: Braz. J. Phys. **33**, 623 (2003)
54. A.O. Sousa: Eur. Phys. J. B **39**, 521 (2004)
55. A. Łaszkiewicz, Sz. Szymczak, S. Cebrat: Int. J. Mod. Phys. **14**, 765 (2002)

Part III

Econophysics

Influence of Information Flow in the Formation of Economic Cycles

J. Miśkiewicz[1] and M. Ausloos[2]

[1] Institute of Theoretical Physics, University of Wrocław, pl M. Borna 9, 50-204 Wrocław, Poland
jamis@ift.uni.wroc.pl
[2] SUPRATECS, B5, University of Liège, B-4000 Liège, Euroland
marcel.ausloos@ulg.ac.be

1 Introduction

Despite the fact that Verhulst's idea [1] of regulated biological populations is 200 years old, it is still very useful since it allows to investigate features of various systems. Here an eight order logistic map is applied in modelling the influence of information flow delay onto the behaviour of an economic system.

The delay of information flow is an internal feature of all economic systems, because continuous monitoring of such systems both on macro and microeconomic scales is either extremely difficult or even impossible. The more so since the data is not easily available nor even reliable as it could be in physics laboratories. Macroeconomic parameters such as Gross Domestic Product, Gross National Product, inflation, demographic data etc. are announced in well defined time intervals (monthly, quarterly or annually). The same situation is observed in the case of various companies. They announce their financial statements about their economic results at specific dates and for given time intervals – according to internal or external rules (usually according to law regulations). Sometimes some "warning" is issued. However the tendency is that intervals between announcements are rather long, e.g. the value of a dividend is announced annually or at various trimester ends. It seems obvious that only very small companies are able to perform continuous monitoring. But even then, the process of collecting information from a significant (on a macroscopic scale) number of such companies inhibits or makes it impossible to perform continuous monitoring. In view of the data collecting procedure it is clear that every economic decision is based on some information describing a past situation. It is also important to notice that the time delays between information gathering, decision taking, policy implementation, and subsequent data gathering are not constant, nor are a fortiori continuous variables, as that was considered in [2]; indeed the information about the system is updated at the end of discrete time intervals.

Therefore econophysics-like modelling of such features encounters some difficulty, surely at the testing level. Recently a microscopic-like approach has been presented, through a model [3–5] including some measure of a company fitness with respect to an external field, and a birth-death evolution, according to some business plan, and the local company close range environment. The information flow was however considered to occur instantaneously.

In order to investigate the discrete information flow time delay and its effect, a model, hereby called the ACP model [3–5], has been modified by splitting the information about the system into two parameters. One is monitored continuously (is updated at every iteration step) and is known to the system itself; the second, like official statements of the system, is announced at the end of discrete time intervals and is used by companies for calculating their strategies. Therefore the strategy of a company depends on the delay time information and the information itself. As it is shown in Sect. 3 the length of the time delay (t_d) influences quite strongly and in a nontrivial way the behaviour of the overall system.

Detailed description of the ACP model is given in Sect. 2 and the properties of the system as a function of time delay and initial concentration are investigated (Sect. 3) in the case of short, medium and long time delays.

2 ACP Model

For the sake of clarity the basic ingredients of the ACP model are recalled here below. The main problem was to simulate the behaviour of economic systems in spatio-temporally changing environmental conditions, e.g. political changes and destruction of economic barriers. The model was set in the form of a Monte Carlo simulation. Notice that the ACP model [3–5] contains among its variants an adaptation of the Bak–Sneppen model and was built in order to answer economic questions[3]. The model consists of

1. **space** – a square symmetry lattice,
2. **companies**, which are initially randomly placed on the lattice, in an
3. **environment** characterized by a real field $F \in [0, 1]$ and a selection pressure sel,
4. each company (i) is characterized by one real parameter $f_i \in [0, 1]$ (its so-called **fitness**).

The following set of actions was allowed to companies:

1. companies survive with the probability

$$p_i = \exp(-sel|f_i - F|) \tag{1}$$

2. companies may move on the lattice horizontally or vertically, one step at a time, if some space is available in the von Neuman neighbourhood.
3. if companies meet they may

[3] Let us recall that the Bak–Sneppen model was originally built in order to investigate the coevolution of populations [6]

a) either merge with a probability b,
b) or create a new company with the probability $1 - b$.

The ACP model may be described in a mean field approximation [7, 8] by introducing the distribution function of companies $N(t, f)$, which describes the number of companies having a given fitness f at time t. The system is then additionally characterized by the concentration of companies $c(t)$.

The present report of our investigations is restricted to the case of the best adapted companies ($f = F$), so that the selection pressure has no influence on the survival of companies. So the only factor which could alter the number of companies is the strategy, i.e. the decision to merge or create a new entity. The ideas behind the mean field approximation [7, 8] are applied here and developed by introducing a strategy depending on the system state and the discrete time of the official announcement about the state of the system.

The introduction of the strategy depending on the state of the system reflects the idea of Verhulst [1], when replacing the constant Malthus grow rate by the function $1 - x$, which introduced a limit for the system to grow. In the present investigation it is assumed that the strategy should depend on the state of the system. Moreover the company board takes its decision knowing information announced about its environment. The generation of new entities is more likely in the case of a low concentration of companies than when this concentration is high. The merging parameter describes the reversed dependency, i.e. merging is more likely to occur in the case of a high density of companies than if the density is low. The simplest function which fulfils this condition is $1 - c$, the same as in Verhulst original work [1].

The additional ingredients to the ACP model are thus

1. the merging parameter b is replaced by a strategy $(1 - c)$,
2. the companies know the value of the concentration c according to official statements announced after the time delay t_d.

The evolution equation of the system with companies, using the state dependent strategy is:

$$c_t = c_{t-1} + \frac{1}{2} c_{t-1}(1 - c_{t-1}^8)(1 - (1 - c_{t-1})^8)(2ST(c(g(t))) - 1) , \qquad (2)$$

where $ST(c) = 1 - c$, $g(t) = k[t/k]$ and $[]$ denotes the procedure of taking a natural number not larger than the one given in the brackets. The time is measured in iteration steps IS.

3 Results

Numerical methods were used in order to investigate properties of the system. Because the coevolution equation (2) is given as an iteration equation the time is discrete and counted in iteration steps (IS). The following features of the system were examined:

1. the **coevolution** of $c(t)$ as a function of the initial concentration,
2. the **stability time** defined as the time required to achieve a unique stable solution; because of numerical reasons the criterium applied here is $|c_{n+1} - c_n| < 10^{-10}$,
3. the **crash time** t_c, such that $c_{t_c} < 0$ (it is understood as the time when all companies are wiped out from the system),
4. the **stability intervals** – the intervals of initial values for which the evolution of the system is longer than a given time t_s
5. the complex Lyapunov exponent

$$\lambda = \lim_{N\to\infty} \frac{1}{N} \sum_{n=1}^{N} \log_2 \left(\frac{\mathrm{d}x_{n+1}}{\mathrm{d}x_n} \right) . \tag{3}$$

The Lyapunov exponent calculated in its complex form (3) gives also some information about the oscillations of the system. Using the properties of logarithm:

$$a < 0 \Rightarrow \log a = \log(-1 \cdot |a|) = \log(-1) + \log|a| . \tag{4}$$

The imaginary part of $\log_2(\mathrm{d}x_{n+1}/\mathrm{d}x_n)$ gives some information on whether the distances between consecutive iterations are monotonic.

The numerical iterations were performed for the initial concentration in the interval $c_0 \in (0, 1)$, at consecutive values distant of 0.02. Therefore 500 histories of evolution were investigated.

It is possible to observe three types of coevolution: a unique, a periodic and a chaotic solution. In the case of a unique solution the system may approach this solution as a "damped" coevolution or "damped oscillation". The damped coevolution happens if $\forall t > 0$: $c(\infty) - c(t) > 0$ or if $\forall t > 0$: $c(\infty) - c(t) < 0$ and $|c(t-1) - c(t)| \geq |c(t) - c(t+1)|$, where $c(\infty)$ is the asymptotic state of the system. This means that the distance between concentration and asymptotic concentration is decreasing in every iteration step and the concentration is either smaller or bigger than the asymptotic concentration. The damped oscillations are observed if $|c(t-1) - c(t)| \geq |c(t) - c(t+1)|$ and $\exists t_0$ such that $\forall t > t_0$: $c(t) > c(\infty)$ and $c(t+1) < c(\infty)$. This means that the distance between consecutive concentrations of companies is decreasing. In the case of a periodic solution for $t > t_0$ there exists a n-tuple of concentrations which is repeated for $t > t_0$, where t_0 is the time required by the system to reach the stable or periodic solution. The length of the n-tuple is defined as the period of oscillations. The system is chaotic if the real part of the Lyapunov exponent is positive: $\mathrm{Re}\,(\lambda) > 0$.

The coevolution of the system is presented either as a function of time (Figs. 1, 4, 7, 10, 13, 15, where the coevolution is plotted for chosen initial concentrations) or as a function of initial concentration (Figs. 2, 5, 8, 11, 14, 16, where the coevolution of the system is plotted in one vertical line so the plot is a set of coevolutions for 500 different initial concentrations.

3.1 Stability Window

The short time delay (t_d) is defined by $2\,\text{IS} \leq t_d \leq 4\,\text{IS}$. In this case the system evolves to the unique stable solution $c = 0.5$. Within this time delay the Lyapunov exponent is equal to zero; no chaotic behaviour is seen.

$t_d = 2\,\text{IS}$

The time delay $t_d = 2\,\text{IS}$ means that the information about the system is updated every two iteration steps. The evolution of the system is presented in Fig. 2 and is plotted as a function of initial iteration. For every 500 initial concentrations 103 iteration steps have been used. The history examples are presented in Fig. 1 as a function of concentration in time. In the case of the shortest time delay considered here the system has a unique solution $c = 0.5$. The stability time as a function of initial concentration is shown in Fig. 3. For a very low initial concentration $0 < c_0 \lesssim 0.01$ a long time ($t_s \geq 47\,\text{IS}$) is needed in order to achieve the stable state. It is also illustrated in Fig. 1, where in the case of low initial concentrations $c_0 = 0.002$ the stability time is quite long (about 100 IS). However except for very small initial concentrations ($c_0 > 0.1$) the stability time is short, i.e. $t_s \in (10\,\text{IS}, 20\,\text{IS})$.

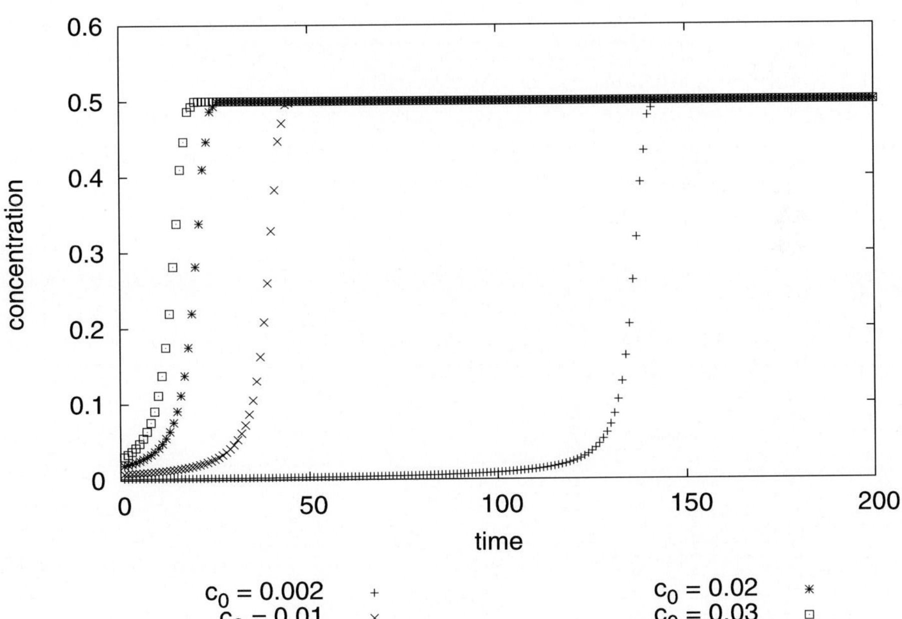

Fig. 1. Coevolution of the system for chosen initial concentrations; delay time $t_d = 2\,\text{IS}$

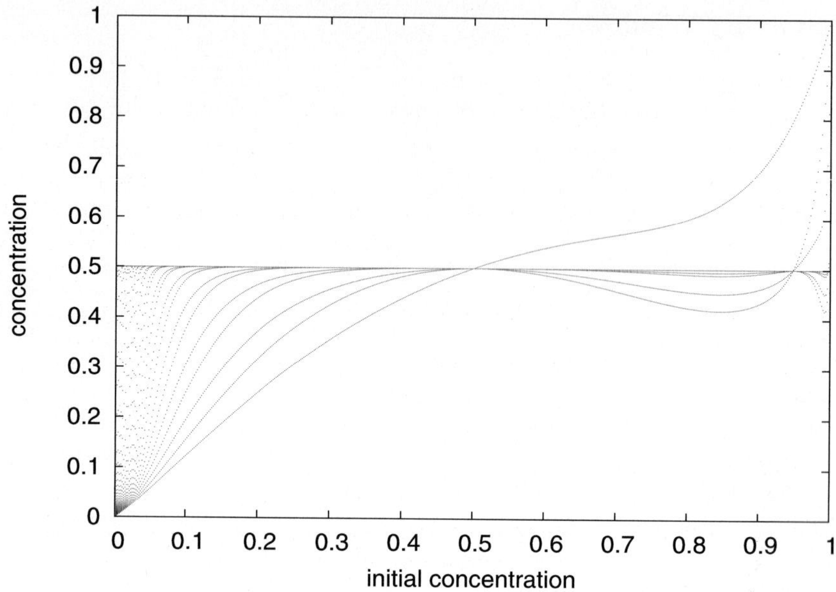

Fig. 2. Coevolution of the system as a function of initial concentration. The coevolution of a system is represented by a vertical series of dots; delay time $t_d = 2\,\mathrm{IS}$

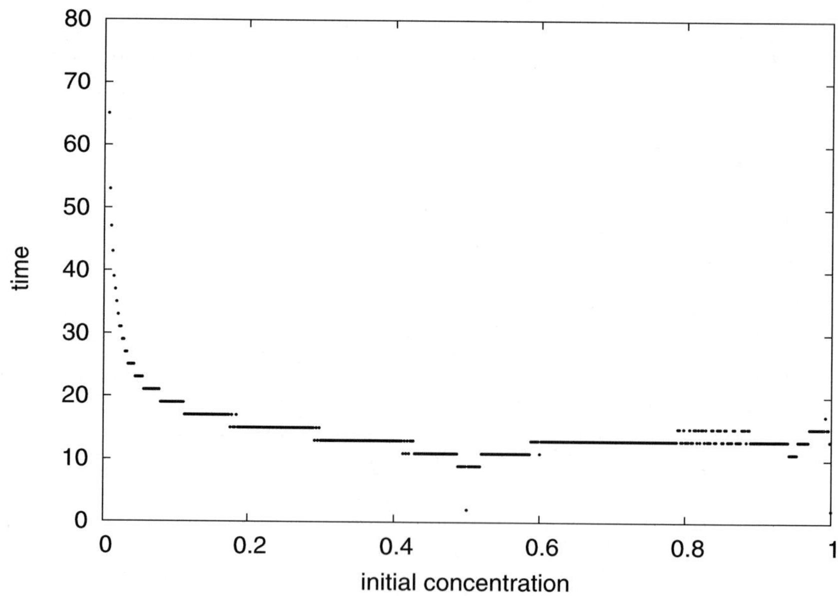

Fig. 3. The time required for the system to achieve a stable concentration as a function of initial concentration; delay time $t_d = 2\,\mathrm{IS}$

$t_d = 3$ IS

Increasing the time delay by one unit, up to three iteration steps, induces important changes in the system. In the evolution of the system, damped oscillations become observable, e.g. for $c_0 = 0.002$ damped oscillations are observed for $t \in (140\,\text{IS}, 155\,\text{IS})$ (Fig. 4 and Fig. 5). The maximum time required for the system to achieve a stable state extends to $t_s \geq 220\,\text{IS}$ as compared with $t_s \geq 47\,\text{IS}$ for $t_d = 2\,\text{IS}$. For most initial concentrations ($c_0 > 0.05$) the stability time is in the interval $t_s \in (70\,\text{IS}, 100\,\text{IS})$. Therefore the system requires a longer time to achieve a stable state. However there are some "stability points" for which the system achieves a stable state markedly faster. These can be found on Fig. 6; these points are: $c_0 = 0.074$, $t_s = 76\,\text{IS}$; $c_0 = 0.136$, $t_s = 73\,\text{IS}$; $c_0 = 0.284$, $t_s = 61\,\text{IS}$; $c_0 = 0.5$, $t_s = 1\,\text{IS}$; $c_0 = 0.826$, $t_s = 58\,\text{IS}$; $c_0 = 0.952$, $t_s = 67\,\text{IS}$.

Comparing the results obtained in the case $t_d = 2\,\text{IS}$ and $t_d = 3\,\text{IS}$ it can be noticed that the stability times is significantly extended and new features become visible (damped oscillations). Therefore we can conclude that the system is very sensitive to the flow of information and that increasing t_d to $t_d + 1$ IS changes the behaviour of the system quite significantly.

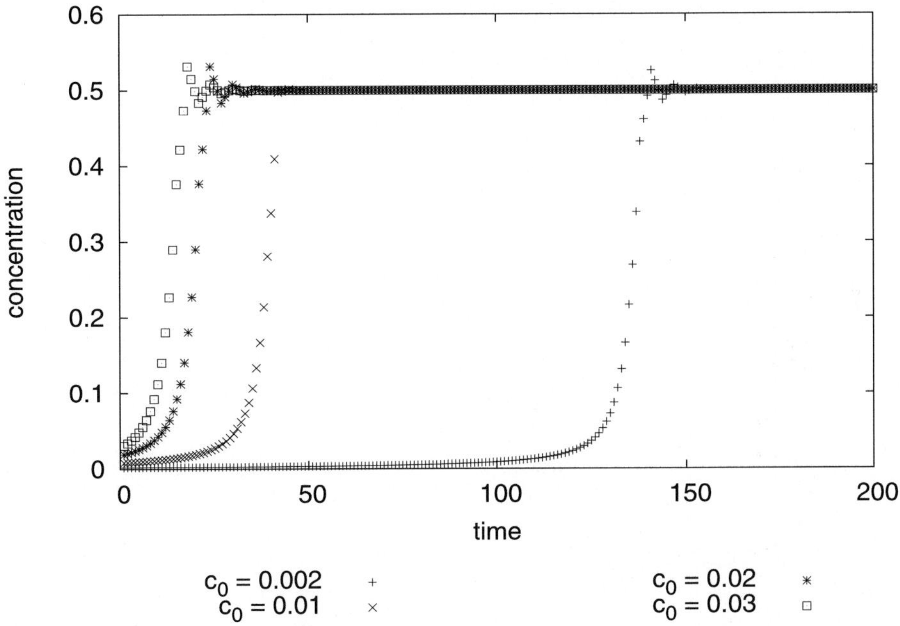

Fig. 4. Coevolution of the system for given initial concentrations; delay time $t_d = 3\,\text{IS}$

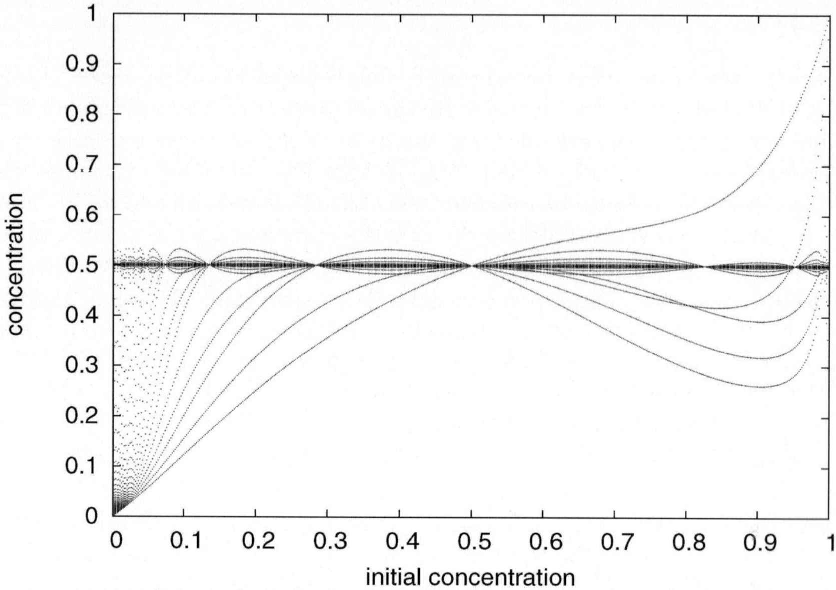

Fig. 5. Coevolution of the system as a function of initial concentration. The coevolution of a system is represented by a vertical series of dots; delay time $t_d = 3\,\mathrm{IS}$

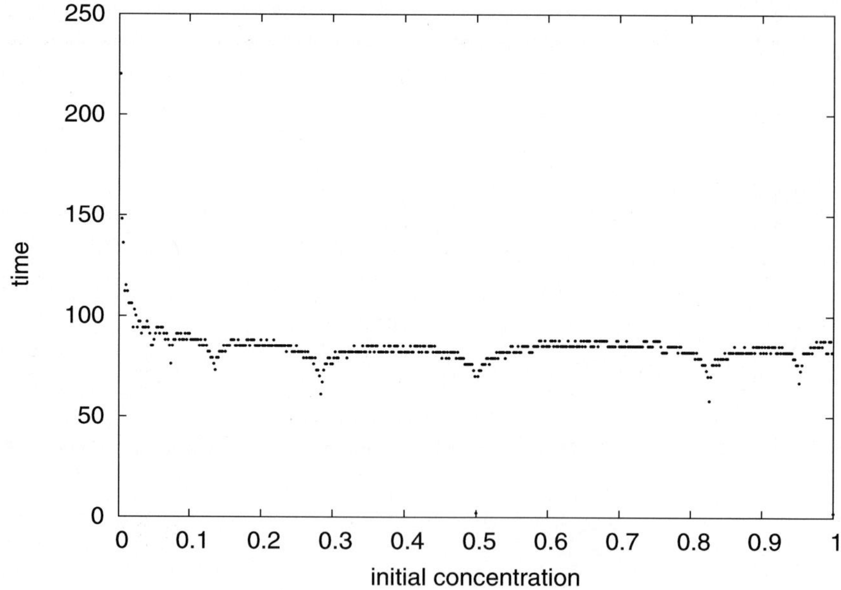

Fig. 6. The time required for the system to achieve stable concentrations as a function of its initial concentration; delay time $t_d = 3\,\mathrm{IS}$

$t_d = 4\,\text{IS}$

For a time delay $t_d = 4\,\text{IS}$, the features seen above (damped oscillations) are also present as it can be observed on both figures showing the coevolution for the considered initial concentrations and for chosen histories presenting explicitly the time evolution of the system – Fig. 5 and Fig. 8 respectively. It is worth noticing that the damping of the oscillations is much weaker than in the case $t_d = 3\,\text{IS}$ (compare Figs. 5 and 8). The oscillation amplitude is decreasing significantly more slowly for the case $t_d = 4\,\text{IS}$ than for $t_d = 3\,\text{IS}$. However in all considered cases $t_d = 2\,\text{IS}$, $t_d = 3\,\text{IS}$, $t_d = 4\,\text{IS}$, the system has one stable solution, but the stability time depends on the delay time; it is the longest in the case $t_d = 4\,\text{IS}$ ($3200\,\text{IS} \leq t_s \leq 4200\,\text{IS}$). The time required for the system to achieve a stable state is presented in Fig. 9. As in the previous case $t_d = 3\,\text{IS}$ there are initial concentrations for which the system reaches the stable state significantly quicker, e.g. $c_0 = 0.23$.

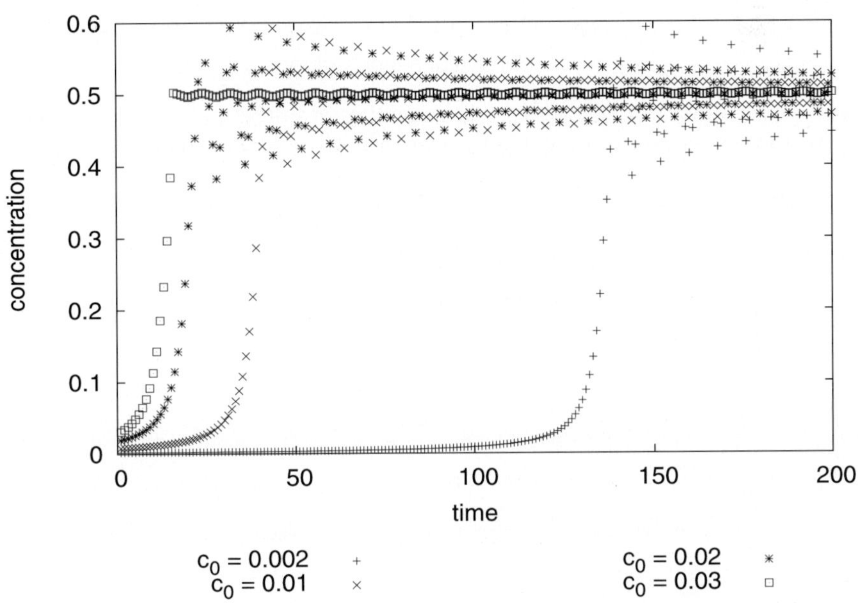

Fig. 7. Evolution of the system for given initial concentrations; delay time $t_d = 4\,\text{IS}$

3.2 Medium Time Delay

$t_d = 5\,\text{IS}$ and $t_d = 6\,\text{IS}$

The five iteration step delay time ($t_d = 5\,\text{IS}$) is very interesting, because this is the shortest time for which cycles of concentration can be observed. For

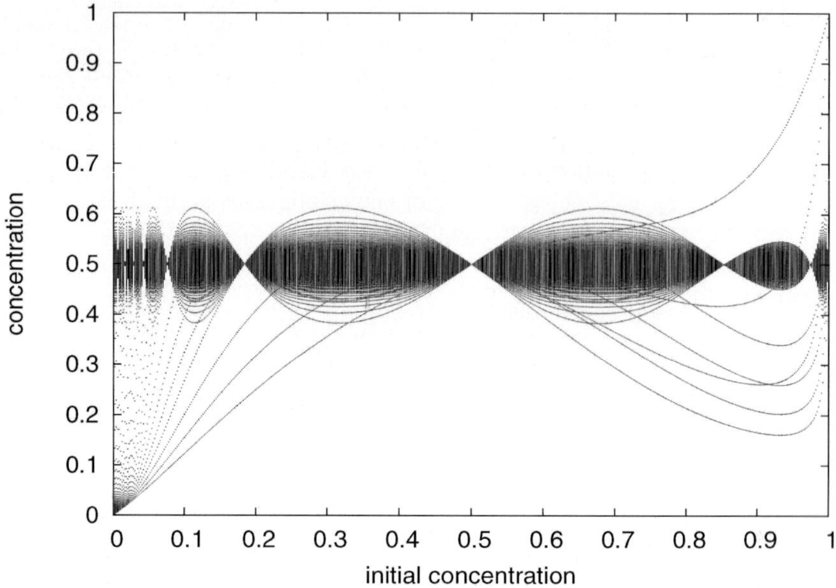

Fig. 8. Coevolution of the system as a function of initial concentration. The coevolution of a system is represented by a vertical series of dots; delay time $t_d = 4\,\mathrm{IS}$

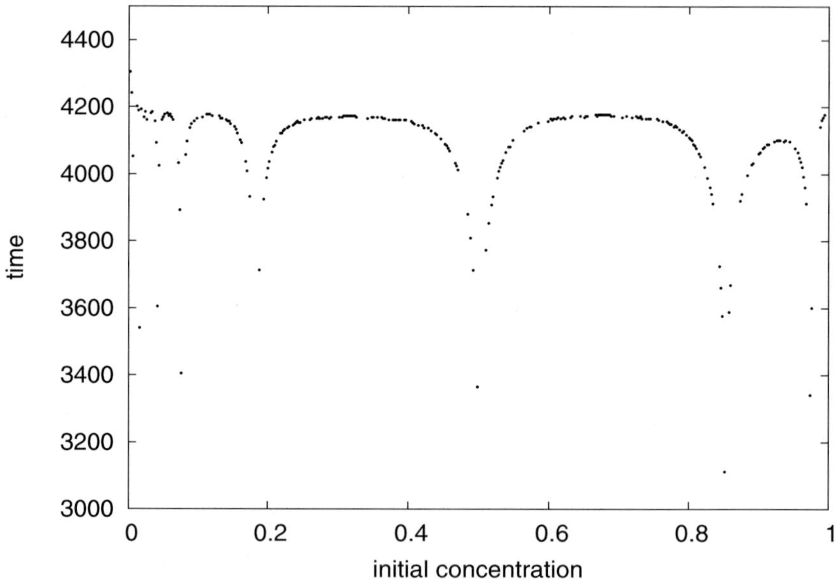

Fig. 9. The time required for the system to achieve stable concentrations; delay time $t_d = 4\,\mathrm{IS}$

this time delay the system has several solutions. Despite the fact that the real part of the Lyapunov exponent is still equal to zero, its imaginary part is not. According to (4) this shows that the system has a periodic solution. These solutions can be seen in Fig. 11 and Fig. 10. In the case of Fig. 11 the evolution is shown as a function of its initial concentration, whereas Fig. 10 presents the system evolution as a function of time for chosen initial concentrations.

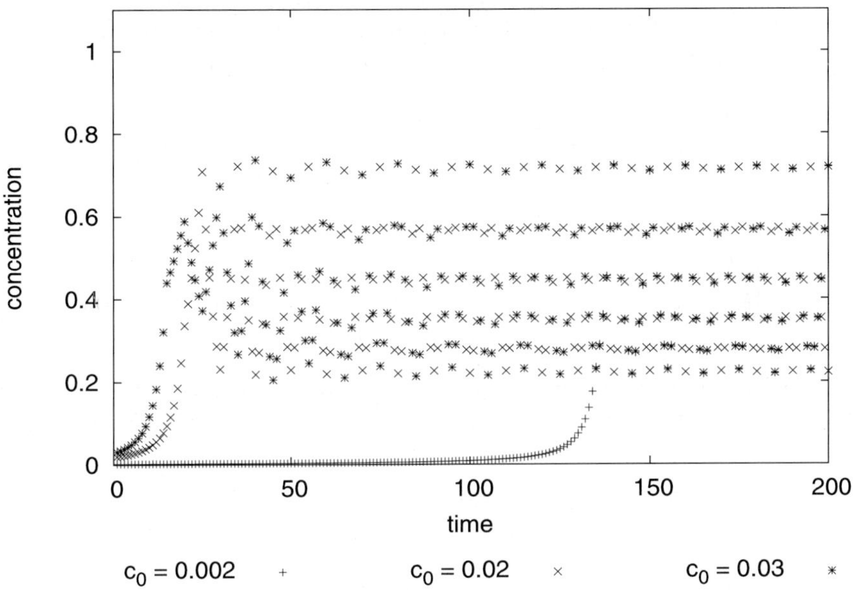

Fig. 10. Evolution of the system for chosen initial concentrations; delay time $t_d = 5\,\text{IS}$

Oscillating solutions can also be found in the case $t_d = 6\,\text{IS}$; the imaginary part of the Lyapunov exponent, as in the previous case ($t_d = 5\,\text{IS}$) is negative (Fig. 12).

3.3 Long Time Delay

$t_d \geq 7\,\text{IS}$

Extending the delay time above six iteration steps leads to a possible collapse of the system. For $t_d \geq 7\,\text{IS}$ the system may crash. The crash is defined when the concentration of companies becomes negative or zero. Examples of such evolutions which lead to a crash are presented in the cases $t_d = 12\,\text{IS}$ and $t_d = 15\,\text{IS}$ in Fig. 14 and Fig. 16 respectively. The crash of the system is presented in such plots as a white band containing very few points in the vertical direction. It is also seen in Fig. 17, where for several intervals on

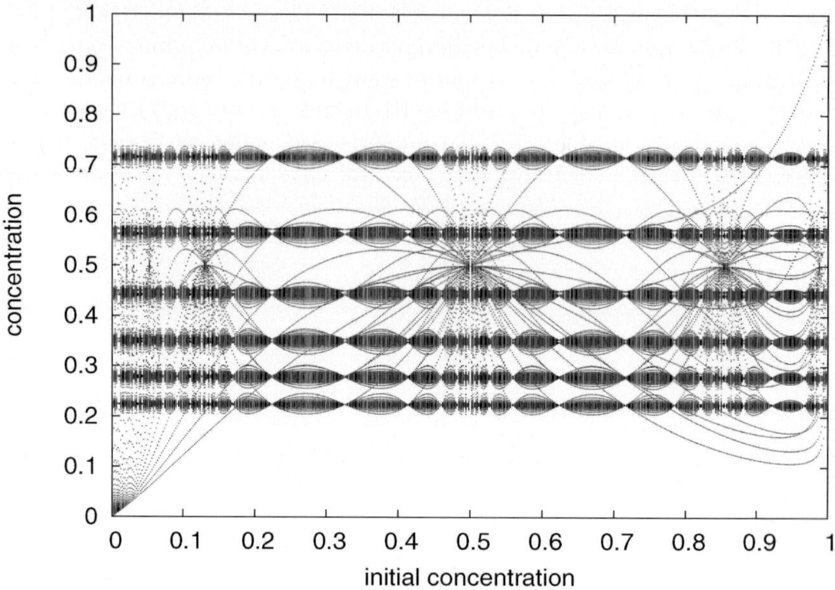

Fig. 11. Coevolution of the system as a function of initial concentration. The coevolution of a system is represented by a vertical series of dots; delay time $t_d = 5\,\mathrm{IS}$

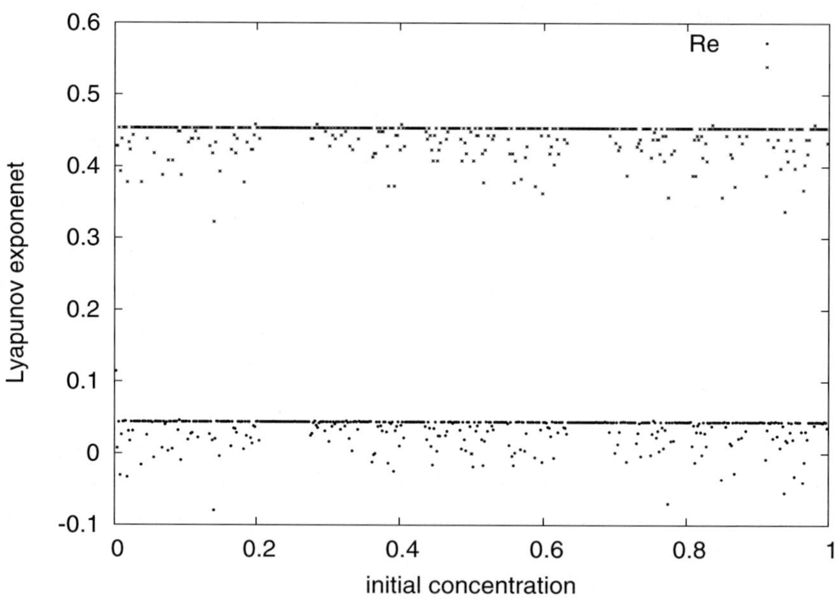

Fig. 12. Lyapunov exponent for $t_d = 6\,\mathrm{IS}$

the initial concentration axis, e.g. $t_d \in (0.15; 0.2) \cup (0.34; 0.36) \cup (0.53; 0.61)$ the crash of the system occurs very quickly. However there are some initial concentrations for which the evolution of the system before crash time is quite long (up to 400 IS). Additionally in the case $t_d = 15\,\mathrm{IS}$, the system may evolve toward a stable state, with a full occupation of the environment by companies. Examples of such an evolution as a function of time for given initial concentrations are presented in Fig. 13 and Fig. 15 for the cases $t_d = 12\,\mathrm{IS}$ and $t_d = 15\,\mathrm{IS}$ respectively.

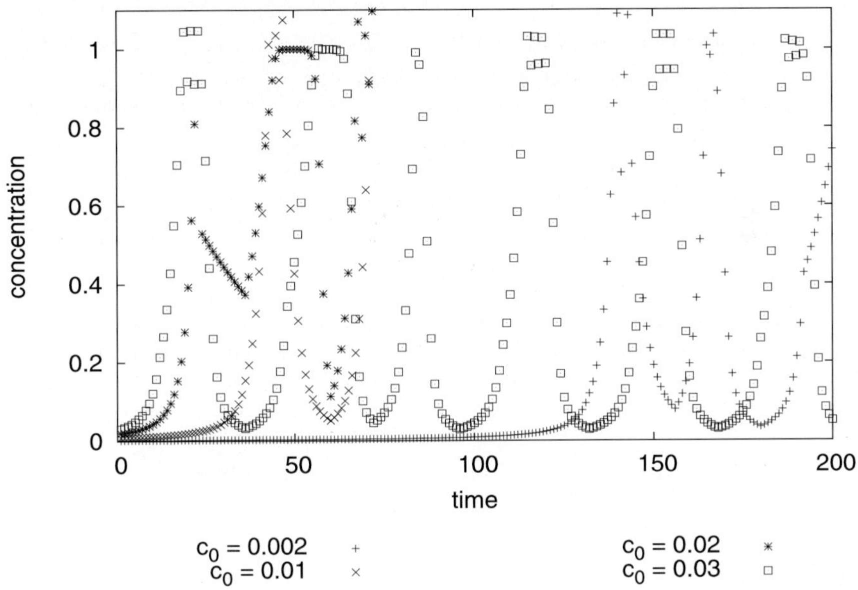

Fig. 13. Coevolution of the system as a function of time for chosen initial concentrations; $t_d = 12\,\mathrm{IS}$

4 Conclusions

Economic cycle causes and occurrences are fascinating and relevant subjects of interest in many economic questions [9, 10]. The problem has also been studied by means of sociology techniques [11], showing that opinions about recession or prosperity undergo drastic changes from one equilibrium to another, both having fluctuations in stochastically resonant systems. In the present investigation, an information flow, typical of economic systems, has been incorporated into the ACP model [3–5]. This has led to observe different forms of so-called cycles, through concentration oscillations. In the case of a *short* delay time $t_d \in (2\,\mathrm{IS}, 4\,\mathrm{IS})$, between data acquisition and policy

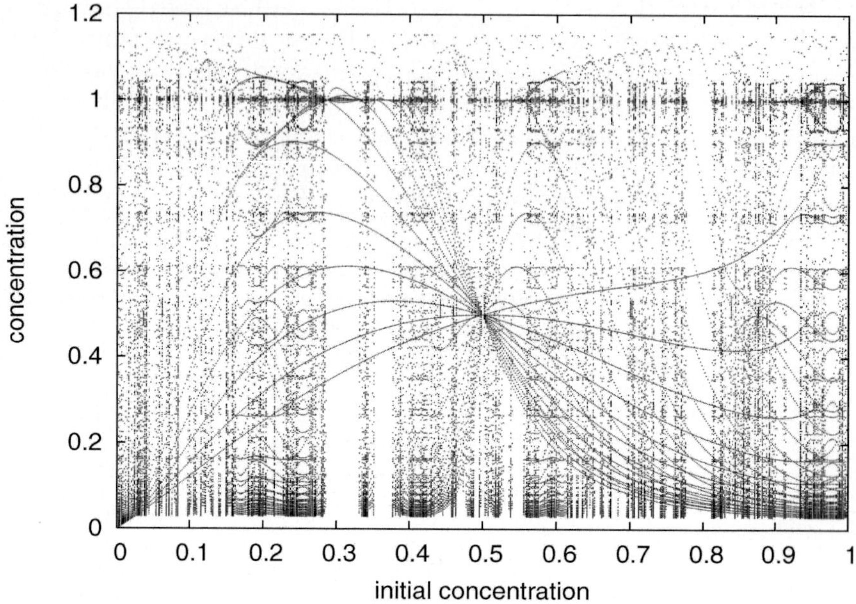

Fig. 14. Coevolution of the system as a function of its initial concentration. The coevolution of a system is represented by a vertical series of dots; $t_d = 12\,\mathrm{IS}$

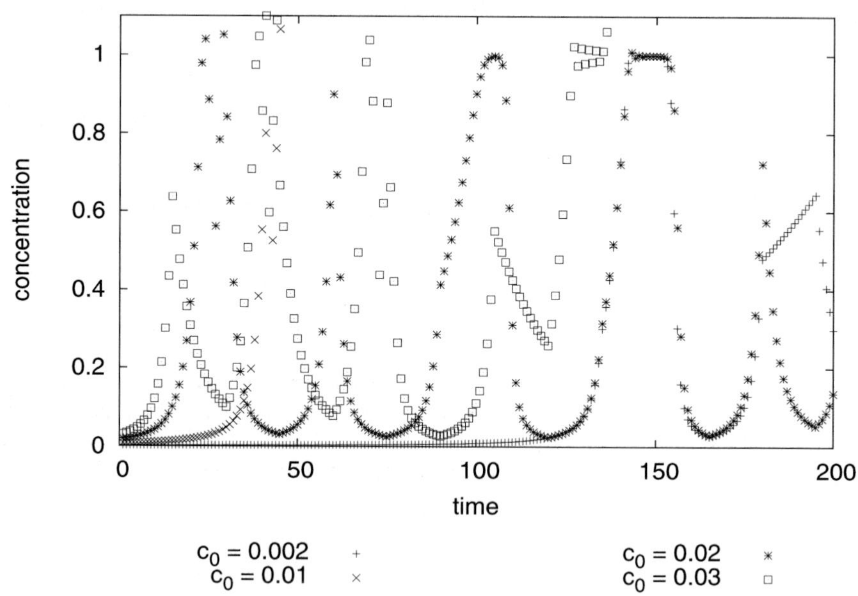

Fig. 15. Coevolution of the system as a function of time for chosen initial concentrations; $t_d = 15\,\mathrm{IS}$

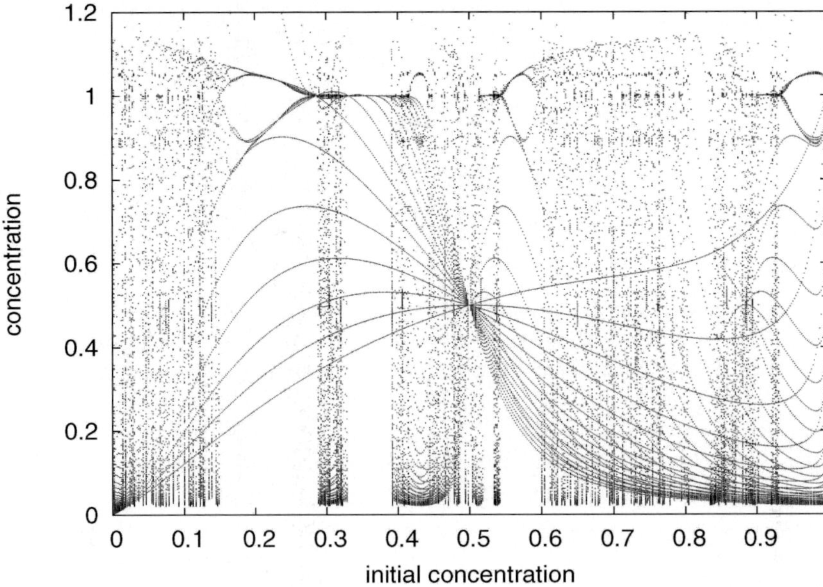

Fig. 16. Coevolution of the system as a function of its initial concentration. The coevolution of a system is represented by a vertical series of dots; $t_d = 15\,\mathrm{IS}$

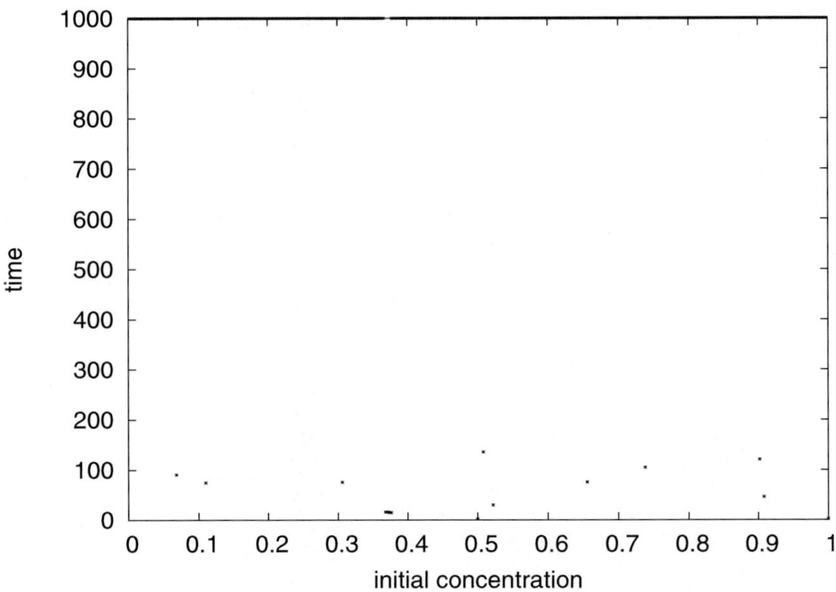

Fig. 17. The crash time of the system as a function of its initial concentration; $t_d = 15\,\mathrm{IS}$

implementation by a company, the system evolves toward a unique stable equilibrium state. This situation can be highly welcomed in some economic systems. Indeed this indicates that, through an information control, a system can insure the existence of a high number of companies, whence not threatening the system of a collapse.

In the case of *medium* size delay times $t_d = 5\,\mathrm{IS}$ or $t_d = 6\,\mathrm{IS}$, the system undergoes oscillations: stable concentration cycles appear in the system. This form of evolution is often observed in economy, e.g. agricultural markets, where without external control the level of agricultural production oscillates between over- and underproduction. Since the enlarging of the delay time leads to the possibility of the system to crash, such a system may require some external (governmental) control, for its stability. In reality, the delay of information flow and policy implementation may also fluctuate. For long information flow delay times, $t_d \geq 7\,\mathrm{IS}$, the systems may crash for most initial concentrations. However, despite the frequent possibility of the system to crash the situation is not hopeless because the crash time in many cases is long enough to allow for some particular control and to avoid the collapse of the company concentration. It is also possible to observe an "economic resonance" where despite a long delay time the system evolves for a long time or can even reach a stable state, which insures its existence. This latest observation is especially interesting for market control purposes, because it points to the existence of initial conditions for which the system may evolve during a very long time, which is vital for the possibility of creating and applying some control procedures.

5 Acknowledgement

JM stay in Liège was partially supported by a STSM grant funded by the COST P10 Physics of Risk program.

References

1. P.-F. Verhulst: Nouveaux Mémoires de l'Académie Royale des Sciences et Belles-Lettres de Bruxelles **18**, 1 (1845)
2. D. Huber, L.S. Tsimring: Phys. Rev. Lett. **91**, 260601 (2003)
3. M. Ausloos, P. Clippe, A. Pękalski: Physica A **324**, 330 (2003)
4. M. Ausloos, P. Clippe, A. Pękalski: Physica A **337**, 269 (2004)
5. M. Ausloos, P. Clippe, A. Pękalski: Physica A **332**, 394 (2004)
6. P. Bak, K. Sneppen: Phys. Rev. Lett. **71**, 4083 (1993)
7. J. Miśkiewicz, M. Ausloos: Physica A **336**, 206 (2004)
8. M. Ausloos, P. Clippe, J. Miśkiewicz, A. Pękalski: Physica A **344**, 1 (2004)
9. M. Kalecki: Econometrica **3**, 327 (1935)
10. M. Aoki: Macroeconomic Dynamics **4**, 487 (2000)
11. M. Hohnisch, S. Pittnauer, S. Solomon, D. Stauffer: Physica A **345**, 646 (2005)

Logistic Function in Large Financial Crashes

G. Rotundo

Department of Business, Technological and Quantitative Studies, Faculty of Economics, University of Tuscia, Viterbo, Italy 01100
Giulia.Rotundo@uniroma1.it

Summary. This chapter examines the signatures that arise after large financial crashes in order to evidence the presence of logistic growth in indices of markets that validates the hypothesis of symmetry on the trend before and after the crash. It is also shown how the probability meaning of the logistic function can be exploited in order to set up a bayesian analysis model. The large crash in the NASDAQ 100 composite index which occurred in April 2000 is presented as a case study.

1 Introduction

The aim of this chapter is to show how the logistic function can occur in financial markets and to explore its relevance when dealing with large crashes produced by the collapse of speculative bubbles due to endogenous causes. The theory about crashes has already assessed the validity of a scaling property in the proximity of large financial crashes ([5,7–9,11,19–22,24–26]). The studies examine also the behaviour of financial data after the occurrence of a large drawdown. The analysis of signatures in financial indices that undergo some large crash is important in order to estimate the time necessary for the market to exit from the descent that follows the fall and thus for driving decisions about investment policies. Empirical evidences show that the symmetry between pre and post crash behaviours can be established only in the presence of log-periodic corrections to scaling [18]. After the occurrence of a crash a market experiences an oscillatory deflating. The time of the exit of the market from the anti bubble and its return to the fundamental values can be obtained from the symmetry with respect to the rise of the bubble. However the mirroring is not perfect and the analysis of deviations from symmetry can be useful also for the estimate of the expected deflating behaviour of prices.

A logistic sigmoidal law [1] can be detected by suppressing the oscillations through the patterns that arise after the occurrence of a crash. This is based on symmetry hypotheses. The case of the large crash during April 2000 of the NASDAQ index is deeply studied here.

The outline of the chapter is the following: the next section resumes the main results of the theory of large financial crashes. Section 3 examines the logistic function for a description of post crash behaviour. In Sect. 4 the large

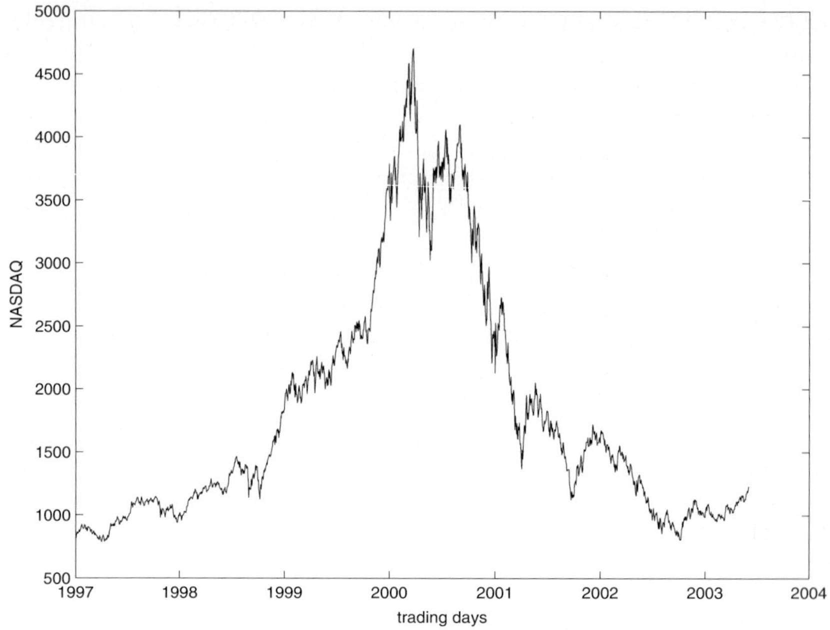

Fig. 1. The NASDAQ 100 composite index from January 1st, 1997 to June 5th, 2003

crash which occurred during April 2000 in the NASDAQ 100 composite index is examined. In the last section it is shown how the detection of the logistic function allows to set up a bayesian analysis model for crashes.

2 Large Financial Crashes Models

A series of papers about speculative bubbles proposes an explanation of large financial crashes due to causes endogenous to the market. Similarities with critical phenomena like earthquakes [4] and the sound emission in materials close to the rupture point led the research to the detection of cooperative underlying phenomena evidenced through discrete scale invariance in financial data close to large crash [15] that can thus be well described as critical points [7]. Market indices have been widely studied through the literature.

Let F be a market index and τ the interval of time to the crash. Then a general form for an evolution equation for F close to the crash time is obtained by the Landau expansion of F around $\tau = 0$:

$$\frac{\mathrm{d}F(\tau)}{\mathrm{d}\log\tau} = \alpha F(\tau) + \beta \mid F(\tau) \mid^2 F(\tau) \cdots \tag{1}$$

where in general the coefficients may be complex [10]. In this expansion only terms that are invariant with respect to a change of phase $\phi \to \phi + C$ where

$F \equiv Be^{i\phi}$ are kept because a phase translation corresponds to a change of time units.

Equation (1) is independent from the sign of τ and can be useful for both the description of the ascent of a bubble and the relaxation dynamics of after crashes market prices. Thus before the crash $\tau = t_c - t$ $(t < t_c)$, where t is the time and t_c is the time of the crash, and (1) describes the bullish behaviour of a bubble; after the crash $\tau = t - t_c$ $(t > t_c)$ and (1) characterizes the bearish phase of an anti bubble.

The research of foreshock and aftershock patterns through the same equation relies on the hypothesis that self-aggregation phenomena of the market are suitable for its description through hierarchical structures that drive the system to the crash. A further taxonomy that distinguishes the causes that generate crashes can be found in [13] and confirms the detection of log-periodic power law signatures as a hallmark of speculative bubbles due to causes endogenous to the market. An important remark is that the crash time t_c is the expected time for the end of the bubble. It corresponds to the most probable time for the occurrence of a crash [15], but it could either occur before, albeit with small probability, or do not happen at all, leading the bubble to deflate smoothly. If the amplitude of the crash is proportional to the bubble part of the crash then $F(t)$ actually is the value of the index at time t. However if the amplitude is proportional to the total price then the best quantity to model is the logarithm of the index value [12, 27].

2.1 First Order Solution

The maintenance of only the first term in the r.h.s. of (1) gives the following evolution equation

$$F(t) \simeq A + B(t_c - t)^m + C(t_c - t)^m \cos[\omega \ln(t_c - t) - \phi] \text{ where } t < t_c \,. \quad (2)$$

t_c is the crash time and A, B, C, m , ω, ϕ, t_c are parameters to be estimated via numerical optimization, for example by a least square method.

In the limit $m \to 0$ and under the hypothesis $\omega \ln(1 - t/t_c) \ll 1$ (2) becomes [15]

$$F(t) \simeq A + B \ln(t_c - t) + C \ln(t_c - t) \cos[\omega \ln(t_c - t) - \phi] \text{ where } t < t_c \,. \quad (3)$$

Setting $C = 0$, (3) becomes

$$F(t) \simeq A + B \ln(t_c - t) \text{ for } t < t_c \,. \quad (4)$$

The symmetry hypothesis leads to the following aftershock behaviour

$$F(t) \simeq A + B(t - t_c)^m + C(t - t_c)^m \cos[\omega \ln(t - t_c) - \phi] \text{ where } t > t_c \,. \quad (5)$$

Hence in the limit $m \to 0$ and under the hypothesis $\omega \ln(1 - t/t_c) \ll 1$, (5) becomes

$$F(t) \simeq A + B\ln(t-t_c) + C\ln(t-t_c)\cos[\omega\ln(t-t_c)-\phi)] \text{ where } t > t_c \quad (6)$$

leading to the following approximation

$$F(t) \simeq A + B\ln(t-t_c) \text{ where } t > t_c \quad (7)$$

if $C = 0$. Empirical studies available through the literature state that the values of m and ω belong to a narrow range of values [13]. Such a behaviour is a robust feature of the bubbles that collapse into a large crash and can be obtained by models of cooperative behaviour among investors [15]. As an alternative to ω the values of $\lambda = e^{2\pi/\omega}$ have been studied. The value of λ describes the difference equation of successive either maxima or minima in the oscillations and thus it is of immediate use. Available results about the range of λ are consistent with the ones about ω [25].

2.2 Second Order Solution

The second order solution of (1) is obtained by keeping the two first terms in the r.h.s.:

$$F(t) \simeq A' + \frac{\tau^\alpha}{\sqrt{1+(\frac{\tau}{\Delta t})^{2\alpha}}}\left\{B' + C'\cos\left[\omega\log\tau + \frac{\Delta_\omega}{2\alpha}\left(1+\left(\frac{\tau}{\Delta t}\right)^{2\alpha}\right) + \phi'\right]\right\} \quad (8)$$

where $\tau = \mid t - t_c \mid$. This equation doesn't change form for the description of pre and post crash data.

3 The Logistic Function

The logistic function, also known as sigmoidal function because of its shape,

$$\sigma : R \to R, \ \sigma(x) = \frac{1}{1+e^{-a(x)}}, \quad x \in R$$

was born for the description of population dynamics [2], but is now widely used in probability theory, statistics and neural networks. The aim of this section is to show how it can occur through functions that describe large financial crashes.

3.1 The First Order Solution

The first order solution can lead to the logistic function for the description of after crashes behaviour. In fact from (7):

$$\frac{t_c}{t} \simeq \frac{1}{1+e^{[F(t)-A]/B}} \,. \quad (9)$$

Thus the after crash behaviour can be described through the logistic function $\sigma(\cdot)$

$$\frac{t_c}{t} \simeq \sigma[F(t)] , \tag{10}$$

where $a(x) = -ax + b$, with $a = -1/B$ and $b = A/B + \log t_c$. Therefore when using the first order solution the sigmoidal function is suitable for the description of data after the crash, but not before it.

3.2 The Second Order Solution

When setting $C' = 0$ (8) can be approximated by

$$F(t) \simeq A' + B' \frac{\tau^\alpha}{\sqrt{1 + (\frac{\tau}{\Delta t})^{2\alpha}}} \tag{11}$$

$$\left(\frac{F(t) - A'}{B'}\right)^2 \left(\frac{1}{\Delta t}\right)^{2\alpha} \simeq \sigma(\log \tau) \text{ where } a(x) = 2\alpha(\log \tau) - 2\alpha \log \Delta t \, . \tag{12}$$

This equation does not change form for the description of pre and post crashes data $\tau =\mid t - t_c \mid$. In this case the logistic function relies over the logarithm of the time and describe the rescaled quantity of interest.

3.3 Probabilistic Approach

The logistic function is well known in probability and especially in bayesian analysis. The probabilistic meaning of the logistic function allows to provide an interpretation of the r.h.s. of (10) as a probability estimate for the occurrence of the crash at time t_c, that can be calculated relying on the level of $F(t)$. Moreover, the approximation states that this probability decays like the inverse of the time to crash, under an appropriate rescaling of the time. This allows to remark the analogy with Omori formula [23] for a law for the rate $Q(t)$ of the shocks following a main earthquake, that decays from the main shock as $Q(t) \propto 1/t^p$ where p is close to 1.

The sigmoidal function occurring in (10) has the interesting feature to be well suited for the description both before and after a crash. Moreover it depends on the logarithm of time, thus keeping trace of the log-periodic structure embedded in the market.

4 Numerical Results

The crash of the index NASDAQ 100 composite which occurred in April 2000 has been deeply examined. The data set collects index values of each trading day since January 1st, 1997 to September 26th, 2002.

As also pointed out in Sect. 2 a rational bubble model proposed in [14] shows that if the magnitude of the crash is proportional to the increase of values exclusively produced by the rise of the bubble [12] then the right quantity to study is the data time series, but if the depth of the crash is proportional to the values, then it is necessary to study the logarithm of data.

As evidenced in [13], in the case of NASDAQ data, the logarithm of the index values must be chosen and thus the data $\{y(t)\}_t$ used here below are the logarithm transformation of the NASDAQ 100 composite index.

The time is expressed as a fraction of the year, according to the series of papers about crashes. Different unit measures for the time lead to different numerical values of parameters. One year contains approximately 260 trading days, depending on the overlap between holidays and weekends, but it is normalized to 100 days. As an example 2000.18 is March 10th, 2000.

A minimum least square method was used in order to perform the fit of functions (2–11) to data. Constraints are mandatory over the scope of the logarithm function ($t_c > t$ before the crash and $t > t_c$ after the crash). Moreover lower and upper bounds for each parameter were fixed across all the examples by relying on the literature for the fit of data before the crash [11], [13]. In (2) parameters A, B, C, ϕ represent the rescaling of measure units, whilst parameters m and ω contain structural information about the market dynamics because they quantify the acceleration in the price increase and the oscillations, respectively [13]. In the fit of (2, 3, 5, 6, 8) the optimal value of ω is searched by taking into account the narrow range across the series of speculative bubbles due to endogenous causes. The same remark holds for the exponent m. In fact the values of m and ω were found to belong to a short range through several analyses of speculative bubbles observed on a wide variety of markets. As an example in [13] for more that thirty crashes on the major financial indices $\omega \approx 6.36 \pm 1.56$ and $m \approx 0.33 \pm 0.18$. The bounds on $\lambda \in [2.3, 2.5]$ are consistent with the ones on ω [25]. Thus the bounds used in the optimization procedures related to the fit of (2) are $\omega > 4.5$ and $\omega < 8.5$. Another bound over the time is obtained from the empirical evidence that the estimate of the most probable crash time is in a range of approximately at most three months from the actual date of the beginning of the index drop [16]. The end of the bubble is defined as the date at which the market reached its maximum. The bounds for each fit after the crash are obtained from the ones stated before the crash by using the symmetry hypothesis.

The optimal set of parameters is very sensitive to the choice of the initial point. Thus optimization procedures were performed by choosing a grid of 100 initial points for each parameter, equally spaced in the interval between the related bounds. For each fit the value of the parameters that gave rise to the lowest value of the residuals without active constraints is reported in Tables 1 and 2.

Table 1. Analysis of NASDAQ 100 composite precursory patterns of the crash of April 2000. Data set contains daily index values from January 1st, 1997 to March 10th, 2000

Formula	Parameters	Figure
(13)	$A = 9.36$, $B = -1.60$, $t_c =$ April 11th, 2000, $m = 0.26$.	(2)
(15)	$A = 9.36$, $B = -1.60$, $t_c =$ April 11th, 2000, $m = 0.26$, $C = 0.05$, $\omega = 6.32$, $\phi = -0.83$.	(3)
(4)	$A = 7.91$, $B = -0.54$, $t_c =$ July 4th, 2000.	(4)
(3)	$A = 7.93$, $B = -0.60$, $t_c =$ August 10th, 2000, $\phi = 2.96$, $\omega = 6.26$, $C = -0.07$.	(5)
(18)	$A = -18.60$, $B = 34.65$, $t_c =$ June 21st, 2000, $\frac{1}{\Delta t} = 144.01$, $\alpha = -0.034$.	(10)
(19)	$A' = -18.60$, $B' = 34.65$, $t_c =$ June 21st, $\frac{1}{\Delta t} = 144.01$, $\alpha = -0.03$, $C' = 0.08$, $\omega = 7.32$, $\Delta\omega = 1.02$, $\phi' = 0.10$.	(11)

Table 2. Analysis of NASDAQ 100 after crash signatures. Crash of April 2000. Data set contains daily index values from September 13th, 2000 to September 26th, 2002

Formula	Parameters	Figure
(16)	$t_c =$ August 8th, 2000 $A = 10.57$, $B = -3.00$, $m = 0.11$.	(6)
(17)	$t_c =$ August 8th, 2000, $A = 10.57$, $B = -3.00$, $m = 0.11$, $C = 0.097$, $\omega = 9.713$, $\phi = -1.82$.	(7)
(7)	$A = 7.62$, $B = -0.40$, $t_c =$ July, 20th, 2000.	(8)
(6)	$A = 7.58$, $B = -0.33$, $t_c =$ July 31st, 2000, $\phi = -2.47$, $\omega = 8.38$, $C = 0.10$.	(9)
(18)	$A' = 9.17$, $B' = 21.85$, $t_c =$ July 8th, 2000, $\frac{1}{\Delta t}$ =179.58, $\alpha = -0.03$.	(12)
(19)	$A' = 9.17$, $B' = 21.85$, $t_c =$ July 8th, 2000, $\frac{1}{\Delta t}$ =179.58, $\alpha = -0.03$, $C = 0.13$, $\omega = 6.86$, $\Delta\omega = 4.62$, $\phi = 3.99$.	(13)

The next paragraphs describe the methods and the results both for the fit to first and second order solutions, each of them before and after the crash. The data set used in order to study the ascending speculative bubble ending to the crash uses index data available for the trading days since January 1st, 1997 till March 10th, 2000, for a total amount of 833 points [11]. The first point can be identified as the lowest values of the index prior to the onset of the bubble while the last point is that of the all time high of the index. Moreover this data set gives rise to the best estimate of the crash time when using (4) [16]. For the post crash study data range since September 13th, 2000 till September 26th, 2002. Data between March 10th, 2000 and September 13th, 2000 are not considered in order to allow the market to overcome its most unstable phase and to start a descent with the same slope as that of the last month before the crash. The results are collected in Tables 1, 2.

4.1 The First Order Fit

Data $\{y(t)\}_t$ were at first fitted by

$$y(t) \simeq A + B(t_c - t)^m \tag{13}$$

by the least square estimate, whence the residuals

$$\frac{y(t) - A - B(t_c - t)^m}{(t_c - t)^m} \tag{14}$$

should be a pure cosine function

$$\frac{y(t) - A}{(t_c - t)^m} - B = C \cos[\omega \ln(t_c - t) - \phi] \tag{15}$$

and were fitted again by least square methods. This procedure splits the optimization of (2) into two phases and allows avoid numerical problems due to the number of variables.

Before the Crash

These fits are necessary in order to get the values of the parameter in order to compare them with the after crashes ones. The results of the fit are collected in Table 1. The first function fitted is (13). The lower bound for (A, B, t_c, m) is $(5, -10, 2000.19, 0)$, the upper is $(10, -0.1, 2001.21, 10)$, and the optimal values are consistent with the ones reported in [13] (see Fig. 2). The fit of (13) is necessary in order to proceed with the fit of (15). The bounds for (C, ω, ϕ) are $(-10, 4, -1)$ and $(10, 8, 1)$, so they are not tight. The results obtained are consistent with the one reported in [11] (see Fig. 3). The next function to fit is (4) and the bounds for (A, B, t_c) are $(7, -1, 2000.19)$ and $(9, -0.1, 2000.52)$. The procedure used for the best fit of (2) evidences its limits when fitting (3). In this case it gave rise to poor fitting and it was necessary to perform a simultaneous optimization over all the parameters. The bounds on (C, ω, ϕ) given by $(-10, 1, -1)$ and $(10, 10, 0)$.

After the Crash

The data set contains NASDAQ data since September 13th, 2000 to September 26th 2002. The best results with no active constraints are collected in Table (2). Formula (5) was tested starting with the best fit of

$$y(t) \simeq A + B(t - t_c)^m \,. \tag{16}$$

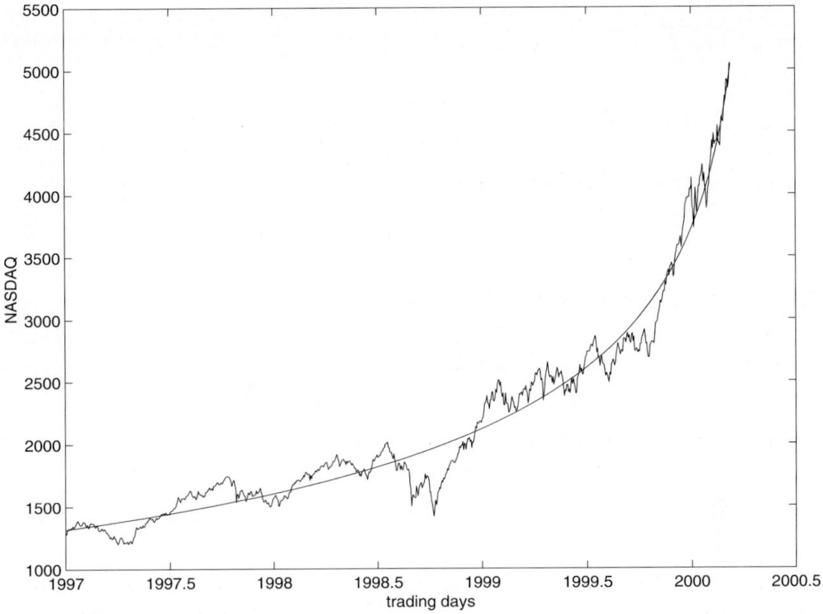

Fig. 2. NASDAQ 100 before the crash. The continuous curve is the best first order fit according to (13)

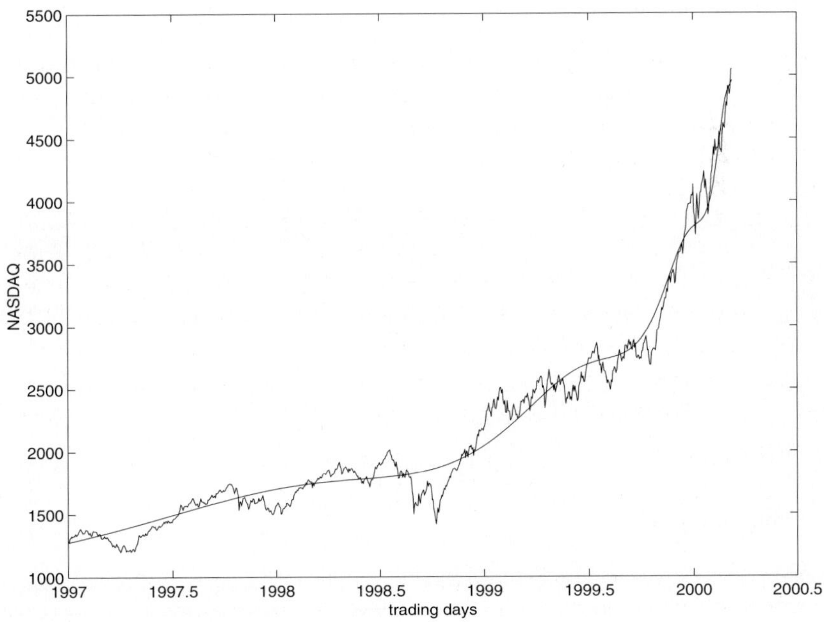

Fig. 3. NASDAQ index before the crash. The continuous curve is the best first order fit according to (15)

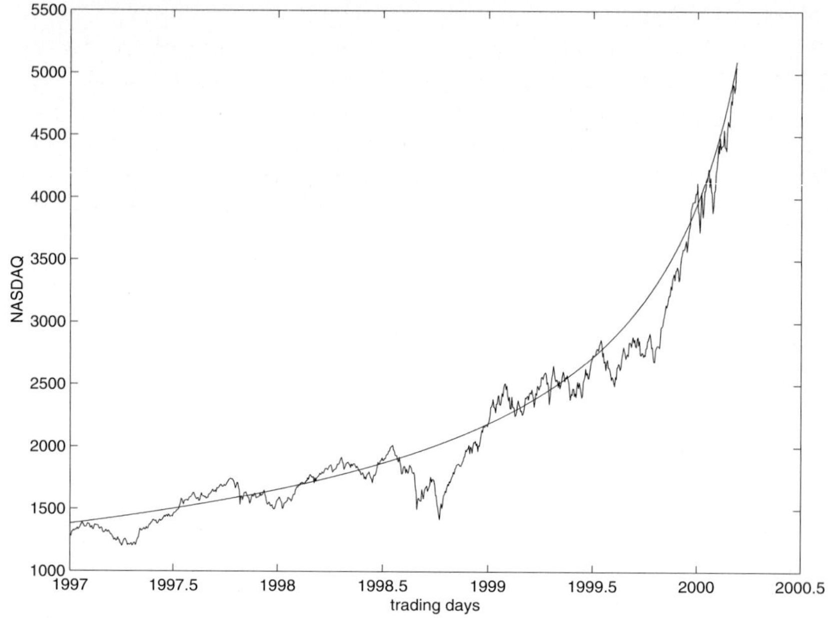

Fig. 4. NASDAQ index before the crash. The continuous curve is the best first order fit according to (4)

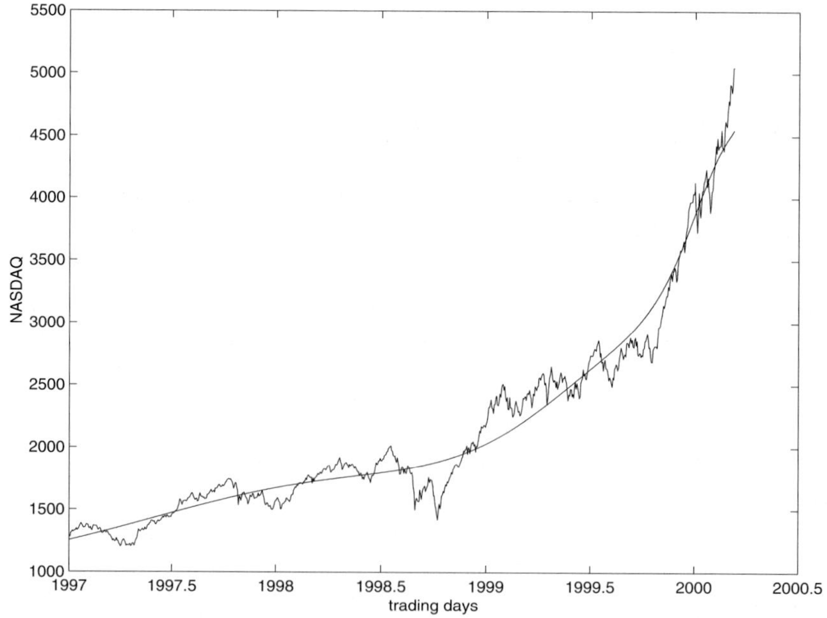

Fig. 5. NASDAQ index before the crash. The continuous curve is the best first order fit according to (3)

The bounds for (A, B, t_c, m) are $(9, -3, 2000.06, 0)$ and $(20, -1, 2000.70, 1)$. The bounds are chosen relying over the results got for data that precede the crash they are not tied with respect to these values, in order to allow a wide search for the minimum value. A remark can be made about the estimate of t_c: for (16) the estimate of the crash time is approximately 20 days after the last day used for the fit and in (16) t_c is approximately 20 days before the initial day. This gives a confirmation about the symmetry of the process.

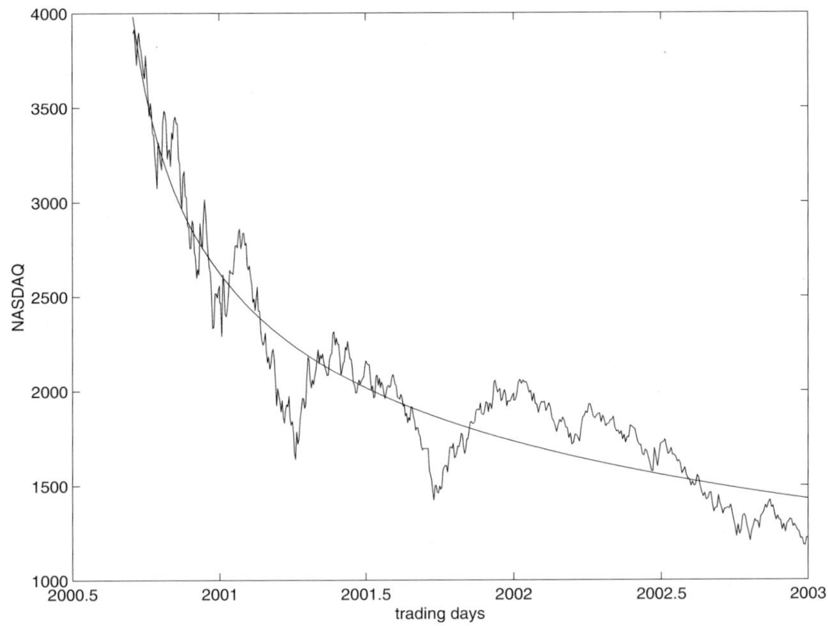

Fig. 6. NASDAQ index after the crash. The continuous curve is the best first order fit according to (16)

Formula (5) was fitted by using the values of A, B, t_c, and m already got with the fit of (16) and by using the least squares method over the residuals

$$\frac{y(t) - A}{(t - t_c)^m} - B = C \cos[\omega \ln(t - t_c) - \phi] \tag{17}$$

for the estimate of C, ω, ϕ. The bounds for (C, ω, ϕ) are $(-1, 3, -20)$, $(1, 10, 20)$.

There is a wide basin of attraction also for $(C, \omega, \phi) = (0.06, 4.04, -4.94)$, but no convergence has been evidenced to values of ω around 2π. The bounds for (17) are again not strict $(-1, 3, -2) \leq (C, \omega, \phi) \leq (1, 10, 0)$ and the fit of the residuals leads to the estimate of (5).

The estimate of (6) was made over all the variables simultaneously, as done for (3)

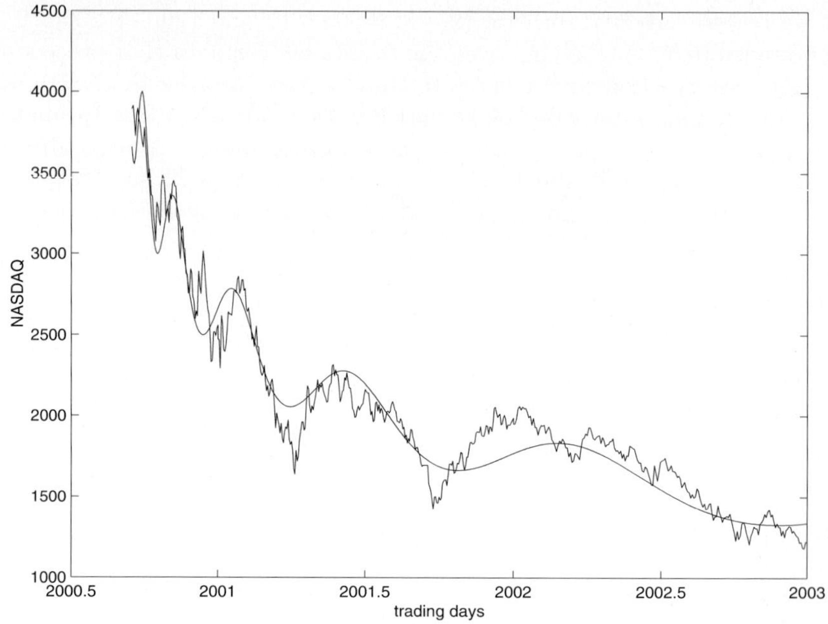

Fig. 7. NASDAQ index after the crash. The continuous curve is the best first order fit according to (17)

4.2 The Second Order Fit

The fit of the (8) has also been performed in two steps. At first the parameters A', B', t_c, Δ_t were estimated by the following approximation:

$$y(t) \simeq A' + B' \frac{\tau^\alpha}{\sqrt{1 + (\frac{\tau}{\Delta t})^{2\alpha}}} \ . \tag{18}$$

Then searching for the minimum of the residuals has been performed by using

$$\frac{y(t) - A'}{\frac{\tau^\alpha}{\sqrt{1+(\frac{\tau}{\Delta t})^{2\alpha}}}} - B' \simeq C' \cos \left\{ \omega \log \tau + \frac{\Delta_\omega}{2\alpha} \left[1 + \left(\frac{\tau}{\Delta t} \right)^{2\alpha} \right] + \phi' \right\} \tag{19}$$

Before the Crash

The best fit of (18) was made with lower and upper bounds for $(A, B, t_c, 1/\Delta_t, \alpha)$ given by $(-100, 0, 2000.19, -10, -20)$ and $(100, 100, 2000.70, 1000, 0)$, respectively. The results are given in Table 1.

After the Crash

The bounds are the same ones used for the best fit reported in the previous paragraph.

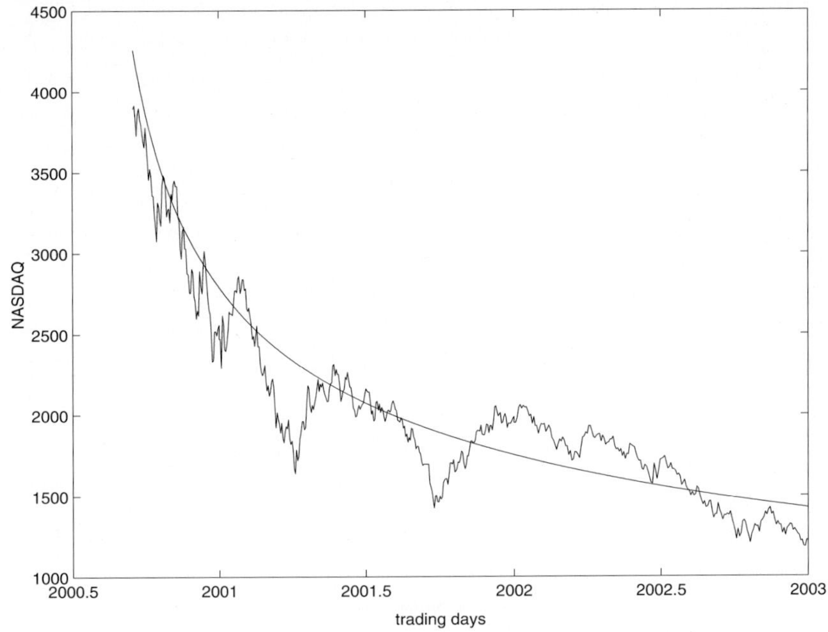

Fig. 8. NASDAQ index after the crash. The continuous curve is the best first order fit according to (7)

4.3 The Logistic Approach

As shown in Sects. 3 and 4 from the approximations to first and second order solutions it is possible to get the logistic function. The values of the parameters of the logistic can be calculated immediately from the fits in Tables 1 and 2. Some important remarks must be made. The first one is about the fact that this kind of aftershock behaviour relies over the structure of the market that is detected through the log-periodic oscillations. Thus it is necessary to perform the fit of the formula that contains oscillations, too, in order to test for the presence of the periodic part. Moreover the description of the after crash behaviour is interesting in order to give an estimate of the deflate of the bubble whilst it is still at its beginning. An empirical strategy that can be held by investors can be the one of performing the data fit as soon as possible. However a raw application of this procedure can lead to huge instabilities and high volatility over the parameter set because the overall function is not stationary.

An important remark is that the residuals on the left hand side of each (15) after the crash, and (17) before and after the crash are stationary log-periodic processes with zero mean. Also (7) that gives rise to (10) can be fitted from (6) taking into account the residuals

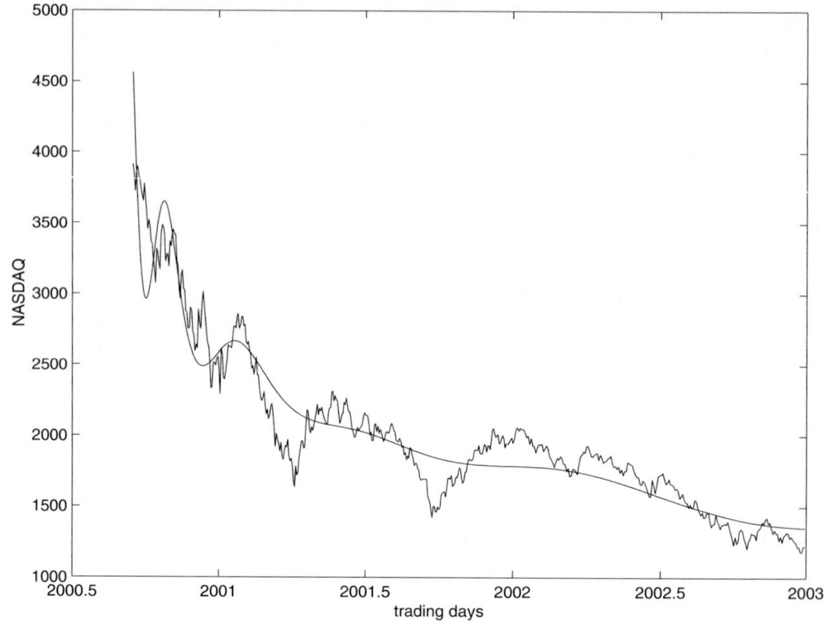

Fig. 9. NASDAQ index after the crash. The continuous curve is the best first order fit according to (6)

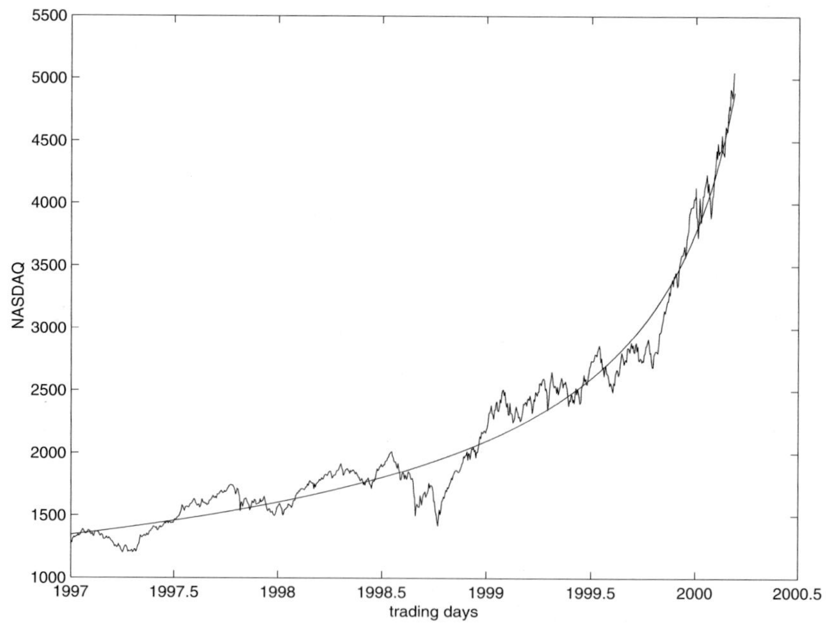

Fig. 10. NASDAQ index before the crash. The continuous curve is the best second order fit according to (18)

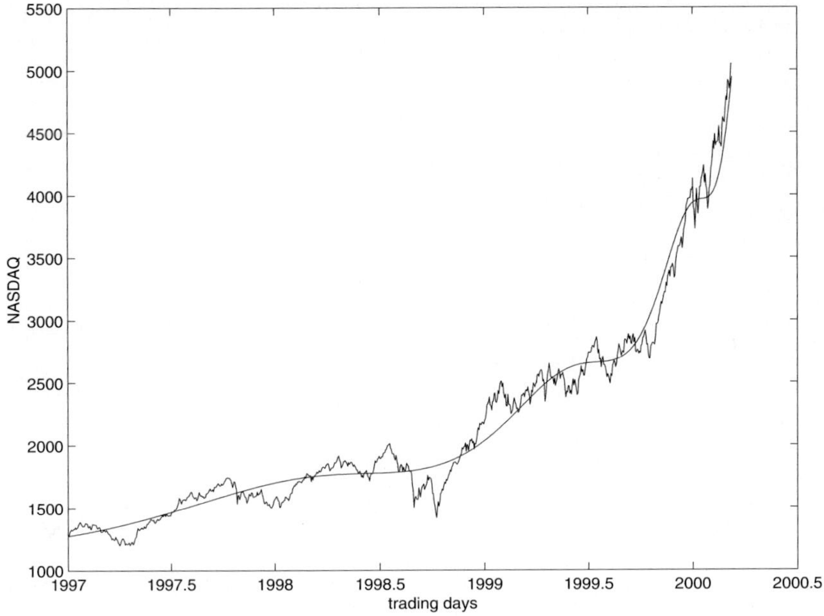

Fig. 11. NASDAQ index before the crash. The continuous curve is the best second order fit according to (19)

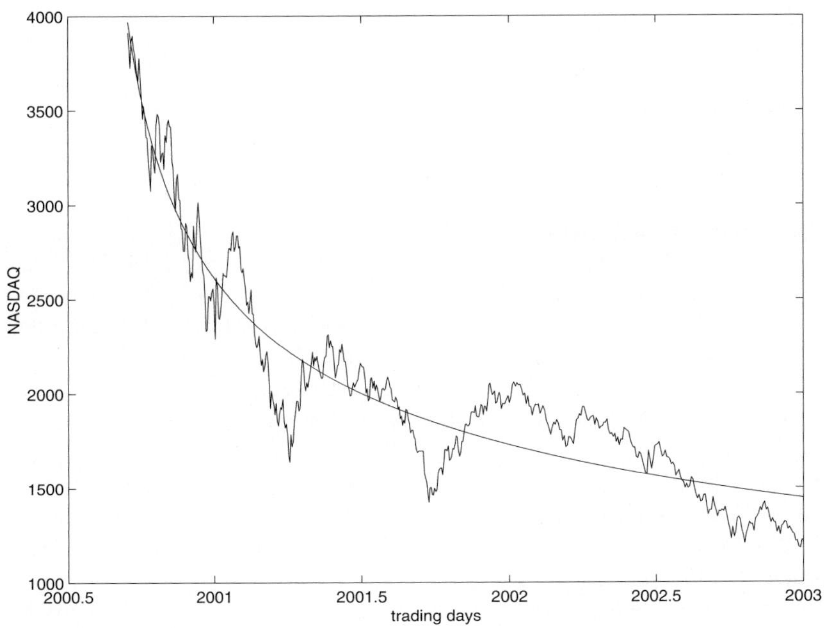

Fig. 12. NASDAQ index after the crash. The continuous curve is the best first order fit according to (18)

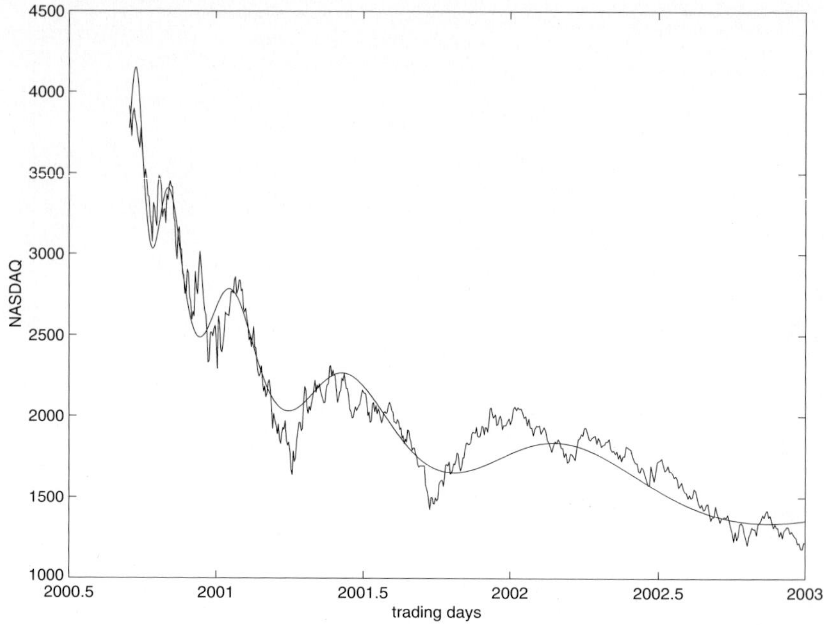

Fig. 13. NASDAQ index after the crash. The continuous curve is the best first order fit according to (8)

$$\frac{F(t)-A}{\ln(t-t_c)}+B \simeq C\cos[\omega\ln(t-t_c)-\phi] \text{ where } t>t_c \tag{20}$$

that exhibits the usual log-periodic structure.

This remark provides a guideline for the estimate of the parameters of the left hand side that allows to improve the stability of their estimate. For trading purposes the optimization task can be performed by adding subsets of data, as soon as they are available and thus depending on the time. In this dynamic setting it is necessary to take into account the accelerating oscillations $\cos[\omega\ln(t_c-t)-\phi]$ which lead to local periods converging to zero according to a geometrical progression with scale factor $\lambda \approx e^{2\pi/\omega}$.

Therefore a dynamic fitting procedure that can ensure the best stability over the parameters is a mixed online and batch one where data points are added only when a sequence as long as one period at least is available. The length of sequences changes because the period changes as the logarithm of the time. When more than one frequency is detected over the data before the crash then the best choice is to use the theoretical results and to try the theoretical value $\omega \simeq 2\pi$ for the after crash data.

Because of the fact that the function is periodic with the logarithm of time then the length of the sequences to be added is not fixed, but goes as the logarithm of the time. Thus when working with respect to the variable

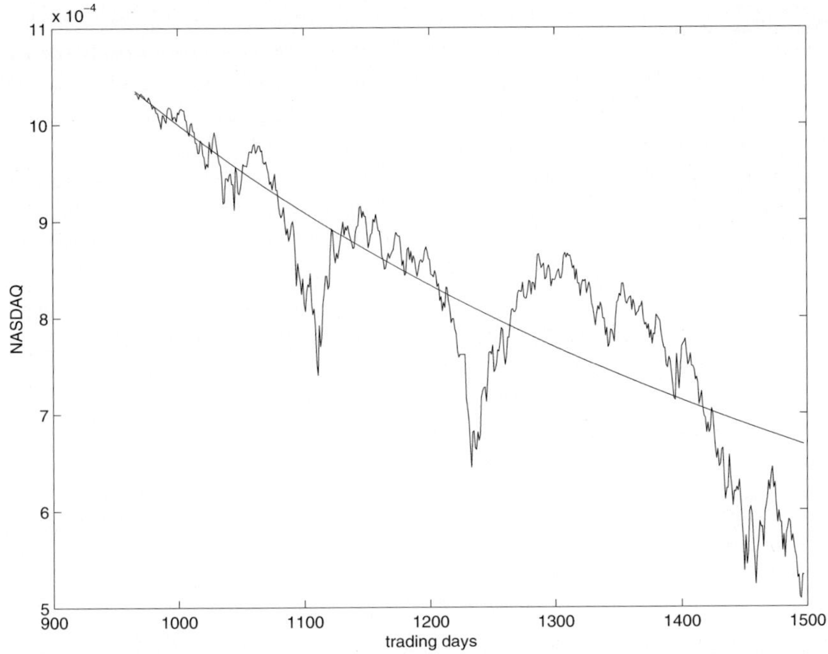

Fig. 14. NASDAQ index after the crash, rescaled through (10). The continuous curve is a detail of the best first order fit according to (10). Please refer to Table 2, line corresponding to the (7) for the values of parameters

"time" it is not possible to work over vectors of fixed size, but it is necessary to change their length.

The point symmetric to January 1st, 1997 with respect to the initial crash time is June 5th, 2003. At that time the market has stopped its descent and it is going to start a longer rise, but at that point it is still in the bounds of log-periodic oscillations.

5 Bayesian Analysis

Large financial crashes can be classified on the causes that generate them. Only endogenous crashes exhibit scaling. It is interesting to understand how strong is the mirroring between the progressively accelerating power law up to the crash and the subsequent approximately symmetrical power law decay, that is lower for exogenous peaks, in order to detect the end of the deflating period of the anti-bubble and the restart of the market. The parameters from the fit of either the first or the second order solutions or both of them can be used in order to write the proper form of the logistic functions. Thus it is possible to get an immediate estimate of them. An interesting question is to understand how the parameters of the logistic function that describe the pre crash growth have to change in order to describe the after crash decay. In order to perform the addressed tasks it is possible to rely over some bayesian analysis results that can be useful for the fit of the parameters of the logistic function [3]. Let $\bar{t}_c$ be the crash time estimated over data before the crash. Define a binary variable $I_\delta(t_c)$. $I_\delta(t_c) = 1$ if $|t_c - \bar{t}_c| < \delta$, and 0 otherwise. It is a measure of the stability of the estimate of t_c. The estimate of the crash is correct when $I_\delta(t_c) = 1$. The value of δ that describes the range of the estimate of t_c can be chosen from numerical results over the data before the crash. The estimate of parameters depends on data $\boldsymbol{y}_t = \{y_t\}_t$. It is possible to fix an a priori probability distribution over the parameters that occur in the logistic function in $a = a(\boldsymbol{y}_t)$. Usually the prior distribution over them is given as a gaussian one, centered on the expected values of parameters. In this case it is possible to use as expected values of the parameters the ones obtained from the best fits before the crash, thus relying on the symmetry hypothesis. The variance can be fixed and gives the spread of the numerical instabilities. With the usual notations of conditional probability it is possible to define

$$p(I_\delta(t) = 1|\boldsymbol{y}, a(\boldsymbol{y})) = \frac{1}{1 + \exp^{-a(\boldsymbol{y})}} . \tag{21}$$

This probability assignment is consistent with the detection of the logistic law (10). The estimate of the probability to get the crash in the given interval can be performed when more data are available and thus considering $\boldsymbol{y}_{t+T}$ instead of $\boldsymbol{y}_t$. As remarked in the previous paragraph the interval time T must be equal to a period in order to get some stability in the numerical estimates. The best fit of the parameters of the sigmoidal function proceeds via the usual least square method with the observation of successive data $\boldsymbol{y}_{t+T}$. After each observation there is an a posteriori distribution

$$p(I_\delta(t) = 1|\boldsymbol{y}_{t+T}) = \int P(I_\delta(t) = 1|a(\boldsymbol{y}_{t+T}))P(a(\boldsymbol{y}_{t+T})|\boldsymbol{y}_{t+T})\, da(\boldsymbol{y}_{t+T}) \tag{22}$$

where

$$P(a(\boldsymbol{y}_{t+T})|\boldsymbol{y}_{t+T}) = \int P(a(\boldsymbol{y}_{t+T})|\boldsymbol{y}_{t+T})\, da .$$

And the joint a posteriori distribution is given by

$$p(a(\boldsymbol{y}_{t+T})|\boldsymbol{y}_{t+T}) = \frac{1}{Z} P(a(\boldsymbol{y}_{t+T})) \prod_{n=1}^{N} P(I_\delta(t) = 1 \mid a(\boldsymbol{y}_t)) ,$$

where Z is a normalizing constant. Theoretical upper and lower bounds are available [3,6] in order to approximate (22), that provides a framework for the description of the adjustment of the probability when new data are available.

6 Conclusions

This chapter shows the occurrence of the logistic function in the description of large financial crashes due to speculative bubbles. It can be accomplished for both the first and second order approximation to the solution of the Landau equation. In the case of first order expansion its probabilistic meaning is also on line with Omori law. The study of after crashes signatures is important in order to estimate the time necessary to the market to absorb the shock and for taking decisions about after crashes investment policies. Whereas the speculative bubbles exhibit symmetry, the logistic function can be used also for the description of after crash data and for the deflating of bubbles. A further observation is that the use of a logistic function instead of the traditional solutions to the Landau equation allows the use of bayesian learning results that give an estimate over the fit of the function to data. The large crash of the index NASDAQ 100 composite that has occurred during April 2000 is presented as a case study. The null hypothesis is that post crash perfectly symmetric to pre crash, thus giving a description of the deflating of the bubble. As soon as time passes and new data arrives it is possible to detect whether the symmetry is respected, thus confirming the estimated end of the bubble, or allowing a different classification of the post crash behaviour. Further examples can be found in [12]. The Argentina stock market bubble of 1992 is well fitted by (4) and the logarithm of data if its anti-bubble is well fitted by (7). The same happens for the Venezuela bubble ending in 1997 and its related anti-bubble [12]. In these cases however the descent is more rapid, even if the log-periodic structure is maintained and allows to keep the difference with the case of crashes where the values return quickly to their fundamentals looking close to the behaviour of a damped harmonic oscillator [25,26]. As an example in [20] the S&P500 dumps like a single dissipative harmonic oscillator and thus the symmetry after crash is not maintained and it is interesting to study the fastness of the rejection of the after crash symmetry. Moreover it is also interesting to give some bounds [17] in order to understand if the decay is violating the mirroring, thus getting a different classification of the crash and consequently a different estimate for the end of the anti-bubble. A further development of this study could be to investigate a taxonomy among crashes in order to detect the minimal time which is necessary after the occurrence of the crash in order to give a stable estimate for the post crash logistic function and thus to determine some further information about the type of crash towards the forecast of the end of the bearish period and for the start of a new phase of the market.

References

1. C.M. Bishop: *Neural Networks for pattern Recognition* (Oxford University Press, Oxford 1996)

2. B. Delmas: Mathématiques et Sciences humaines **167**, 27 (2004)
3. M.N. Gibbs, D.J.C. MacKay: IEEE-NN **11**, 1458 (2000)
4. Y. Huang, A. Johansen, M.W. Lee, H. Saleur, D. Sornette: J. Geophys. Res. **105**, 25451 (2000)
5. Y. Huang, H. Saleur, C.G. Sammis, D. Sornette: Europhys. Lett. **41**, 43 (1998)
6. T.S. Jaakkola, M.I. Jordan: Computing upper and lower bounds on likelihoods in intractable networks. In: *Uncertainty in Artificial Intelligence (UAI) Proceedings of the Twelfth Conference*, ed by E. Horvitz (1996) pp 340–348
7. A. Johansen, O. Ledoit, D. Sornette: Int. J. Theo. Appl. Finance **3**, 219 (2000)
8. A. Johansen, D. Sornette: Physica A **245**, 411 (1997)
9. A. Johansen, D. Sornette: Physica A **261**, 581 (1998)
10. A. Johansen, D. Sornette: Int. J. Mod. Phys. C **10**, 563 (1999)
11. A. Johansen, D. Sornette: Eur. Phys. J. B **17**, 319 (2000)
12. A. Johansen, D. Sornette: Int. J. Theo. Appl. Finance **4**, 853 (2001)
13. A. Johansen, D. Sornette: Endogenous versus Exogenous Crashes in Financial Markets. In press in Contemporary Issues in International Finance (2004)
14. A. Johansen, D. Sornette: Risk **12**, 91 (1999)
15. A. Johansen, D. Sornette, O. Ledoit: Risk **1**, 5 (1999)
16. G. Rotundo: Physica A **344**, 77 (2004)
17. L.K. Saul, M.I. Jordan: Exploiting tractable substructures in intractable networks. In: *Advances in Neural Information Processing Systems (NIPS)*, vol 8, ed by D. Touretzky, M. Mozer, and M. Hasselmo (MIT Press, Cambridge 1996)
18. D. Sornette: Phys. Rep. **378**, 1 (2003)
19. D. Sornette, A. Johansen: Qu. Fin. **1**, 452 (2001)
20. D. Sornette, A. Johansen, J.-P. Bouchaud: J. Phys. I France **6**, 167 (1996)
21. D. Sornette, W.-X. Zhou: Qu. Fin. **2**, 468 (2002)
22. D. Sornette, W.-X. Zhou: Qu. Fin. **3**, C39 (2003)
23. T. Utsu, Y. Ogata, R. Matsu'ura: J. Phys. Earth **43**, 1 (1995)
24. N. Vandewalle, M. Ausloos, Ph. Boveroux, A. Minguet: Eur. Phys. J. B **9**, 355 (1999)
25. N. Vandewalle, M. Ausloos: Eur. Phys. J. B **4**, 257 (1998)
26. N. Vandewalle, Ph. Boveroux, A. Minguet, M. Ausloos: Physica A **255**, 201 (1998)
27. W.-X. Zhou, D. Sornette: Physica A **329**, 249 (2003)

Agent Based Approaches to Income Distributions and the Impact of Memory

Przemysław Repetowicz, Peter Richmond, Stefan Hutzler, and Eimear Ni Dhuinn

Department of Physics, Trinity College Dublin 2,
repetowp@tcd.ie,
richmond@tcd.ie,
shutzler@tcd.ie,
nidhuine@tcd.ie

Summary. Agent based models have been used to study the dynamics of wealth or income distributions in populations. In this chapter we develop a generic theory of interacting agents where the property of memory may be included. The model may be analysed both analytically and numerically via computer simulation. Particular regimes are analysed and compared with empirical data for both the UK and the Republic of Ireland. We also demonstrate that with the inclusion of memory the predicted wealth distribution can exhibit Pareto tails with values of the exponent that are greater than unity unlike the model without memory where the exponent is always unity.

1 Introduction

Reverend Thomas Malthus [3] is generally recognized as the person who made the first serious attempt to understand and quantify economic growth. However his predictions of everlasting exponential growth were quickly recognized as being incorrect. The issue was resolved by Verhulst [1] in a series of seminal papers [2]. A statistical as opposed to deterministic approach to economic modelling was first made by Italian economist Pareto [4] who examined the distribution of wealth in Italian society. He found that the high end (tail) of the wealth distribution was described by power law, $f_1(w) \sim w^{-1-\alpha}$. Pareto noted at the time that *"The problem of pure economics bears a striking likeness to that of rational mechanics... It therefore appears quite legitimate to appeal also to mathematics for assistance in the solution of the economic problem."*

Recently it has been shown via numerous empirical studies [5–8] that this power-law tail is temporarily stable over a wide range of wealth values. Studies of income data in Sweden in [5], Revenue data concerned with the net capital of individuals at death in the United States [6], income distributions in Japan in [8], the UK [18] and Australia [19]) all support Pareto's initial conclusions. In addition, distributions of sizes, cities, Internet files, biological taxa, gene families and protein family frequencies in [7]) also suggest that Pareto's conclusions are of a general nature and suggest that this high end tail that exhibits the power law is characterized by several multiples or even tens of multiples of the average wealth. Note that in this paper we use words income and wealth interchangeably. This corresponds to the starting assumption that these economic systems are in equilibrium ((quasi-)stationary) and the plausible assertion here that income of an individual is roughly equivalent to wealth in his or her possession in the long term i.e. at death.

Models based on stochastic processes with multiplicative noise (with the updated value of process depending on a product of the present value and a random variable) that yield a power-law distribution of the value of the process have been known for years from studies as diverse as studies of economic growth [12], city populations [13], wealth distributions [14], stochastic renewal processes [15] and of frequency of words in text [11]. Recently Di Matteo [19] published a model based on additive stochastic processes with individuals (agents) interacting through a network. In this model the distribution that is a weighed sum of Gaussians with averages proportional to the connectivity of the network, equal variances, and weights depending on the degree distribution of the network appears to be a power law in the high end and a log-normal distribution in the low end. Other models [20] simulate in a simplistic way the effect of tax. However all these models relate essentially to a Generalized Lotka–Volterra equation that combines a multiplicative stochastic process with a process that redistributes a fraction of the total money to ensure the money possessed by an agent is never zero. The resulting distribution $f_1(v)$ has the form:

$$f_1(v) \sim \frac{\mathrm{e}^{(1-\alpha)/v}}{v^{1+\alpha}} \tag{1}$$

where $\alpha - 1$ is a positive number that is, it is suggested related to social security and some random investments.

A different class of models [21–24] are based by analogy with the kinetic theory of dilute gases (so called wealth exchange models). In these models, agents exchange money via pairwise interactions and the rate of exchange depends on the number of agents. The models yield a power-law tail as a result of the breaking of the conservation of wealth in exchange processes (inelastic scattering).

We have examined in detail elsewhere the basic model due to Patriarca, Chakraborti and Kaski (PCK) [9] and shown that aspects of the model [10] can be solved analytically. The model yields a form for the wealth distribution that appears reasonable for low values of wealth. At the high end, the model admits a power law but the value of α for the model is precisely unity whereas it is known that for empirical data that α may be 1.5 or higher. Here we propose a generalized model (RRH model) that includes the possibility of agents with memory. The model is developed in section (2). We then proceed to examine various limits. The generalization clearly encompasses the PCK model as a special case therefore and continues to fit reasonably well the low end of the wealth distribution. This is now demonstrated in section (5) using data from both the UK and the Republic of Ireland. In the following sections, we examine the possibility that multi point and opposed to pairwise interactions might allow a change in the tail index, α. The result is negative. We finally show that when memory is included, the power law describing the higher end tail can take on values that differ from unity and that are more in agreement with empirical data. This result is demonstrated both analytically and supported via numerical simulation.

2 Kinetics of Wealth Distributions

Elsewhere [16] we developed the solution of the PCK model where agents have no memory. Here we consider a more generalized model that includes the possibility of agents with a memory. As with the simple PCK model the time evolution of the distribution of wealth is described through dynamical rules that involve exchanges between two agents (two-agent exchanges) in two consecutive times t and $t + \delta t$, viz.:

$$
\begin{aligned}
w_i\,(t+\delta t) &= \lambda_i \hat{w}_i(t) + \epsilon\,[(1-\lambda_i)\hat{w}_i(t) + (1-\lambda_j)\hat{w}_j(t)] \\
w_j\,(t+\delta t) &= \lambda_j \hat{w}_j(t) + \epsilon_1\,[(1-\lambda_i)\hat{w}_i(t) + (1-\lambda_j)\hat{w}_j(t)]
\end{aligned} \tag{2}
$$

where λ_i and λ_j describe the amount of money saved during the exchange process by agents i and j respectively (saving propensities). The random number $\epsilon \in [0,1]$ is uniformly distributed with $\langle\epsilon\rangle = 1/2$ and $\epsilon_1 = 1-\epsilon$. The variables $\hat{w}_i(t)$ and $\hat{w}_j(t)$ are computed from values of wealth at time t and at some random time $q \leq t$ that is sampled from a distribution $p_t(q)$, that we call the memory horizon distribution, prior to the start of the dynamical process.

$$\hat{w}_i(t) = w_i(t) + \gamma w_i(q) \tag{3}$$

$$\hat{w}_j(t) = w_j(t) + \gamma w_j(q) \tag{4}$$

Thus our process is not Markovian and is measured by some decay rate of the distribution $p_t(q)$ (to be specified in next paragraph). Notice that with this definition of memory, the agents tend to accumulate money in the long term with a rate that is specified by the parameter γ (the money accumulation parameter).

Since the model involves two-agent exchange processes at two times it is convenient to describe it in terms of a four-agent distribution function $f_4(v,w,v',w';t,q)$. This function specifies the probability density of an event in which two randomly chosen agents have wealth values $V_t = v$ and $W_t = w$ at time t and values $V_q = v'$ and $W_q = w'$ at time $q < t$. This means that $f_4(v,w,v',w';t,q) := P\,(V_t = v, W_t = w, V_q = v', W_q = w')$. The two-agent function $f_2(v,w;t)$ is obtained by integrating the four-agent function over values $v', w' \geq 0$ at time q. The kinetic equation for the two-agent distribution function is constructed by counting all states of agents (v', w') at time t and (v'', w'') at time q, states that give rise to a predefined state (v, w) at time $t + \delta t$. We assume that the waiting times between wealth exchanges are distributed exponentially with some rate r. Therefore conditioning on the occurrence of the exchange process in time interval $[t, t + \delta t]$ for small δt yields:

$$\begin{aligned} f_2(v, w; t + \delta t) &= (1 - r\ \delta t) f_2(v, w; t) \\ &+ (r\ \delta t) \int_{R_+^4} dv' dw' dv'' dw''\ f_4(v', w', v'', w''; t, q) \\ &\times P\left(V_{t+\delta t} = v, W_{t+\delta t} = w \,|\, V_t = v', W_t = w', V_q = v'', W_q = w''\right) \end{aligned} \quad (5)$$

where the integration in (5) runs from zero to infinity, the conditional probability in the integrand in (5) is expressed via a delta function (no dissipation of wealth assumption) as follows:

$$\begin{aligned} &\delta\left(v - \lambda(\hat{v})\hat{v} - \epsilon\left\{[1 - \lambda(\hat{v})]\hat{v} + [1 - \lambda(\hat{w})]\hat{w}\right\}\right) \\ &\times \delta\left(w - \lambda(\hat{w})\hat{w} - \epsilon_1\left\{[1 - \lambda(\hat{v})]\hat{v} + [1 - \lambda(\hat{w})]\hat{w}\right\}\right), \end{aligned} \quad (6)$$

where

$$\hat{v} = v' + \gamma v'' \quad \text{and} \quad \hat{w} = w' + \gamma w'' \quad (7)$$

and the saving propensities $\lambda(v)$ may depend on the values of wealth v. In order to change (5) into an algebraic equation we make use of a generalized Laplace transform:

$$\begin{aligned} \mathcal{L}^{(\lambda,\gamma)} : f(v_1, v_2) &\rightarrow \tilde{f}^{(\lambda,\gamma)}(\boldsymbol{x}) := \mathcal{L}^{(\lambda,\gamma)}_{v_1,v_2}\left[f(v_1, v_2)\right](\boldsymbol{x}) \\ &:= \int_0^\infty dv_1 \int_0^\infty dv_2\ e^{-(x_1 + \lambda(v_1 + \gamma v_2)x_2)(v_1 + \gamma v_2)} f(v_1, v_2) dv \end{aligned} \quad (8)$$

where $\boldsymbol{x} = (x_1, x_2)$. Note that (8) depends on the functional form $\lambda(\hat{v})$. For $\lambda(\hat{v}) = \lambda = \text{const}(\hat{v})$ the generalized transform $\tilde{f}^{(\lambda,\gamma)}(\boldsymbol{x}) = \mathcal{L}_{v_1,v_2}[f(v_1, v_2)]$ $(x_1 + \lambda x_2, \gamma(x_1 + \lambda x_2))$ and in general $(\lambda(\hat{v}) \neq \text{const}(\hat{v}))$, as we will see in next paragraph, the generalized transform is expressed via partial derivatives of the Laplace transform itself. Multiplying both sides of (5) by $\exp(-xv - yw)$ and integrating over $v', v'' \in [0, \infty]$ and $w', w'' \in [0, \infty]$ we get:

$$\begin{aligned} &\frac{\tilde{f}_2(x, y; t + \delta t) - \tilde{f}_2(x, y; t)}{r\ \delta t} + \tilde{f}_2(x, y) \\ &\qquad = \tilde{f}_4^{(\lambda,\gamma)}\left(\begin{pmatrix} \epsilon x + \epsilon_1 y \\ \epsilon_1(x - y) \end{pmatrix} \begin{pmatrix} \epsilon x + \epsilon_1 y \\ -\epsilon(x - y) \end{pmatrix}; t, q\right) \end{aligned} \quad (9)$$

where $\tilde{f}_2$ and $\tilde{f}_4^{(\lambda,\gamma)}$ are the Laplace and generalized Laplace transforms of the two-agent and the four-agent functions respectively. Note that the first and the second vector on the right hand side of (9) corresponds to values of wealth of the first agent and of the second agent at times t and time q respectively.

In order to solve (9) we need to express the generalized Laplace transform through the Laplace transform. We assume that values of wealth of different agents at the same time are independent whereas values of wealth of the same agent at times t and q are not. This is a plausible assumption since in real societies densities of agents are small but agents possess a memory horizon even though its distribution decays fast with the time lag. This means that the four point function in (9) factorizes as follows:

$$\tilde{f}_4^{(\lambda,\gamma)}(\boldsymbol{x}_1, \boldsymbol{x}_2; t, q) = \tilde{f}_2^{(\lambda,\gamma)}(\boldsymbol{x}_1; t, q)\tilde{f}_2^{(\lambda,\gamma)}(\boldsymbol{x}_2; t, q) \tag{10}$$

We assume the saving propensity $\lambda(v)$ to be a periodic function $\lambda(v) = \lambda(v+2L)$ of the wealth v. This means that $\lambda(v)$ can be expanded in a Fourier series:

$$\lambda(v) = \lambda_0 + \sum_{n=1}^{\infty} \lambda^{(n)} \sin(\frac{n\pi}{L} v) \tag{11}$$

Inserting (11) into the definition of the generalized Laplace transform (8), factorizing the exponential in the integrand into two parts that depend on x_1 and on x_2 only and expanding the exponential with x_2 in a Taylor series we obtain:

$$\tilde{f}_2^{(\lambda,\gamma)}(\boldsymbol{x}) = \mathcal{L}_{v_1,v_2}^{(\lambda,\gamma)}\left[f_2(v_1, v_2)\right](\boldsymbol{x}) = \left[1 + \mathcal{A}_{x_2, s(\boldsymbol{x})}\right] \tilde{f}_2(s(\boldsymbol{x}), \gamma s(\boldsymbol{x})) \tag{12}$$

where $s(\boldsymbol{x}) = x_1 + x_2\lambda_0$ and

$$\mathcal{A}_{x_1,x_2} := \sum_{n=1}^{\infty} \frac{(-x_1)^n}{n!} \left(\sum_{m=1}^{\infty} \lambda^{(m)} \partial_{-x_2} \sin(\frac{\pi}{L} m \partial_{-x_2}) \right)^n \tag{13}$$

is a pseudo-differential operator that is defined by means of a Taylor expansion in partial derivatives of higher order. Inserting (12) into the kinetic equation (9) we obtain in the limit $\delta t \to 0$ the following pseudo-differential equation:

$$\begin{aligned} &\left[r^{-1}\partial_t + 1\right] \tilde{f}_2(x, y; t) = \\ &\quad \left[1 + \mathcal{A}_{\epsilon_1(x-y), \mathcal{P}_1(x,y)}\right] \tilde{f}_2(\mathcal{P}_1(x, y), \gamma\mathcal{P}_1(x, y); t, q) \\ &\quad \times \left[1 + \mathcal{A}_{-\epsilon(x-y), \mathcal{P}_2(x,y)}\right] \tilde{f}_2(\mathcal{P}_2(x, y), \gamma\mathcal{P}_2(x, y); t, q) \end{aligned} \tag{14}$$

where

$$\mathcal{P}(x, y) = ((\lambda_0 + \epsilon\lambda_1)x + \epsilon_1\lambda_1 y, (\lambda_0 + \epsilon_1\lambda_1)y + \epsilon\lambda_1 x) \tag{15}$$

we defined a vector whose components $\mathcal{P}_1(x, y)$ and $\mathcal{P}_2(x, y)$ are invariant with respect to interchanging (ϵ, x) with (ϵ_1, y).

A further analysis of (14) is left for future work.

In the following we now assume $\lambda(v) = \text{const}(v) = \lambda$ and consider the following particular cases:

- Agents that have no memory ($\gamma = 0$) and the saving propensity λ is the same fixed number for all agents
- An extension of the model to include three-agent exchange processes.
- Agents have a memory ($\gamma > 0$) and the savings propensity for agents may be different and characterized by a distribution $\rho_\Lambda(\lambda)$. Note that in this case agents can accumulate wealth during an exchange process.

3 Lack of Memory and Equal Savings

In the absence of memory i.e. $\gamma = 0$, wealth is conserved during the exchange processes and the two-agent distribution function reaches a stationary solution at large times. The model has been solved numerically by Chakrabarti [17], who found that to a good approximation the distribution of wealth could be fitted well by a heuristic function:

$$f_1(v) = \frac{n^n}{\Gamma(n)} v^{n-1} \exp(-nv) \tag{16}$$

where $\Gamma(n)$ is the gamma function and the parameter n is related to the saving propensity, λ as follows:

$$n(\lambda) = 1 + \frac{3\lambda}{1-\lambda} \tag{17}$$

This relation (17), in effect a conjecture by Patriarca, Chakraborti and Kaski (PCK) [10], can in fact be derived analytically. Since the PCK conjecture is concerned with the one-agent distribution we may assume that its explanation can be achieved using the mean field approximation

$$\tilde{f}_2(x, y; t) = \tilde{f}_1(x; t)\tilde{f}_1(y; t) \tag{18}$$

using this approximation, (30) takes the form:

$$\tilde{f}_1(x) = \left\langle \tilde{f}_1((\lambda + \epsilon\lambda_1)x)\tilde{f}_1((\epsilon\lambda_1)x) \right\rangle_\epsilon = \frac{1}{\lambda_1 x} \int\limits_0^{\lambda_1 x} \tilde{f}_1(\lambda x + \phi)\tilde{f}_1(\phi)\, \mathrm{d}\phi \tag{19}$$

where the average in the first equality in (19) is over $\epsilon =$ Uniform (0,1) and in the second equality we substituted for $\epsilon\lambda_1 x$. In this case a solution $\tilde{f}_1(x) = \sum_{n=0}^{\infty}(-1)^n m_n x^n$ that is analytic in x exists. The moments $\langle v^n \rangle = m_n \cdot n!$ satisfy recursion relations:

$$m_p = \sum_{q=0}^{p} m_q m_{p-q} \tilde{C}_q^{(p)}(\lambda) \quad \text{where} \quad \tilde{C}_q^{(p)}(\lambda) = \frac{\int\limits_0^{(1-\lambda)} (\lambda + \eta)^q\, \eta^{p-q} \mathrm{d}\eta}{1-\lambda} \tag{20}$$

where the constants $\tilde{C}_q^{(p)}$ are computed from following recursion relations:

$$\tilde{C}_{q+1}^{(p)} = \frac{(1-\lambda)^{p-q-1} - (q+1)\tilde{C}_q^{(p)}}{p-q} \quad \text{with} \quad \tilde{C}_0^{(p)} = \frac{(1-\lambda)^p}{p+1} \tag{21}$$

Solving the equations of moments (20) with initial conditions $m_0 = 1$ and $m_1 = 1$ recursively, i.e. expressing, via the pth equation, m_p as a function of λ and all previous values of m one obtains:

$$m_2 = \frac{\lambda+2}{2(1+2\lambda)} \qquad m_3 = \frac{\lambda+2}{2(1+2\lambda)^2} \tag{22}$$

$$m_4 = \frac{72 + 12\lambda - 2\lambda^2 + 9\lambda^3 - \lambda^5}{24(1+2\lambda)^2(3+6\lambda-\lambda^2+2\lambda^3)} \tag{23}$$

The first three moments m_1, m_2 and m_3 coincide with the moments m_1^{conj}, m_2^{conj} and m_3^{conj} of PCKs function (16) if the relation between the parameters n and λ is given by (17). Indeed the coefficients $m_i^{\text{conj}}(n)$ of a series expansion of the Laplace transform

$$\tilde{f}_1(x) = \left(\frac{n}{x+n}\right)^n = \sum_{j=0}^{\infty} (-1)^j m_j^{\text{conj}}(n) x^j \tag{24}$$

of the function (16) agree with moments (23) up to the third order, subject to (17) being satisfied. This is shown in a nice way in Fig. 1. The deviation $\Delta\tilde{f}_1(x)$ between the exact solution of (20) and the Ansatz (16) has a leading fourth order:

$$\Delta\tilde{f}_1(x) = \frac{(n-1)(n+1)(n+8)}{8n^3(10n^3+30n^2+45n-4)} x^4 + O\left(x^5\right) \tag{25}$$

It is hard to say whether a more general class of functions than (16) would fulfill (20) to higher expansion orders.

4 Three-Agent Exchange Processes

Elsewhere [16] we have shown that the two-agent exchange process in the mean field approximation does not provide a full explanation of the model. Obviously many-agent distribution functions $f_m(x_1, \ldots, x_m)$ may not be produced correctly within this approach but more importantly the value of α for the high end wealth distribution where the savings are chosen according to a random distribution is always unity. Might this change if we include m-point interactions:

$$w_i(t+\delta t) = \lambda_i w_i(t) + \epsilon_i \left[\sum_{j=1}^{m} (1-\lambda_j) w_j(t)\right] \tag{26}$$

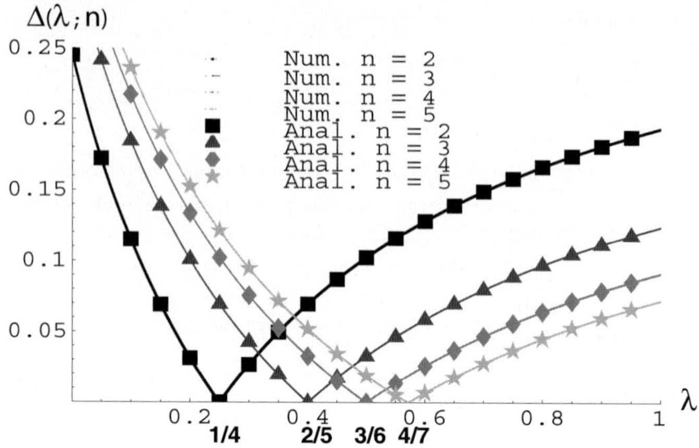

Fig. 1. Deviations $\Delta(\lambda; n) = \sum_{p=0}^{10} \left|(m_p(\lambda) - m_p^{\mathrm{conj}}(n))/p!\right|$ of the exact moments m_p of the wealth distribution from the moments m_p^{conj} derived from the conjecture plotted as a function of λ for $n = 2, 3 \ldots, 9$. *Solid lines (dot symbols)* correspond to analytical (numerical) solutions of the moment (20). We see that the minima $\lambda = (n-1)/(n+2) = \{1/4, 2/5, 3/6, 4/7, 5/8, 6/9, 7/10, 8/11\}$ of the deviations do correspond to the PCK conjecture (17)

where $\epsilon_i \in [0, 1]$ and $\sum_{i=1}^{m} \epsilon_i = 1$. This means that at every time step, exchange processes involving any number of agents can happen with a certain likelihood. We perform the analysis for $m = 3$ in order to establish the type of mathematical difficulties we might encounter. Now the kinetic equation for the two-agent distribution function in the Laplace domain is expressed through three-agent distribution functions and takes the form:

$$\begin{aligned}
&\frac{1}{r}\partial_t \tilde{f}_2(x, y; t) + \tilde{f}_2(x, y; t) \\
&= (1-\sigma)\left\langle \tilde{f}_2\left(\lambda x + \lambda_1\left[\epsilon x + (1-\epsilon)y\right], \lambda y + \lambda_1\left[\epsilon x + (1-\epsilon)y\right]; t\right)\right\rangle \\
&+\frac{\sigma}{2}\left\langle \tilde{f}_3\left(\lambda x + \lambda_1\left(\epsilon_1 x + \epsilon_2 y\right), \lambda y + \lambda_1\left(\epsilon_1 x + \epsilon_2 y\right), \lambda_1\left(\epsilon_1 x + \epsilon_2 y\right); t\right)\right\rangle \\
&+\frac{\sigma}{2}\left\langle \tilde{f}_3\left(\lambda_1\left(\epsilon_2 x + \epsilon_3 y\right), \lambda x + \lambda_1\left(\epsilon_2 x + \epsilon_3 y\right), \lambda y + \lambda_1\left(\epsilon_2 x + \epsilon_3 y\right); t\right)\right\rangle \quad (27)
\end{aligned}$$

where $\lambda + \lambda_1 = 1$, $\epsilon + \epsilon_1 \leq 1$ and σ and $(1-\sigma)$ denote likelihoods of three-agent and two-agent exchange processes respectively. The first (second and third) term(s) on the right-hand side in (27) account(s) for two-(three-)agent exchange processes respectively. Setting $y = 0$ and $\sigma = 0$ we obtain the kinetic equation of two-agent exchanges (19) Setting $x = y = 0$ we obtain an identity because $f_2(0, 0) = f_3(0, 0, 0) = 1$.

Now the transcendental equation derived from the kinetic equation (27) has the following form (compare with (41)):

$$
\begin{aligned}
1 = (1-\sigma)\left(\langle(\lambda+\epsilon\lambda_1)^\alpha\rangle + \langle(\epsilon\lambda_1)^\alpha\rangle\right) \\
+\frac{\sigma}{2}\left(\langle(\lambda+\epsilon_1\lambda_1)^\alpha\rangle + \langle(\lambda+\epsilon_2\lambda_1)^\alpha\rangle + 2\langle(\epsilon_1\lambda_1)^\alpha\rangle + 2\langle(\epsilon_2\lambda_1)^\alpha\rangle\right)
\end{aligned} \tag{28}
$$

where $\epsilon, \epsilon_1, \epsilon_2 =$ Uniform(0,1) and $0 < \epsilon_1 + \epsilon_2 < 1$ are sampled independently from ϵ. Note that since $\langle\epsilon\rangle = 1/2$ and $\langle\epsilon_1 + \epsilon_2\rangle = 2/3$ both sides of (28) coincide for $\alpha = 1$. For $\alpha > 1$ the right hand side is a strictly decreasing function of α because it is a sum of powers functions a^α where $a < 1$.

Thus $\alpha = 1$ is once again the only solution of this equation for arbitrary saving propensity distributions $\rho_A(\lambda)$ and for any likelihood $\sigma \in [0, 1]$ of three-agent exchange processes. This result is confirmed by numerical simulations.

5 Comparison of the Model to Empirical Data

We have analysed distributions of yearly incomes in the United Kingdom within 1993–2002 and in the Republic of Ireland within 2001–2004 and fitted the Patriarca, Chakraborti and Kaski model with a non-random saving propensity to the data (see Fig. 2). The fitted values of saving propensities for both countries are plotted versus time in Fig. 3. Since the number of years-data available is small it is hard to see any trend in the saving propensities as a function of time. However, the accuracy of the fit, in terms of the error bars, is reasonable. Most of the data that we analysed were characterized by an exponential decay of the distribution of wealth in the high end of the distributions. Hallmarks of power laws were vaguely visible in few data sets and set in only starting from a couple of multiples of the average income. This may be due to the fact that high-income data are underestimated due to the fact that rich people (tax-evaders) are not willing to disclose their full earnings to the revenue commissioner.

6 Presence of Memory and Random Savings

For $\lambda(v) = \text{const}(v)$ the pseudo-differential operators on the right hand side in (14) are equal to zero. Therefore (14) takes the form:

$$
\begin{aligned}
&\left[r^{-1}\partial_t + 1\right] \tilde{f}_2(x, y; t) \\
&= \tilde{f}_2(\mathcal{P}_1(x, y), \gamma\mathcal{P}_1(x, y); t, q)\tilde{f}_2(\mathcal{P}_2(x, y), \gamma\mathcal{P}_2(x, y); t, q)
\end{aligned} \tag{29}
$$

Setting $y = 0$ in (29) and using $\tilde{f}_2(x, 0) = \tilde{f}_2(0, x) = \tilde{f}_1(x)$ we obtain an equation for the one-agent function

$$
\begin{aligned}
&\left[r^{-1}\partial_t + 1\right] \tilde{f}_1(x; t) \\
&= \tilde{f}_2(\mathcal{P}_1(x, 0), \gamma\mathcal{P}_1(x, 0); t, q)\tilde{f}_2(\mathcal{P}_2(x, 0), \gamma\mathcal{P}_2(x, 0); t, q)
\end{aligned} \tag{30}
$$

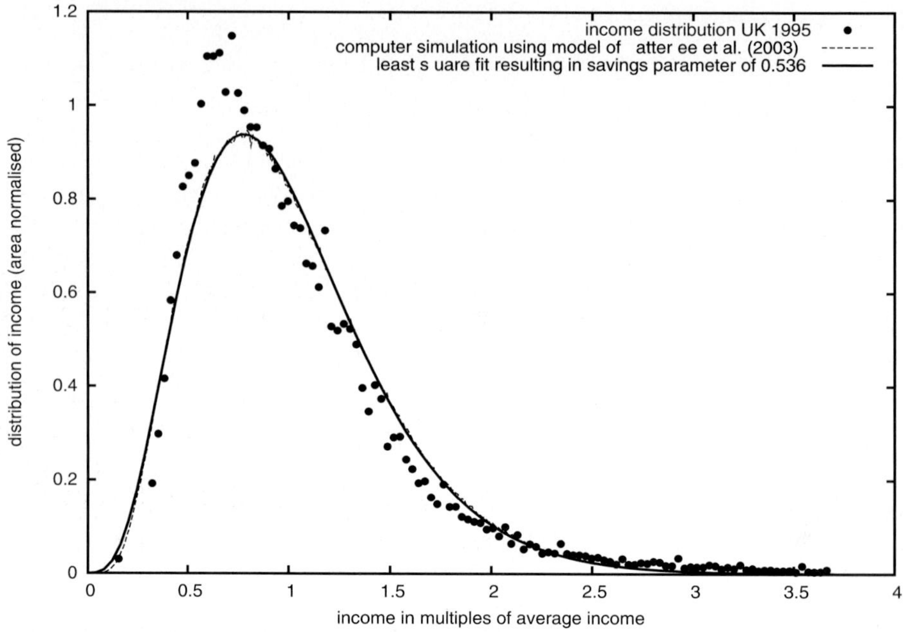

Fig. 2. A fit of the PCK function (16) to the distribution of yearly incomes in £ in the United Kingdom in 1995 for a saving propensity $\lambda = 0.54 \pm 0.05$ overlayed with numerical simulations of the two-agent wealth exchange lack of memory model (2) for the same value of λ

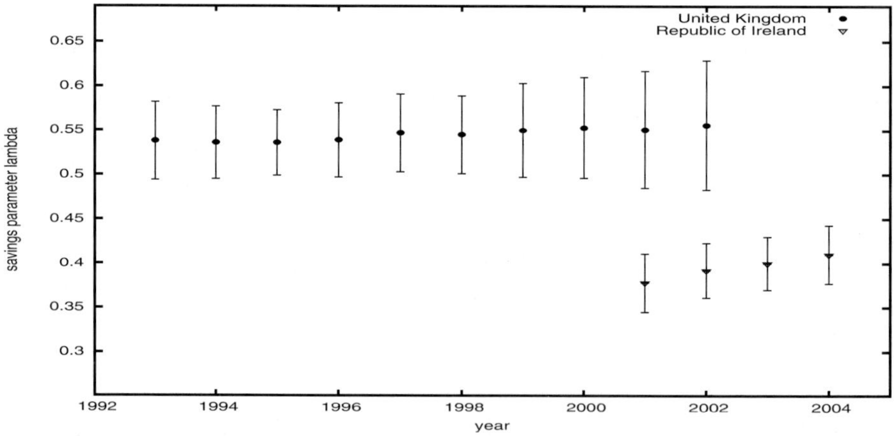

Fig. 3. Yearly saving propensities λ for the United Kingdom and for the Republic of Ireland derived from the PCK model. The error bars are derived from the least square deviations fit errors

We differentiate (30) with respect to x at $x = 0$ and obtain the time dependence of the mean wealth

$$\langle v \rangle (t) := -\left.\partial_{x_1} \tilde{f}_2(x_1, 0; t)\right|_{\boldsymbol{x}=\mathbf{0}} = -\left.\partial_{x_2} \tilde{f}_2(0, x_2; t)\right|_{\boldsymbol{x}=\mathbf{0}} \tag{31}$$

We have:

$$\left[r^{-1}\partial_t + 1\right] \langle v \rangle (t) = (\lambda + 2\epsilon\lambda_1)\left(\langle v \rangle (t) + \gamma \langle v \rangle (q)\right) = \langle v \rangle (t) + \gamma \langle v \rangle (q) \tag{32}$$

where in the second equality in (32) we have averaged over ϵ using $\langle \epsilon \rangle = 1/2$. Now we assume that the memory horizon distribution $p_t(q) = p(t-q)$ is homogenous in time. This allows us to solve (32) as follows:

$$\langle v \rangle (t) = \mathcal{L}_s^{-1}\left[\frac{\langle v \rangle (0)}{s - \gamma r \tilde{p}(s)}\right](t) \tag{33}$$

where $\tilde{p}(s) := \mathcal{L}_t\left[p(t)\right](s)$ is the Laplace transform of $\langle v \rangle (t)$ with respect to time t. We assume for simplicity that the memory horizon decays exponentially:

$$p(t) = \theta \mathrm{e}^{-\theta t} \tag{34}$$

with a mean $1/\theta$ (memory horizon), and that $\langle v \rangle (0) = 1$. This yields

$$\langle v \rangle (t) = \frac{s_1 + \theta}{s_1 - s_2}\exp(s_1 t) - \frac{s_2 + \theta}{s_1 - s_2}\exp(-|s_2| t) \tag{35}$$

where

$$r\gamma = s_{1,2}\left(1 + \frac{s_{1,2}}{\theta}\right) . \tag{36}$$

Since the wealth is not conserved we assume that the two-point function has a quasi-stationary solution of the form:

$$\tilde{f}_2(x, y; t, q) = \Phi_2(x\hat{v}_{t,q}, y\hat{v}_{t,q}) , \tag{37}$$

where $\hat{v}_{t,q} := \langle v \rangle (t) - C \langle v \rangle (q)$ and C is a constant that will be determined later. Inserting (37) into (30) and replacing x by $x/(\hat{v}_{t,q})$ we obtain:

$$\left[r^{-1}\langle s_1 \rangle x\partial_x + 1\right] \Phi_2(x, 0) = \Phi_2((\lambda + \epsilon\lambda_1)x, \gamma(\lambda + \epsilon\lambda_1)x)\Phi_2(\epsilon\lambda_1 x, \gamma\epsilon\lambda_1 x) , \tag{38}$$

where

$$\begin{aligned}\langle s_1 \rangle &:= \left\langle \frac{\dot{\hat{v}}_{t,q}}{\hat{v}_{t,q}} \right\rangle_q = s_1 \int_0^t \frac{\theta \mathrm{e}^{-\theta\tau}\mathrm{d}\tau}{1 - C\mathrm{e}^{-s_1\tau}} \\ &= s_1\left[1 + \sum_{n=1}^{\infty} C^n \frac{1 - \mathrm{e}^{-(\theta + n s_1)t}}{(1 + n\frac{s_1}{\theta})}\right] \underset{t\to\infty}{=} s_1\left[1 + \sum_{n=1}^{\infty}\frac{C^n}{(1 + n\frac{s_1}{\theta})}\right]\end{aligned} \tag{39}$$

For small values of x and y we expand the function

$$\Phi_2(x, y) = 1 - x - y + A(x^\alpha + y^\alpha)$$

in a series, insert it into (38), and compare coefficients at powers of x. This yields:

$$x\colon\ 1 + r^{-1}\langle s_1 \rangle = 1 + \gamma \tag{40}$$

$$x^\alpha\colon\ 1 + \alpha r^{-1}\langle s_1 \rangle = \left(\langle(\lambda + \epsilon\lambda_1)^\alpha\rangle + \langle(\epsilon\lambda_1)^\alpha\rangle\right)(1 + \gamma^\alpha) \tag{41}$$

where the average on the right hand side in (41) is over the distributions of λ and of ϵ. We insert (39) into (40) and we realize that (40) is equivalent to (36) if $C = r\gamma/\theta$ and $C \lll 1$.

From (40) we see that (41) is fulfilled for $\alpha = 1$ and we seek for a different solution $\alpha \in [1, 2]$. Since for $\gamma > 1$ the derivative of the right hand side of (41) with respect to α changes sign from minus to plus for some $\alpha \geq 1$ there will be another solution for $\alpha \neq 1$ of (41). The analysis of these solutions and their comparison to the numerical calculations is left for future work.

We emphasize that this analysis is valid in following limits:

(a) The money accumulation parameter per wealth exchange $r\gamma$ is much smaller than the inverse memory horizon θ.
(b) The money accumulation parameter $\gamma > 1$.
(b) The time in the simulation is large enough $t \to \infty$.

We have also performed numerical simulations with the rate of wealth exchanges $r = 1$, and with $\gamma > \theta$. Results of these simulations, shown in Fig. 4, clearly show that the distribution has a slope different from unity in the high end and the corresponding value of α can take on values that conform with empirical values.

7 Conclusions

We have presented a generalised model of agent interactions that incorporates agent memory and shown that both analytic and numerical solutions for the subsequent wealth distributions may be obtained. The model includes the approach of PCK as a special case. Using data from both the Republic of Ireland and the UK we show that the model describes reasonably well the low end of the wealth distribution. Furthermore contrary to the PCK model our generalised model including agent memory can account for the observed values of the Pareto exponent α that characterises the high end wealth distribution. A more detailed analysis of these results and other aspects will be given elsewhere.

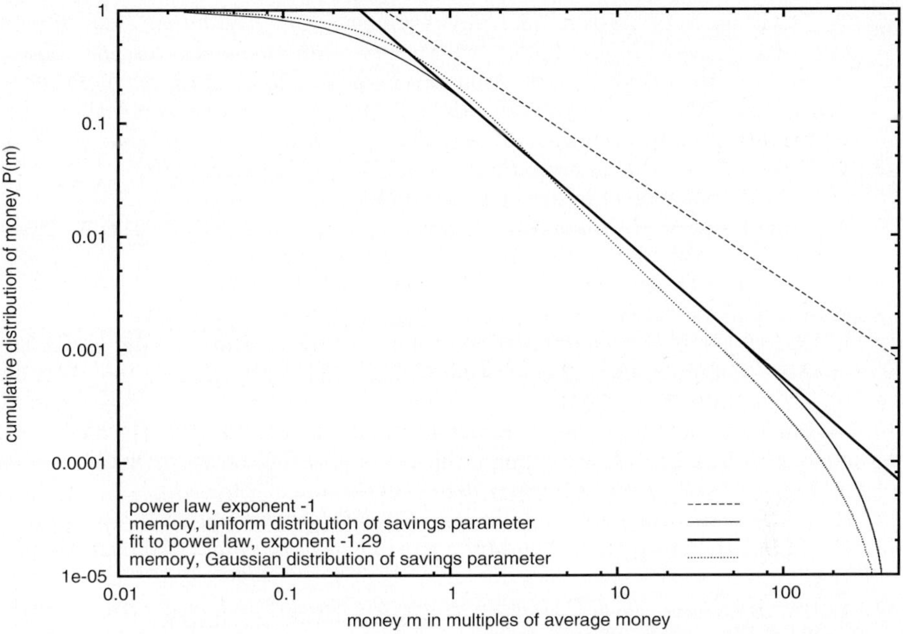

Fig. 4. Numerical calculations of the distribution of wealth with presence of memory and with random saving propensities. Here the rate of exchanges of wealth $r = 1$, the money accumulation parameter $\gamma = 1$ and the inverse memory horizon $\theta = -\log(0.63)$. It is readily seen that the slope of this distribution $|\alpha|$ becomes larger than unity once memory is present

Acknowledgments

The authors are grateful to Science Foundation Ireland for financial support via Basic Award 04/BR/P0251). E. Ni Dhuinn in particular acknowledges support via UREKA Supplement 04/BR/PO251UR04. Participation "Verhulst 200" at the Royal Military Academy in Brussels, BELGIUM was aided by COST Action P10 "Physics of Risk". Irish income data was provided by the Irish Revenue Commissioners http://www.revenue.ie. We are grateful to Geoff Willis for giving us the income data from the United Kingdom data.

References

1. A. Quetelet, Pierre-François Verhulst, *Annuaire de l'Académie royale des sciences de Belgique* **16**, 97 (1850)
2. P.-F. Verhulst: A note on the law of population growth, 1835. In: D. Smith, N. Keyfitz, *Mathematical Demography. Selected Papers* (Springer Verlag, 1977); P.-F. Verhulst: Mem. Acad. R. Belg. **18**, 1 (1845)

3. T. Malthus: *An Essay on the Principle of Population, as it Affects the Future Improvement of Society with Remarks on the Speculations of Mr. Godwin, M. Condorcet, and Other Writers*, LONDON, PRINTED FOR J. JOHNSON, IN ST. PAUL'S CHURCH-YARD, 1798, WWW:http://www.ac.wwu.edu/~stephan/malthus/malthus.0.html
4. V. Pareto: *Cours d'Economie Politique* (F Rouge, Lausanne 1897)
5. M. Levy, S. Solomon: Physica A **242**, 90 (1997)
6. A. Dragulescu, V.M. Yakovenko: Physica A **299**, n 1-2, 213 (2001); A. Dragulescu, V.M. Yakovenko: Eur. Phys. J. B **20**, 585 (2001)
7. W.J. Reed, B.D. Hughes: Phys. Rev. E **66**, 067103 (2002)
8. H. Aoyama, W. Souma, Y. Fujiwara: Physica A **324**, 352 (2003)
9. A. Chatterjee, B.K. Chakrabarti, S.S. Manna: Phys. Scripta T **106**, 36 (2003); preprint cond-mat/0311227; A. Chatterjee, B.K. Chakrabarti, S.S. Manna: Physica A **335**, 155 (2004)
10. M. Patriarca, A. Chakraborti, K. Kaski: Phys. Rev. E **70**, 016104 (2004)
11. June-Yule Lee, Signal processing of chaotic impacting series, *IEE Colloquium on Signals Systems and Chaos*, p 7/1-6 (1997)
12. R. Gibrat: *Les Inégalités économiques* (Librairie du Recueil Sirey, Paris 1931)
13. G.K. Zipf: *Human Behavior and the Principle of Least Effort* (Addison-Wesley, 1949)
14. Y. Ijiri, H. Simon: *Skew Distributions and the Sizes of Business Firms* (North-Holland, New York 1977)
15. H. Kesten: *Random Difference Equations and Renewal Theory for Products of Random Matrices*, Acta Mathematica, CXXXI: 207-248, 1973
16. P. Repetowicz, S. Hutzler, P. Richmond: preprint cond-mat/0407770, submitted to Physica A
17. A. Chatterjee, B.K. Chakrabarti: Physica Scripta T **106**, 36 (2003) and cond-mat/0311227; A. Chatterjee, B.K. Chakrabarti, S.S. Manna: Physica A **335**, 155 (2004)
18. T. Cranshaw: poster presentation at EPS meeting Applications of Physics in Financial Analysis (APFA) 3, London December (2003)
19. T. Di Matteo, T. Aste, S.T. Hyde: preprint cond-mat/0310544
20. S. Solomon et al: Phys. Rev. E **66**031102 (2002); S. Solomon, P. Richmond: Eur. Phys. J. B **27**, 257 (2002); P. Richmond, S. Solomon: Int. J. Mod. Phys. C **12**, 333 (2001)
21. F. Slanina: Phys. Rev. E **69**, 46102 (2004)
22. D. Ben-Avraham, E. Ben-Naim, K. Lindenberg, A. Rosas: preprint cond-mat/0308175
23. E. Ben-Naim, P.L. Krapivsky: Phys. Rev. E **61**, R5 (2000)
24. P.L. Krapivsky, E. Ben-Naim: J. Phys. A: Math. Gen **35**, L147 (2002)

Part IV

Condensed Matter

Agglomeration/Aggregation and Chaotic Behaviour in d–Dimensional Spatio–Temporal Matter Rearrangements Number–Theoretic Aspects

Adam Gadomski[1] and Marcel Ausloos[2]

[1] U.T.A. Bydgoszcz, Institute of Maths & Physics, Bydgoszcz PL–85796, Poland
agad@atr.bydgoszcz.pl
[2] University of Liège, SUPRATECS, Liège B–4000, Euroland
marcel.ausloos@ulg.ac.be

Summary. Matter gets organized at several levels of structural rearrangements. At a mesoscopic level one can distinguish between two types of rearrangements, conforming to different close-packing or densification conditions, appearing during different evolution stages. The cluster formations appear to be temperature- and space-dimension dependent. They suffer a type of Verhulst-like saturation (frustration) when one couples the growing (instability) and mechanical stress relaxation modes together. They manifest a chaotic behaviour both in space and time domains. We pretend to offer a comprehensive and realistic picture of a material or mega-cluster formation in d dimension.

1 Introduction

Matter organisations at a mesoscopic (molecular–cluster) level typically manifest a multitude of microstructural rearrangements. Cluster–cluster aggregations of proteins and/or colloids, phase separations, flocculation–coagulation phase transformations, sol-gel systems, (wet) sand or rice piles, etc., are manifestations of loosely-packed rearrangements, typically occurring under moderate or high temperature conditions. In contrast ripened polycrystals, sintered powders, soap froths and bubbles, and other cellular systems, constitute a type of rearrangement that usually emerges in a (relatively) low temperature limit and under certain ("field dependent") matter close–packing constraints. Beside such an agglomeration, fracture, desaggregation, desorption, dissolution, and alike, can be thought to be the "inverse process", finding its place in the opposite part of the relevant phase diagram [1].

In all of them spatial as well as temporal signatures of chaotic behaviour, due to matter reorganisations, can be detected: They are temperature and space-dimension dependent. In particular, one can show rigorously that in the limit of the spatial dimension going to infinity loosely-packed agglomerations become non–chaotic by suppressing totally their instability growing mode since it is related to the nonequilibrium agglomerate's entropy [2], while their closed-packed counterparts are not. When the growing mode is coupled to a mechanical stress relaxation mode as a power law via some phenomenological relation of Hall–Petch–Griffith (H-P-G) [3] type (an Onsager-type conjecture [4] of the present study), certain marks of Bethe–lattice frustration, related to a spatial overcrowding of the Cayley–tree branches, appear in the (mean-field) approach – a kind of frustration qualitatively of very similar type than that observed in Verhulst-type systems in an adequate time and parametric zone.

In the specific cases discussed in this review, however, by increasing the space dimension, d, we automatically induce some increase of the possible number of degrees of freedom in the system. Thus, when taking into account the coupling of the late-time growing and relaxation modes, say, in a fairly synchronized viz. power-law way of H-P-G type, one obtains that even though the material's relaxation goes slower than in the case when such a coupling is proposed in an unsynchronized (Debye-relaxation involving, i.e. rapid) way, one is, however, able to establish or restore an apparent dynamic microstructural order within the system the nonequilibrium (chaotic) measures of which are proposed below. There is, unfortunately, no way of establishing such an order when the coupling fails the power-law type synchronization requirement [3, 4].

Thus, when an ample space amongst the clusters is recovered by the system at its mature growing stage, we consider that the system successfully tries to avoid a chaotic matter organisation in space. Note that temperature may markedly help in surmounting the activation-energy barrier of the agglomeration, especially when it is raised appropriately, whence when not "damaging' a possibly smooth evolution of the system. Full success is, however, guaranteed when the limit of $d \to \infty$ is reached. If there is no chance for recovering the ample space, the late-time growing stage is realized in a moderately chaotic way. The mechanical stress relaxation, in turn, enters a readily chaotic regime, since the (nonequilibrium) entropy of the system diverges to plus infinity. The overall scenario resembles, in general, a formation of large (fractal) colloid aggregates that typically occurs with and without temperature and/or space-dimension dependent gravity factor domination, like as if imposing some limits to gelation of colloids [5].

The paper is arranged as follows. In Sect. 2, we define both the closely-packed as well as loosely-packed agglomerations, calling the latter the aggregation throughout. In Sect. 3, we list some qualitative signatures of chaos in matter-agglomerating systems, and refer briefly to different definitions as well as meanings of chaos. In Sect. 4, we present quantitative measures of chaos signatures in systems of interest, whereas in Sect. 5 we unveil number-theoretic measures, featuring a chaotic spatio-temporal behaviour of them. In Sect. 6, according to some suggestions given in [6], on which much of our report is based, in order to see which agglomerations behave orderly or non-chaotically, we explore the limit of $d \to \infty$, and arrive at a certain interesting (perhaps, surprising) conclusion, favouring aggregation of matter, or some structural loosely-packed, and typically high-temperature, matter rearrangements – in contrast to those emerging under close-packing low-temperature conditions. We close the paper by offering a concluding address in Sect. 7.

2 Agglomeration vs Aggregation of Matter – a Model Description

Following [7], throughout the present study, we wish to distinguish between the notions of agglomeration and aggregation of matter. By the former we mean an assembly of grains or molecular clusters, kept together by relatively strong forces (e.g. ionic), so that there is no easy possibility of taking the clusters apart, or destroying them. For the latter, because of the appearance of weak bonds between clusters, such as Van der Waals or hydrogen types, the possibility of cluster separation becomes an observed tendency of the matter rearrangement due to their weak bonds. For a schematic explanation of the difference between both matter arrangements, see Fig. 1.

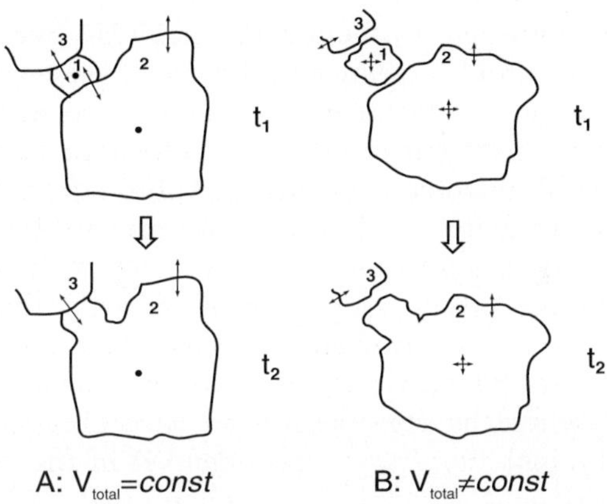

Fig. 1. Typical cluster-merging (three-grain) scenario for closely-packed (left, denoted by A) and loosely-packed (right, denoted by B) agglomerations. Two consecutive time steps t_1 and t_2 are shown. The former usually goes by a scenario with the preservation of the total agglomerate's volume (though in a more irregular way, when its logarithmic speed is measured, cf. Sects. 3–6), whereas the latter does not [8]. In the former, the clusters do not perform a translational motion but their boundaries may fluctuate in time and space, even though they are quite strongly confined by their neighbourhood. In the latter, an almost opposite situation in the time-and-space domain is typically observed. Some void is left behind a loosely packed system

2.1 Basic System of Equations Describing Model Matter Agglomeration

As in previous work [8, 9] we begin with a local continuity equation

$$\frac{\partial}{\partial t} f(v,t) + \frac{\partial}{\partial v} J(v,t) = 0 \,, \tag{1}$$

preferentially supplemented by the corresponding initial (of delta-Dirac type as a first attempt) and boundary (typically, of absorbing type) conditions (IBCs).

In a few subsequent studies a thermodynamic-kinetic description [8–10] of model complex matter agglomeration has been worked out. For the current[3] in the space of cluster volumes

$$J(v,t) = -\left[B(v)\frac{\partial}{\partial v}\Phi\right] f(v,t) - D(v)\frac{\partial}{\partial v} f(v,t) \,, \tag{2}$$

has been used [9], where $f(v,t)$ is the distribution of clusters of volume v: this means, that $f(v,t)dv$ is the (relative) number of clusters with size in the infinitesimal volume interval $[v, v+dv]$; t is the time; Φ represents the physical potential, equivalent to the free energy of the system (see [9] for an explanation of the term). It is assumed to be one of the most relevant drivers of the agglomeration process at the mesoscopic level, assuring its nonequilibrium character. In fact, the current (2), in the form presented above, comes from a rigorous derivation, starting from the Gibbs equation for the entropy production [10]. It has quite strong foundations anchored in nonequilibrium thermodynamics [9]. It is worth mentioning that upon inserting (2) into the continuity equation, (1), one gets a second-order partial differential equation of the Fokker–Planck–Kolmogorov (F-P-K) type [11].

For the dynamics of such a system some routes to chaotic behaviour have been sketched elsewhere by considering the (in)stability of Markov semi-groups in [12]. The mobility $B(v)$ is also defined in the configurational v–space and reads [9]

$$B(v) = \frac{D}{k_B T} v^\alpha, \qquad \alpha = \frac{d-1}{d} \,, \tag{3}$$

where D is a diffusion reference constant. Realize that the principal role of D is to scale the time variable; k_B represents the Boltzmann constant. Note that the mobility $B(v)$ is related to the Onsager coefficient, $L(v)$, that appears in the derivation of the matter flux equation (2) under a set of assumptions, mostly based on the locality of the Kramers-type process [10, 13], etc. $L(v)$, and hence $B(v)$, could be measured by comparing the current and the thermodynamic force [10]. The quantity $D(v) = Dv^\alpha$ is to be inferred

[3]For a method to derive diffusion currents for different types of systems one is encouraged to look into [10]

from the Green–Kubo (G-K) formula [14], so that there is some quite strong suggestion for deriving $B(v)$ both, experimentally as well as theoretically [9].

There is a debate about a possible violation of the G-K formula [14, 15]. For instance, it is proved that for a gas of charged particles subjected to an external electric field, the mean mobility of a charged particle, based on the G-K formula, is reliably well estimated for suitably small values of the external field. Moreover, at a microscopic scale one observes a nonlinear (or chaotic) behaviour of the particles, which is, unfortunately, not reflected by the macroscopic (mobility) measure. In our case, we assume algebraic correlations in v–space, for a phenomenological formula. The assumption seems to be as natural as possible: $D(v) \propto v^{\alpha}$; that means that both the diffusivity $D(v)$ and the mobility $B(v)$ are proportional to the cluster hypersurface, $s^{(d)} := R^{d-1}$. It should be underlined that it is, in our opinion, the common physical case in clustering phenomena, and is working properly at the mesoscopic level considered in our approach.

Notice right here that exactly the same assumption has been used to model in a F-P-K way the formation of surface nanostructure arrays [16]. Therein, an experimentally-observed passage between direct curvature-dependent ripening of matter nano-islands (our densely-packed agglomeration), and inverse ripening, with an elastic-field caused contraction of growing quantum dots [17] (our sparsely-distributed agglomeration of matter), has been presented.

There are, however, matter agglomerations, for a given T, that do not conform usually to

$$D(v) \propto B(v) \sim v^{\alpha} \simeq s^{(d)} \,. \tag{4}$$

To them belong both some physical-metallurgical transformations [18], such as martensitic, and presumably, also certain phase orderings of non-diffusive kind, emerging in model biosystems, such as those occurring in lipid biomembranes [19]. Other than algebraic types of correlations in the hyperspace can likely be expected for these. If a power law of the type given by (4) can be kept for further modelling, some additional correlations in time must complete a more comprehensive correlational proposal, cf. [20]. Other types of correlations in the hyperspace, even if they allow to get a general solution to the problem, may not accommodate the boundary conditions [9], so that one would expect either to be left with an unsolved specific problem or to encounter anomalous or irregular behaviour of the agglomerating system [8, 19, 20]. In such a case another type of finite, instead of infinite boundary boundary conditions [20], can sometimes give a remedy for the problem [21]. Here, under the term *infinite boundary conditions* [8] we typically understand the boundary conditions of absorbing (Dirichlet) type

$$f(v = 0, t) = f(v = V_{\rm clust}, t) = 0 \tag{5}$$

in which the single cluster volume is taken at infinity, $V_{\rm clust} = \infty$, whereas in case of the finite boundary conditions it assumes a finite value, $0 < V_{\rm clust} <$

∞, cf. [21], and a discussion therein. Although the latter unquestionably seems more physical the former is more frequently used to reveal the evolutions in matter-agglomerating systems [22] – this resembles to some extent a situation in statistical-thermodynamical systems undergoing an equilibrium phase transition: As such they are typically considered in the so-called thermodynamic limit (here, with a number of subunits going to ∞) under the mentioned agglomeration-oriented, e.g. condensation conditions, and the analogy would presumably extend over the examined nonequilibrium evolutions too [9,21].

2.2 Thermodynamic Potentials Driving Matter Agglomerations

In previous work [9] the analytic form of a (so-called) compaction potential was obtained, i.e.

$$\Phi(v) = \Phi_0 \ln (R/R_0) \; , \tag{6}$$

where Φ_0, R_0 – constants, and R stands for some cluster radius. Because

$$v \equiv v^{(d)} \sim R^d, \qquad d = 1, 2, 3, \dots \; , \tag{7}$$

one gets also $\Phi(v) \propto \ln (v/v_0)$, where v_0 is a constant.

The logarithmic potential assures the emergence of rather compact and curvature-involving structures, whence the name of "compaction potential" [22]. It should be noted that Φ is an entropic potential [9]. Thus, it can be a cause of some desaggregation, or matter-influenced impingement effects, occurring within the overall aggregation space.

In a previous study on the phase transformation kinetics for loosely packed "diffusive" agglomerates [8] we have written the matter flux of a purely diffusive nature prescribed in configurational space as follows

$$J(v,t) = -D(v) \frac{\partial}{\partial v} f(v,t) \; . \tag{8}$$

(The diffusion function $D(v) = Dv^{\alpha}$ is proportional to the cluster (grain) surface.)

Both closely-packed and loosely-packed agglomerations follow from the general form (2). Indeed, the loosely-packed case is obtained when the first (drift) term in r.h.s. of (2) can be neglected. Formally, $B(v) \to 0$ when $T \to \infty$. From the physical point of view, it corresponds to sufficiently high temperatures $T \geq T_{\text{pass}} > 0$, where T_{pass} can be treated as a cross-over temperature[4] above which the agglomeration takes place exclusively by yielding loosely-packed microstructures. However, the drift term in r.h.s. of (2) depends both on v and T. It tends *uniformly*, which means independently of v, to zero at the high temperature limit if

[4] Borrowing from the nomenclature of phase transitions and critical phenomena one might sometimes opt for calling it the threshold temperature

$$B(v)\frac{\partial \Phi(v)}{\partial v} \simeq \mathrm{C} = \mathrm{const}\ . \tag{9}$$

Then for a given system, temperature T_{pass} does not depend on v and looks consistently defined.

Some additional argumentation can be provided that such a constant (limit) C exists and is well defined. Namely, when applying both (3) and (7) one sees with sufficient accuracy that

$$\mathrm{C} \propto \frac{1}{\Delta R} \times \frac{\Delta\Phi(v)}{k_B T}\ . \tag{10}$$

This means that C is essentially determined by a product involving two contributions: a certain curvature-like term, $\kappa = 1/ \mid \Delta R \mid$, and some dimensionless energetic argument, $\varepsilon_E = \mid \Delta\Phi(v) \mid / k_B T$. The above claimed high-temperature limit, with the cross-over temperature T_{pass} as a reference temperature characteristic of a system of interest, would naturally demand $0 < \varepsilon_E \ll 1$ while, because of approaching the mature growing stage any change in the cluster radius must be small, $0 <\mid \Delta R \mid\ll 1$, and therefore, its inverse would tend to some big value, i.e. $\kappa \gg 1$. Thus, C will take on a finite value. It is believed that for certain agglomerations under readily high-temperature conditions it will eventually acquire a small value[5], that means, $0 < \mathrm{C} \ll 1$ naturally holds. It is a case when the potential

$$\Phi(v) \propto v^{1-\alpha} \simeq v^{1/d}\ . \tag{11}$$

In [16] a condition of setting the current equal to zero, $J = 0$, has been chosen to balance diffusive and non-diffusive terms in the F-P-K type description, cf. [10], aiming at getting a proper behaviour of the metastable nanostructure [17] arrays. We are of the opinion that such a proposal is legitimate in the relatively low-temperature domain. When the temperature is raised, but agglomeration is still allowed to occur, the proposal may fail. Thus, the above is a possible solution for the high-temperature limit. A type of localization of the Gaussian distribution, characteristic of the inverse ripening (a metastable state of the nanostructure evolution) can also be obtained within the present modelling, cf. [8]. This is the case of (8) when in a (readily) mature growing stage, since the single volume v of the cluster does not change much. As a matter of fact, there is no small-cluster population available for merging (Fig. 1), i.e. $D(v) \rightarrow$ *const.*, which nearly corresponds to the high-temperature criteria of (9), or equivalently (10). In so doing, (8)

[5]Such a belief comes undoubtedly from the fact that we offer our approach for systems evolving in an overdamped regime, such as those of biopolymeric type. For them the Reynolds number is typically of the order of 10^{-3}, i.e. very low, so that the mobility per se, even for a single biomolecule but also for a molecular cluster, must clearly be of negligible value [13], regardless of whether we measure it in the v–space or, what is usually done, in a position space

represents the 1st Fick law in its standard form. Upon inserting it into (1) one immediately arrives at the 2nd Fick law (in the configurational space) with its standard Gaussian solution, the metastable case being emphasized in [16].

The above potential form (11), designed for loosely-packed agglomeration, seems legitimate here: Note that the 'force' $F_{c-c} \propto \partial\Phi(v)/\partial v$ behaves like

$$F_{c-c} \propto \frac{1}{v^{\alpha}} \simeq \frac{1}{s^{(d)}} , \tag{12}$$

because $v \sim R^d$. Thus, F_{c-c} acts as the inverse of the area of the cluster hypersurface, $s^{(d)}$, which implies that the smaller the area is, the bigger the force acting on the cluster can be, this way impeding the formation of new clusters, which would contribute to an aggregate's density increase. Qualitatively, a similar dependence is found for the closely-packed matter agglomeration: from (6) one gets, as above, for the 'force'

$$F_{c-c} \simeq \frac{1}{v^{(d)}} . \tag{13}$$

Here, F_{c-c} acts as the inverse of the hypervolume of the cluster, $v^{(d)} := R^d$, which makes a clear difference between closely-packed and loosely-packed agglomerations, presumably leading to a certain relaxation of the surface tension conditions for loosely-packed clusters-containing systems [22].

Referring further to (9) and using the similarity relation, (7), one gets

$$\Phi(R) \propto \Phi(R_0)\frac{R}{R_0} , \tag{14}$$

where

$$\Phi(R_0) = \frac{k_B T}{D_\alpha} R_0, \qquad T \geq T_{\mathrm{pass}} , \tag{15}$$

and consequently, $\Phi(R) \propto R$. Moreover,

$$D_\alpha = D(1-\alpha) . \tag{16}$$

R_0 can now be specified to be the initial cluster radius. Note that D_α is a d–dependent quantity.

2.3 Cluster Volume Fluctuations as Reliable Characteristics of Matter Agglomeration

Aggregations and agglomerations emerge in a fluctuating changing medium. Therefore, any reasonable quantitative attempt on resolving the fluctuation impact on their speed is worth examining here. In what follows, let us propose an evaluation of the reduced variance

$$\sigma^2(t) = \frac{\langle v^2(t)\rangle - \langle v^1(t)\rangle^2}{\langle v^1(t)\rangle^2} \equiv \frac{\langle v^2(t)\rangle}{\langle v^1(t)\rangle^2} - 1\,, \tag{17}$$

as a direct measure of the cluster volume fluctuations.

The notation used in (17) refers to the statistical moments

$$\langle v^n(t)\rangle = \int_0^\infty v^n f(v,t)\, dv \qquad n = 0, 1, 2, \dots \tag{18}$$

of the stochastic process, where the matter agglomeration is usually described by the local continuity equation, (1).

The explicit solutions, $f(v,t)$–s, have been presented elsewhere [8, 9, 20], and Refs. therein. The zeroth moment, $\langle v^0(t)\rangle$, is related to the average number of molecular clusters in the system, and usually shows an algebraic decrease with time [8]. The first moment, $\langle v^1(t)\rangle$, is related to the total volume which is a constant value for closely-packed agglomerations [22] and an increasing function of time for loosely-packed agglomerations [8], cf. Fig. 1. From the expressions of both moments, it follows that the average cluster radius, $R_{\text{av}}(t)$, behaves as a power law in time, with a growth exponent $1/(d+1)$ that apparently contains some signature of random close-packing of matter by having included the super-dimension $d+1$ [9, 23]. ($d+1$ tells us something about the minimum number of non-overlapping neighbours of a given cluster in a d–dimensional space.) These constitute the main characteristics of the model agglomeration/aggregation process in its late-stage ($t \gg 1$) limit.

The question remains about asymptotic values of the moments $< v^n(t) >$ that must be known when applying formula (17). For closely-packed agglomerations, the moments are found to obey a power law [22]

$$\langle v^n(t)\rangle \sim t^{(n-1)/(2-\alpha)} \qquad (n = 0, 1, 2), \quad t \gg 1\,, \tag{19}$$

whereas for matter aggregation one finds another power law [8]

$$\langle v^n(t)\rangle \sim t^{[(n-1)+\alpha]/(2-\alpha)} \qquad (n = 0, 1, 2), \quad t \gg 1\,. \tag{20}$$

Notice, that for $\alpha = 0$ ($d = 1$) both power laws above approach the same form, namely $\langle v^n(t)\rangle \sim t^{(n-1)/2}$. When utilizing (17) and (19) it appears that for closely-packed agglomerations, $\sigma^2(t)$ can be fully identified with the inverse of $\langle v^0(t)\rangle$ (the average number of clusters), cf. [22] for details, what because of the constancy of $\langle v^1(t)\rangle$, leads to $\sigma^2(t) \propto V_{\text{sp}}(t)$, where $V_{\text{sp}} \equiv V_{\text{sp}}(t) \simeq \langle v^1(t)\rangle / \langle v^0(t)\rangle$, and can be termed the mean specific volume of the tightly-packed agglomerate, being equivalent to the inverse of its mean number density. The specific volume fluctuations read

$$\sigma^2(t) \propto t^{d/(d+1)}\,, \tag{21}$$

and if $d \to \infty$, $\sigma^2(t) \simeq V_{\text{sp}} \propto t$.

When using (17) and (20), however, it turns out that for loosely-packed agglomerations $\sigma^2(t)$ is a quantity equivalent to the average cluster radius $R_{\rm av}(t)$, see [8,9]. They behave in time as

$$\sigma^2(t) \propto t^{1/(d+1)} \,. \tag{22}$$

When $d \to \infty$, $\sigma^2(t) \simeq R_{\rm av}(t) \to const$, which means, that on average the system ceases to grow. Note that the standard diffusional regime, is always characterized by the one-half exponent, is achieved exclusively for the aggregations in $d = 1$ because the only linear characteristic is $R_{\rm av} \equiv R_{\rm av}(t)$: Note that $V_{\rm sp}(t)$ is not a linear characteristic, since $V_{\rm sp}(t) \propto [R_{\rm av}(t)]^3$ usually holds. Here the $d = 1$–case must clearly be disqualified as standard diffusional, cf. (21).

Commenting on the last relations, (21) and (22), one might furthermore conclude that they reflect a well-known Onsager conjecture that the fluctuations in a system undergo the same type of changes as the corresponding macroscopic dynamic variables [4]: Here one may think of the specific volume of the agglomerate and the grain radius, and their behaviour in the late-time domain, respectively.

2.4 Coupling the Instability (Growing) and Mechanical Stress Relaxation Modes of Matter Agglomeration

Poisson was likely the first who recognized that viscoelastic properties of fluids and solids can reasonably be compared in a suitable, mostly short-time domain, though the specification of the domain must be more precise for specifying the systems of interest. Maxwell successfully followed the ideas of his famous French predecessor, arriving at his well-known, in general non–Markovian, model of relaxation [2]. In what follows we present our Maxwell–model-based ideas on how to distinguish between the two agglomerations under study, and how to switch on a kind of coupling between the (late-stage) growing and relaxational modes in the viscoelastic d–dimensional matrix that we investigate. The existence of the coupling seems to be experimentally justified, see [5,15,16], and involves generically the viscoelastic nature of the mega-cluster late-stage formation [2].

Thus, the afore presented rationale toward quantifying the fluctuations of the system can be strengthened with a supporting phenomenological argumentation. The idea comes from a "coupled" diffusion-relaxation picture that appears in such a complex system. In any diffusion-migration growing process, the mechanical strain-stress fields play such a role as well. In our case, such a situation can be safely expected in the temperature domain $T \le T_{\rm pass}$. Another type of relaxation of the stress field, say σ_m, is expected to prevail when the closely-packed agglomeration conditions are met. A different behaviour may be observed when the closely-packed agglomeration conditions are lost for the first time, that is, at $T = T_{\rm pass}$, when the loosely-packed context appears. In both temperature regimes, the relaxation of $\sigma_m(t)$ over the course of time, is very likely to go in a way essentially described by the current (2). This is expected to occur [22] presumably under (nearly) homogeneous strain conditions, $\varepsilon_m \approx const$, for $t \gg 1$. For an additional motivation of coupling matter agglomeration and stress relaxation picture, related to fracture phenomena, see [1, 22].

From [22] it can be learned, that in the absence of non-Arrhenius or fractal type kinetics, seemingly modifying the diffusion coefficient $D(v)$ [20], one expects the Maxwell dashpot-and-spring model to reflect properly the relaxation behaviour. We wish to set up here a phenomenological picture, showing that both agglomeration and mechanical stress relaxation, where the stress relaxation takes place under slow growth conditions, proper of a mature growing stage in a viscoelastic multiphase medium [13,15], are coupled processes [5, 16]. To work out the problem quantitatively, we will represent one of the two contiguous and matter-exchanging clusters in the agglomerate, say cluster (grain) 1, as an expanding one, equivalent to the spring, growing at the expense of its neighbour, to be named cluster (grain) 2, i.e the dashpot, to which, according to the Maxwell model, the contracting action should be assigned, cf. Fig. 2 for details; see [22].

For the system with non-wide gaps, the Maxwell model conditions are almost satisfied, so that the two-cluster action can be extended over all pairs of contiguous clusters until the expanding (growing) eventually survive. In a next step, the same kind of competition appears as in the well argumented Laplace–Kelvin–Young scenario suitable for cellular systems [9]. This picture holds in the closely-packed context.

In the loosely-packed context (a system with wide gaps), we may have qualitatively almost the same picture [2] but with several differences which implies that cluster expansion would not be likely so vigourous. Since the corresponding gap is wider, therefore untight, the fluid leakage might be more pronounced. Thus, the fluid response against the piston wall is weaker, and the Maxwell type stress relaxation no longer applies, cf. the caption of Fig. 2.

The stress relaxation can be described by introducing an exponent χ in the Maxwell-like quasi-fractional model presented here below. This exponent should be, in general, d-dependent, and points to a difference when comparing with the classical Maxwell model [2]. Here, we offer a coupled matter diffusion and stress relaxation picture, but for a random walk performed in the configurational space [22]. As is known, the Maxwell stress relaxation picture leads to an exponential decay of the stress:

$$\sigma_m(t) \sim \exp(-t/\tau_M) \; , \tag{23}$$

where τ_M is a reference time for the concentrated clusters [2] to be eventually inferred from the Einstein–Stokes-like formula [13, 22]. This behaviour holds for $T < T_{\text{pass}}$. As mentioned above, for $T \geq T_{\text{pass}}$ we propose

$$\frac{\mathrm{d}\sigma_m}{\mathrm{d}t} + \frac{\sigma_m^{\chi}}{\tau} = 0 \; , \tag{24}$$

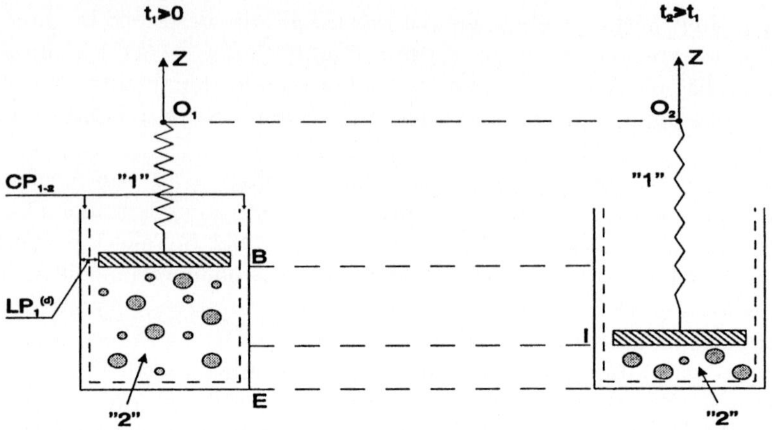

Fig. 2. Maxwell sequential spring-and-dashpot (quasi-fractional) model with narrow (d–independent) and wide (d–dependent) gaps, shown schematically in two subsequent time instants t_1 and t_2, where $t_2 > t_1 > 0$, from left to right, respectively. Grain "1" consists of the spring and the piston's upper wall, to which the second end of the spring is attached, while its first end is mounted either on 0_1 or 0_2, from left to right, respectively. Grain "2" consists of the viscous medium inside the cylinder as well as the inner wall of the piston. The cylinder's walls complete the overall model structure of the viscoelastic grains. The material exchange between "1" and "2" is assured by the existence of the gaps: narrow CP_{1-2} gaps in case of closely-packed agglomeration, and some two wider (here, represented by the left-hand side gap, $\mathrm{LP}_1^{(d)}$) in case of the aggregation. Therefore, the piston-and-cylinder system, containing a viscous fluid, here composed of big and small particles, is either more (densely-packed agglomeration) or less (undensely packed formation) leakproof. The overall material exchange is caused by spring expansion along z axis, which results here in a longitudinal expansion of grain "1" at the expense of grain "2", cf. *http://www.j-npcs.org/abstracts/vol2000no4.html.* Notice formally that: $\mathrm{O_1E} = \mathrm{O_2E}$, and for t_1 one has $z_1(t_1) = \mathrm{O_1E}$, $z_2(t_1) = \mathrm{BE}$ as well as for t_2 one gets $z_1(t_2) = \mathrm{O_2I}$, $z_2(t_2) = \mathrm{IE}$, which results in grain expansion-contraction behaviour, like $\mathrm{O_1B} < \mathrm{O_2I}$ and $\mathrm{BE} > \mathrm{IE}$, when mutually comparing the distances along the z axis at t_1 and t_2, respectively

where the above is usually true when the internal strain field, ε_m is practically constant, see above. When solving (24), one obtains

$$\sigma_m(t) \sim (t/\tau)^{-1/(\chi-1)}, \qquad t \gg 1 , \tag{25}$$

where $\chi = 2d + 3$; about τ, see [22] or [13]. Notice that for $\chi = 1$ and $\tau = \tau_M$ in (24) one gets the solution (23); for $\chi \neq 1$ (25) is the only solution to the relaxational problem as stated. When comparing (23) and (25) one sees that the relaxational response goes slower for the late-time loosely packed aggregational context than for its densely-packed agglomerational counterpart.

3 Qualitative Signatures of Chaos in Matter–Agglomerating System

Let us consider a few qualitative signatures of chaos in matter–agglomerating systems from the literature. Such certain signatures for systems of the type studied in the present work are summarized in Table 1.

Table 1. Qualitative signatures of chaos in a model matter–agglomerating system of interest

Item	Signature	Refs.
One:	Entropic system of molecular-chaotic behaviour	[13, 14, 24]
Two:	Lack of matter - depletion zones around the charged clusters	[13, 22, 25]
Three:	Competition-and-loss effect: (un)tight spring-and-dashpot model	[2, 23, 26]
Four:	IBCs of normal (e.g., Neumann) or abnormal type should be taken appropriately	[6, 20, 21]
Five:	Bethe-type frustration in coupled relaxation and late-growing events	[1, 23, 26]
Six:	Growth viz. instability: random close-packing with its $d+1$–account	[9, 23, 24]
Seven:	Nonequilibrium entropy measures viz. mean-harmonic speeds	[2, 14, 28]
Eight:	Entropic potential(s) assuring nonequilibrium character of the phenomenon	[2, 9, 14]
Nine:	G-K type construction of $D(v)$, $B(v)$, and its consequences	[2, 14, 27]
Ten:	Diffusion-space pre-chaotic (Fibonacci) feature by D_α	[8, 14, 28]

The items stated in Table 1 do not exclude other possible forms to chaos, or its signatures, in matter-agglomerating systems. We do not pretend to describe all of them, or even their majority. For routes to chaos recommended from physical point of view one would usefully consult [27, 28]; which routes, or scenarios of chaos, are recommended by mathematicians, especially when a partial-differential-equation formalism of F-P-K type is effective, can be found in [6], and in Refs. therein.

4 Some Quantitative Measures of Chaos Signatures in Matter–Agglomerating System

In [22] some entropic-like nonequilibrium measures of growth

$$\nu_{\rm sp}^{(d)} = \left(\frac{\ln\left[\sigma^2(t)\right]}{\ln t}\right)_{\text{for } t\gg 1}, \quad d=1,2,3,\dots \qquad (26)$$

as well as for the mechanical stress relaxation evolution

$$\mu_{\mathrm{sp}}^{(d)} = \left(\frac{-\ln[\sigma_m(t)]}{\ln t} \right)_{\text{for } t \gg 1}, \quad d = 1, 2, 3, \dots \tag{27}$$

have been proposed. This seems to be working most appropriately in a growth-and-relaxation synchronization metastable regime

$$\sigma_m \sim R_{\mathrm{av}}^{-1/2} \sim \sigma^{-1} , \tag{28}$$

which represents the H-P-G condition [3]

$$\sigma_m \sim R_{\mathrm{av}}^{-1/2} , \tag{29}$$

appropriate for the fluctuational late-time regime [13] of interest here. Bear in mind that if certain empirical modifications of the formula (29) are applied toward obtaining a specific form, interconnecting (26) with (27), one gets something like

$$\nu_{\mathrm{sp}}^{(d)} = q \mu_{\mathrm{sp}}^{(d)} , \tag{30}$$

where typically $q > 2$. In the classical H-P-G limit $\nu_{\mathrm{sp}}^{(d)} = 2\mu_{\mathrm{sp}}^{(d)}$ holds. However q may also strive for obtaining superplastic effects, i.e. when taking on fractional values, cf. [9], and Refs. therein. This is sometimes termed in physical-metallurgical literature the inverse H-P-G effect.

Because of (22)

$$\nu_{\mathrm{sp}}^{(d)} = \frac{1}{d+1}, \quad d = 1, 2, 3, \dots . \tag{31}$$

Realize that formulae (28) and (29) might again be interpreted in terms of the Onsager conjecture [2, 4], see above. Since the overall exponent in (25) reads

$$\frac{1}{\chi - 1} = \frac{1}{2(d+1)} , \tag{32}$$

which is exactly one half of the growth exponent $\nu_{\mathrm{sp}}^{(d)}$ given in (22), see (31) too, one consequently provides

$$\mu_{\mathrm{sp}}^{(d)} = \frac{1}{2(d+1)}, \quad d = 1, 2, 3, \dots \tag{33}$$

Let us emphasize here that $1/(\chi - 1)$ stands for the so-called Nutting exponent for relaxation, and can be interpreted in terms of the loss tangent, that means, a well-known dissipation factor in the relaxation phenomena, mostly in dielectric (e.g., macromolecular) environments [22].

5 Number-Theoretic Measures of Spatial and Temporal Irregularities in Aggregation-Agglomerating Systems

It is interesting to note here that $\chi = \chi(d)$, i.e.

$$\chi(d) = 2(d+1) + 1 . \tag{34}$$

A certain generator of the Bethe–lattice elements, is recovered starting from the 3-bond (initial) generator for $d = 0$, and continuing with d, upon identifying d as the numbers of emerging bonds in a gelling system [5]. This is a very useful tool for the mean-field description of gels, and other multi bond-containing systems.

In this way, an odd number Bethe–lattice generator for subsequent d–s can be offered, see Fig. 3.

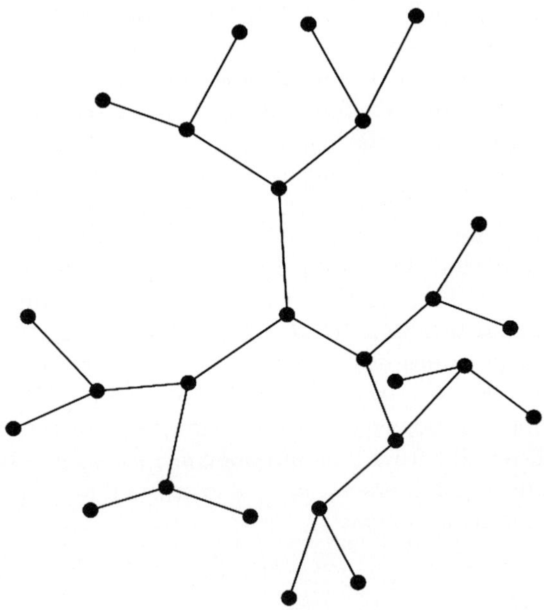

Fig. 3. An example of a not much developed Bethe lattice, which by itself manifests a frustration because of "having problems" with containing all of its nodes in the available d–dimensional space [26], to some extent so as, for example, the population of Verhulst fellow countrymen does in the available Belgium territory [29]

Another, equally interesting observation can be offered, namely

$$\frac{2}{\nu_{\rm sp}^{(2)}} = \frac{1}{\nu_{\rm sp}^{(1)}} + \frac{1}{\nu_{\rm sp}^{(3)}} . \tag{35}$$

This means that for loosely-packed agglomerations the harmonic-mean rule for the growth speed is exactly fulfilled. *Mutatis mutandis*, one can expect the same type of rule, (35), for $\mu_{\rm sp}^{(d)}$. Let us recall that the fluctuations $\sigma^2(t)$ have been proposed as a reliable criterion of differentiating between aggregation and agglomeration, and that an efficiency (harmonic-rule, see (35)) additional criterion, derived from the constructed fluctuational proposal, supports the aggregation in dimension d ($d = 1, 2, 3$), with an emphasis placed on $d = 2$, where 'golden-ratio-like' or harmonic-mean properties are in favor. The mean-harmonic speed implies that the center of mass of a moving body, referred consequently to as the molecular cluster, may not span the same distance, say s, back and forth, during a time period. This leads to a quite realistic quantification of a mean speed on the distance $2s$, and also shows that not an arithmetic mean of the back and forth speeds but a harmonic mean applies here. Such a schedule can likely be extended over the aggregation that essentially relies on random matter attachments and/or detachments of particles [8, 25], in such a way a forward sub-process may essentially go unidimensionally, like in a ballistic motion, whereas its reverse counterpart would explore the whole three-dimensional domain. This situation typically appears in the case of matter desorption, in which detachment occurs part by part from a "reactive" surface spot.

Some other confirmation of (mean) harmonicity, its close relation to the golden rule, and to the Fibonacci sequencing (characterizing well the fractality of "diffusive" microstructures), is hidden in the (macro)ion (or, cluster [22]) diffusion coefficient D_α, in our model, (16), which is also included in the free energy Φ. The label, or the lower index, is simply α, which for $d = 1$ results in $\alpha = 0 \equiv 0/1$, for $d = 2$ gives $\alpha = 1/2$, whereas for $d = 3$ offers $\alpha = 2/3$. The values of the diffusion coefficient (16) are: $D_0 \equiv D/1$, $D_{1/2} \equiv D/2$ as well as to $D_{2/3} \equiv D/3$, respectively. They correspond to the first five-number Fibonacci sequence, composed of the numerators and denominators of α-s, like 0,1,1,2,3, and obey $a_{n+2} = a_{n+1} + a_n$, for the three subsequent Fibonacci numbers a_n, a_{n+1} and a_{n+2}. If so, one can provide the following two recursive formulae

$$\alpha^{(d-1)} = \frac{a_{d-1}}{a_d}, \qquad d = 1 \, , \tag{36}$$

and

$$\alpha^{(d-1)} = \frac{a_d}{a_{d+1}}, \qquad d = 2, 3 \, , \tag{37}$$

where $a_0 = 0$, $a_1 = 1$, $a_2 = 1$, $a_3 = 2$, $a_4 = 3$ are the first five Fibonacci numbers. Since the analogy with gelling systems seems evident [22], this cannot be taken entirely as a surprise. The bonding in gels clearly goes as a branching process, being (as in the case of ultrametric space) quite naturally described geometrically in terms of Fibonacci numbers, thereby involving the notion of fractality [5].

When finishing this section, let us note that both the characteristic chaotic measures, cf. (31) for example, have their random close-packing account $d+1$ involved. This is a landmark of randomness but readily appears as a space-filling action of modelled matter reorganisations. Realize that our rationale may apply just in the same vein to cluters-containing assemblies, evolving in a d–dimensional space, – where a cluster is characterized by its fractal dimension $0 < d_F < d$, cf. [9, 22].

6 Chaos in an Infinite-Dimensional Agglomerating and/or Aggregating System

Consider the case $\lim_{d\to\infty}$. A corresponding chaotic measure for the late-stage growing event in the agglomeration of matter, very reminiscent of nonequilibrium correlational entropy measure [2], reads

$$\nu_{\mathrm{sp}}^{(\infty)} = \lim_{d\to\infty} \left(\frac{\ln[\sigma^2(t)]}{\ln(t)} \right)_{\text{for } t\gg 1} , \tag{38}$$

whereas its counterpart for the relaxation is given by an analogous formula, namely that

$$\mu_{\mathrm{sp}}^{(\infty)} = \lim_{d\to\infty} \left(\frac{-\ln[\sigma_m(t)]}{\ln(t)} \right)_{\text{for } t\gg 1} , \tag{39}$$

holds. They are consistent formally with the so-called correlational entropy (Kolmogorov–type) measure, defined in [28] and follow the rationale presented in [6], in which some measures of chaos in dynamical systems described by partial differential equations have been discussed. For "thermostatic" systems out of equilibrium one has to speak of the so-called generalized fractal dimension formalism, first introduced by Grassberger and Proccacia, see [28], and Refs. therein.

The most attractive reason for introducing such measures arises from the fact that if one evaluates both $\nu_{\mathrm{sp}}^{(\infty)}$ and $\mu_{\mathrm{sp}}^{(\infty)}$, one unambiguously gets for the aggregation

$$\nu_{\mathrm{sp}}^{(\infty)} = \mu_{\mathrm{sp}}^{(\infty)} = 0 , \tag{40}$$

whereas for the close-packed agglomeration one provides

$$\nu_{\mathrm{sp}}^{(\infty)} = 1 , \tag{41}$$

and

$$\mu_{\mathrm{sp}}^{(\infty)} = \infty . \tag{42}$$

Thus, for both cases, (41) and (42), one arrives at a chaotic behaviour in the nonequilibrium system [2, 28] of a densely-packed agglomerate. This is not the case of the aggregation for which the common measure of its chaotic character is zero, cf. (40).

Thus, proceeding consistently with the approach offered in [6] we may conclude that the late-time *aggregation* process develops in an ordered way. The case $d = 2$ appears to be the most efficient since the harmonic-mean rule (35) is applied; for it the nonequilibrium character of the random process should be emphasized [30]. It is intriguing to realize that the system property called the harmonicity throughout is very much related to its nonequilibrium entropic or chaotic characteristic(s).

7 Concluding Address

Based on the above, we are allowed to state the following:

(i) in matter-agglomerating systems chaos is revealed as a complex spatio-temporal and temperature-dependent phenomenon;
(ii) nonequilibrium chaotic measures of any late-stage matter agglomeration modelled can be proposed relying upon the nonequilibrium Kolmogorov-type entropy measure, which makes a reliable (harmonic) quantification of the tempo of the process;
(iii) coupling late-stage matter agglomeration with relaxation of assisting elastic fields via an Onsager-type [4], or, in the parlance of physical metallurgy, H-P-G conjecture [3], leads to several characteristic sub-effects (Bethe-lattice generator, first-five Fibonacci-number signatures, random close-packing $d+1$–criterion [23], etc.) having their rationale in fundamental properties of the entropic or harmonic-mean character of the phenomenon[6];
(iv) as for the formal point of view: The presented mesoscopic system, Sect. 2, serving to describe the matter aggregation can be derived rigorously based on the Gibbs entropy production equation [9, 10, 22], and
(v) its chaotic signatures can be inferred as presented in Sects. 3–6, supported somehow by the ideas contained in [6]; at this point, a general task remains to be done as to connect the type of chaos with the entropy-based scheme [10, 22] used to derive the equations of F-P-K [11], or diffusion, types [8, 9, 27], and how far the proposed measures of chaos (also, the ones used in the present review) are reminiscent of those used conventionally in nonlinear science [2, 14, 27, 28, 30]? Perhaps, the Edwards' entropy measures for slowly moving grains, evolving (bio)polymer- or colloid-type matrices and compacted powders could also contribute to solve the problem [32].

---

[6] As can be for example observed in clays made of an inorganic material known as laponite [31]

Acknowledgement

One of us (A.G.) dedicates this study to Prof. Peter Laggner, Ö.A.W., Graz, Austria, and Prof. Gerard Czajkowski, U.T.A. Bydgoszcz, Poland, in the year of their 60th anniversary. This work is done under 2P03B 03225 (2003-2006) by A.G. Part of M.A. works is in the framework of an Action de Recherches Concertée Program of the University of Liège (ARC 02/07-293). Thanks go to Prof. Peter Richmond (Dublin) for deciding to support the participation of A.G. in the meeting "Verhulst 200 on Chaos", 16–18 September 2004, Royal Military Academy, Brussels, Belgium from the funds of COST P10. Last but not least we are thankful to Prof. Jerzy Łuczka (Katowice) for drawing our attention to Ref. [16] and Prof. Miguel Rubí (Barcelona) for useful comments on the manuscript.

References

1. M. Ausloos: Solid State Commun. **59**, 401 (1986)
2. D. J. Evans, G. P. Morriss: *Statistical Mechanics of NonEquilibrium Liquids* (Academic Press, London 1990)
3. P. O. Hall: Proc. Roy. Soc. B **64**, 747 (1951); N. J. Petch: Phil. Mag. **1**, 186 (1956); A. A. Griffith: Phil. Trans. Roy. Soc. (London) A **221**, 163 (1920)
4. L. Onsager: Phys. Rev. **37**, 405 (1931)
5. S. Manley et al.: Phys. Rev. Lett. **93**, 108302-1 (2004)
6. R. Rudnicki: Math. Meth. Appl. Sci. **27**, 723 (2004)
7. B. H. Kaye: *A Random Walk Through Fractal Dimensions* (VCH Verlagsgesellschaft, Weinheim 1989)
8. M. Niemiec, A. Gadomski, J. Łuczka, L. Schimansky–Geier: Physica A **248**, 365 (1998)
9. A. Gadomski, J. M. Rubí: Chem. Phys. **293**, 169 (2003)
10. J. M. G. Vilar, J. M. Rubí: Proc. Natl. Acad. USA **98**, 11081 (2001)
11. G. M. Zaslavsky: Phys. Rep. **371**, 461 (2002)
12. K. Pichór, R. Rudnicki: J. Math. Anal. Appl. **215**, 56 (1997)
13. I. Derényi: Stochastic processes. In: *Fluctuations and Scaling in Biology* chap. 2.2, ed. by T. Vicsek (Oxford University Press, Oxford 2001) pp. 27–31
14. J. R. Dorfman: *An Introduction to Chaos in Nonequilibrium Statistical Mechanics* (WN PWN Warsaw 2001, *in Polish*) Chap. 6.2
15. J. M. Rubí, I. Santamaría–Holek, A. Pérez–Madrid: J. Phys. C **16**, S2047 (2004)
16. D. E. Jesson, T. P. Munt, V. A. Shchukin, D. Bimberg: Phys. Rev. Lett. **92**, 115503-1 (2004)
17. L. Silvestri, G. Czajkowski, F. Bassani: J. Phys. Chem. Solid. **61**, 2043 (2000)
18. N. Vandewalle, B. Delisse, M. Ausloos, and R. Cloots: Phil. Mag. B **78**, 397 (1998)
19. P. Laggner, M. Kriechbaum: Chem. Phys. Lipids **57**, 121 (1991)
20. A. Gadomski: Phil. Mag. Lett. **70**, 335 (1994)
21. A. Gadomski, J. Łuczka, R. Rudnicki: Physica A **325**, 284 (2003)
22. A. Gadomski, J. M. Rubí, J. Łuczka, M. Ausloos: Chem. Phys. *in press* (2004)

23. R. Zallen: *The Physics of Amorphous Solids* (John Wiley & Sons, New York 1983) Chap. 2
24. J. Schmelzer, G. Röpke, R. Mahnke: *Aggregation Phenomena in Complex Systems* (Wiley–VCH, Weinheim 1999)
25. V. J. Anderson, H. N. W. Lekkerkerker: Nature **416**, 811 (2002)
26. W. Przygocki, A. Włochowicz: *Polymer Physics* (WN PWN, Warsaw 2001, *in Polish*) Chap. 4
27. V. S. Anishchenko, V. V. Astakhov, A. B. Neiman, T. E. Vadivasova, L. Schimansky–Geier: *Nonlinear Dynamics of Chaotic and Stochastic Systems. Tutorial and Modern Developments* (Springer-Verlag, Berlin 2002) Chap. 2
28. H. G. Schuster: *Deterministic Chaos. An Introduction* (VCH Verlagsgesellschaft, Weinheim 1988) Chap. 5.2
29. V. A. Kostitzin: *Biologie Mathématique* (Librairie Armand Colin, Paris 1937)
30. U. Erdmann: Kollektive Bewegung. PhD Thesis, *in German*, Humboldt University of Berlin, Berlin (2004)
31. E. Olivier, E. Pefferkorn: Colloid Polym. Sci. **279**, 1104 (2001)
32. A. Barrat, J. Kurchan, V. Lorreto, M. Sellito: Phys. Rev. Lett. **85**, 5034 (2000)

A Chaos and Fractal Dynamic Approach to the Fracture Mechanics

Lucas Máximo Alves[1] and Rui F.M. Lobo[2]

[1] GTEME - Grupo de Termodinâmica, Mecânica e Eletrônica dos Materiais, lmalves@uepg.br,
Departamento de Engenharia de Materiais, Universidade Estadual de Ponta Grossa - Caixa Postal 1007, Av. Gal. Carlos Calvalcanti, 4748, Campus UEPG/Bloco L - Uvaranas, Ponta Grossa, Paraná,CEP. 84030.000 Brasil.

[2] Grupo de Nanotecnologia e Ciência a Nano-Escala, rfl@fct.unl.pt,
Departamento de Física, Faculdade de Ciências e Tecnologia da Universidade Nova de Lisboa, 2829-516 Caparica – Portugal.

Summary. It is shown that the onset of instabilities observed in the fracture of brittle isotropic materials is a consequence of the mathematical structure of chaos that underlies such phenomena. The straight line crack velocity is written in the form of a logistic map explaining the onset of instabilities observed by Fineberg et al. This approach provides a single and concise tool to study this and other nonlinear aspects presented by dynamic crack growth.

1 Introduction

Dynamic fracture has been stimulating a growing interest not only because of its fundamental importance in understanding fracture processes but also because of the challenges to mathematical analysis and experimental techniques.

Fineberg and co-workers performed experiments on fast crack growth [1,2] for different brittle materials, such as PMMA, soda-lime glass, etc., revealing many new aspects, defiant to the fracture dynamic theory, related to unstable crack growth. They observed the existence of a critical velocity starting from which instability begins and measured the correlation between the fluctuations in crack growth velocity and the ruggedness of the generated surfaces. They found a time delay (of the order of the stress relaxation time) present during the whole fracture process in the materials used by them. In spite of this experimental observation that can be associated to stress relaxation properties in viscoelastic materials, they were unable to relate it physically to the onset of the instability phenomenon itself. Therefore the need for a correct mathematical description of the instability process in crack growth represent one of the interesting challenges in dynamic fracture.

On the other hand, Kostrov–Nikitin [3] and Christensen [4] modelled the quasi-static crack growth case in viscoelastic materials showing that the stress field at the crack tip does not depend on the relaxation properties of the material but that it is determined by the instantaneous elastic modulus at the crack tip. Similarly, analogous considerations have been taken and extended to dynamic case [6, 7, 7]. Analyzing this last situation Freund [8] admits that there is really a paradox because the process zone in front of the crack tip depends strongly on the loading conditions and usually ideal boundary conditions are considered, such as a thin crack tip describing a punctual fracture in a short range effect in the obtaining of asymptotic solutions to the problem. Therefore, such paradox is solved by introducing a separation zone of finite extend in the form of a cohesive flaw zone of some length, say l_o, ahead of the moving crack tip [8, 9] to separate the crack formed from the instantaneous process at the crack tip. Such cohesive zone accompanies the crack growth process at the crack tip, while the crack opens gradually against the resistance of some cohesive stress within this zone. The time, t, required for a crack tip to advance a distance equal to l_o, as the crack grows at crack growth velocity, v_o, introduces a process time, given by $t = l_o/v_o$, that must be compared to a characteristic relaxation time, $\tau \sim t$, of the material to determine whether the process is "fast" or "slow" [8].

Actually, if it is considered that the relaxation time, τ, is much smaller than the time, t, taken by the process zone to move forward the crack, the zone is stable and the relaxation effects don't affect the process as it is shown for the quasi-static case ($v \to 0$) [3] by the results of Kostrov–Nikitin and Christensen [4]. But for fast crack growth the relaxation time τ is very large compared to the time, t, taken by the process zone to move a distance l_o. This happens when the crack growth velocity tends to the Rayleigh waves velocity of the material. In this case the relaxation effects become important and the crack becomes unstable as shown experimentally by Fineberg and co-workers [1, 2] by means of the correlation measured between the fluctuation in crack growth velocity and the ruggedness of the generated surfaces. In these correlation measurements they found a time delay between the oscillations in the crack growth velocity and the surface profile of approximately 3 µs and 1.0 µs for PMMA and soda-lime glass respectively. The critical velocities at which the instability begins are 340 m/s and 1100 m/s, respectively, and both values correspond to 0,34 of the respective Rayleigh surface wave velocity, c_R, in these materials, whose values are 975 m/s and 3370 m/s for PMMA and soda-lime glass respectively. This means that at these velocities the characteristics lengths of the process zone is of the order of $l_o = 3\,\mu\text{s} \times 340\,\text{m/s} \sim 1{,}02\,\text{mm}$ and $l_o = 1.0\,\mu\text{s} \times 1100\,\text{m/s} \sim 1.10\,\text{mm}$ for the two materials cited above, also agreeing with other experimental results found by Fineberg and co-workers.

In view of these results there are two alternatives to be considered: either the fast crack growth process is described by using the stress field modelled with a retardation time that takes into account the effects of relaxation of the material, or the mathematical description of a process zone is used that keeps all the information of delay produced by the relaxation of the material, leaving unsolved the instability process, and the previous case is ignored. Historically the latest option has been the solution proposed by Irwin and other scientists to describe the complex effect of the plasticity and viscoelasticity at the crack tip [10–12]. It is not a surprise that, when intending to study a Mechanics of the Fracture that can be universally applied to any kind of material and to any mode of loading, it should receive another formulation.

In this chapter, in order to explain Fineberg et al. results [1, 2] one makes use of a physical model in which the retardation time is introduced in an explicit way generalizing the classical formulation of Fracture Mechanics. It will also be shown that this mathematical procedure will include the fracture process inside the family of phenomena described by nonlinear dynamic processes with the advantage of using the whole mathematical development so far accomplished for these processes and the classic formulation of Fracture Mechanics generalized by the explicit inclusion of time retardation.

2 Theoretical Development of a Chaotic Model to Dynamic Fracture

2.1 The Fast Crack Growth in the Fineberg–Gross Experiments

Consider a semi-infinite plane plate under Mode-I loading and plane strain in elastodynamic crack growth conditions, as shown in Fig. 1. The experimental configuration of the body under testing is equivalent to a infinite plate condition to avoid the edges effects on the crack growth process. In this experiment a fast crack growth develops as a result of a high loading rate and a high strain rate in the dynamic fracture process. In agreement with

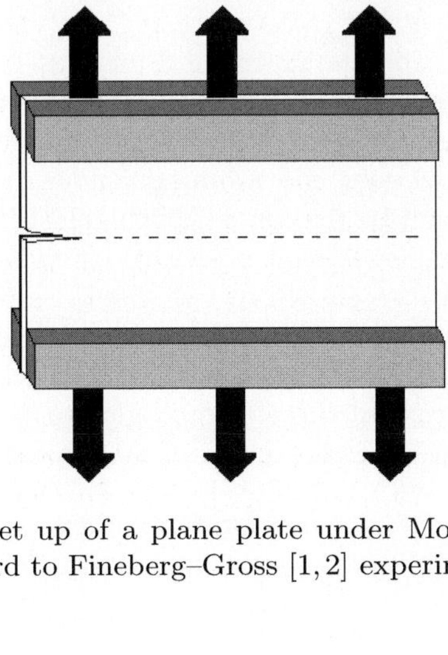

Fig. 1. Experimental set up of a plane plate under Mode I loading in dynamic fracture process in accord to Fineberg–Gross [1, 2] experiments

Fineberg–Gross [1, 2] the aspect of the fracture surfaces is rugged and the crack growth velocity shows a instability from a critical velocity.

In their experiments [1, 2] Fineberg–Gross showed that when the crack speed reaches a critical value a strong temporal correlation between the velocity, $v_o(t)$, and the response in the form of the fracture surface at $A_o(t+\tau)$ takes place (having its notation changed, in the present text, to $L_o(t)$ instead of $A_o(t)$ to designate the fracture surface length). The time delay measured between this two magnitudes presents a value τ of about $\simeq 3\,\mu$s for PMMA and $1.0\,\mu$s for soda-lime glass, for example, showing that there is a given value for each material.

2.2 The Origin of the Time Delay from Fineberg–Gross Experimental Evidences

Crack growth is a complex process that involves many physical phenomena (bond breaking, sound and light emission, heat generation, etc.) which participate in the crack growth process. Also, the relative importance of each one depends on the conditions under which crack growth takes place. It is well known that for any given material a certain time delay exists between the loading and the crack growth, at the beginning of the crack growth, in experiments performed under time-dependent loading conditions [8]. Moreover, this time delay tends to disappear with the crack growth [8]. However, since dynamic fracture involves heat generation it is necessary to take into account the fact that a viscoelastic material such as the one used by Fineberg–Gross [1, 2] shows viscoelastic relaxation phenomena. This is so because the heat developed at the crack tip might have changed the local properties of the material making the hypothesis of viscoelastic relaxation very plausible. In the case of a viscoelastic material, such as PMMA, if a considerable amount of viscoelastic material is formed at the crack tip (size of the order of $l_o \sim v_o\tau$), the time delay appears locally remaining because of the existence of the persistent creep phenomenon that takes place at fast dynamic crack growth, as it is evident by the correlation shown in the Fineberg–Gross experiments [1, 2]. Therefore, this may explain the phenomenology behind the time delay obtained by Fineberg et al. experiments [1, 2]. Taking into account this strong evidence of time delay between the two magnitudes mentioned above the fast dynamic crack growth can be expressed by the elastodynamic energy release rate, G_{oD}, at a given time such as it does not depend on the crack velocity at the same moment [12–15] as shown in Fig. 2 and as it will be shown below.

The time delay obtained by Fineberg and co-workers [1, 2] attributed to the viscoelastic properties of the material at the crack tip, possibly develops in several steps. In the case of the vitreous (glasses) and polymeric materials the viscoelastic properties of the material at the crack tip become evident because at fast crack growth heat develops at crack tip, and due to the poor heat conduction, temperature rises with possible softening of this material, (in front of the crack tip) as explained above, even at room temperature. In

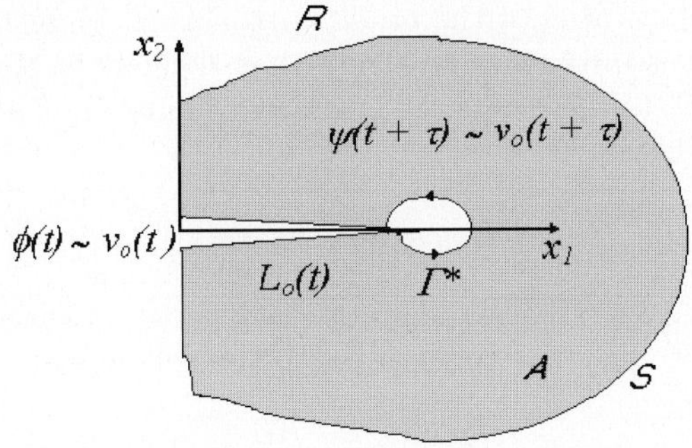

Fig. 2. Schematic crack tip region with time delay between the input flux energy, $\phi(t)$ and output flux energy, $\psi(t+\tau)$

the case of metals, due to higher heat conduction, the viscoelastic properties only appear when the sample is tested at temperatures greater than thirty percent of their absolute melting temperatures [11].

"There are two competing mechanisms involved in the crack growth that characterize the creep deformation. The blunting of the material in front of the crack tip relaxes the crack tip stress field and tends to retard crack growth. The other mechanism results in an accumulation of creep damage in the form of microcracks and voids that enhance crack growth as they coalesce" [11].

2.3 The Foundations of Quasi-Static and Dynamic Fracture Mechanics

For quasi-static crack growth the usual elastic energy release rate, G_o, is given by [12]

$$G_o \equiv \frac{\mathrm{d}(F - U_{Lo})}{\mathrm{d}L_o} , \tag{1}$$

where F is the work performed by external forces on the sample, U_{Lo} is the change in elastic strain energy caused by the introduction of a crack with length, L_o, into the sample. L_o is the distance between two points of the crack (or projected crack length) and the subscript zero denotes the plane projected magnitudes.

The Griffith–Irwin energy balance approach for stable fracture requires that [10, 16]

$$G_o \geq R_o , \tag{2}$$

where R_o is the crack resistance per unit thickness defined as

$$R_o \equiv \frac{\mathrm{d}U_{\gamma o}}{\mathrm{d}L_o} , \tag{3}$$

and $U_{\gamma o}$ is the product of the specific elastic surface energy of the material, $\gamma_{o(\mathrm{eff})}$, by the projected surface area of the crack (two surfaces, length L_o), i.e., the plane surface area of the crack with actual (rugged) crack length L and unit width. It is important to stress that L refers to the actual (rugged) crack length and L_o to the projected crack length of L, i.e., a distance between two points [17]. Therefore

$$U_{\gamma o} = 2L_o \gamma_{o(\mathrm{eff})} \ , \tag{4}$$

and according to Alves [17] the constant crack growth resistance, $R_o = 2\gamma_{(\mathrm{eff})}$ needs to be corrected by the ruggedness in the follow way

$$R_o = 2\gamma_{\mathrm{eff}} \frac{\mathrm{d}L}{\mathrm{d}L_o} \ , \tag{5}$$

where the mathematical term $\mathrm{d}L/\mathrm{d}L_o$ was called the local ruggedness of the fracture surface [17]. From the quasi-static Griffith criteria, the condition to trigger the beginning of crack growth, is given by $G_o = R_o$. Gao [18] introduced the role of the surface roughening and branching instabilities in dynamic fracture proposing a so-called "wavy-crack model" motivated by experimental observations where rapidly moving cracks develop roughened fracture surfaces. The essence of his model consists in separating the microscopic crack-tip motion with local velocity, from the macroscopically observable crack motion with apparent velocity.

Similarly for the stationary case the elastodynamic energy release rate, G_{oD}, is given by [12]

$$G_{oD} \equiv \frac{\mathrm{d}[F - (U_{Lo} + T_o)]}{\mathrm{d}L_o} \ , \tag{6}$$

and T_o is the kinetic energy of the crack growth which can be written as:

$$T_o = \frac{1}{2}\rho v_o^2 \int\int \left[\left(\frac{\mathrm{d}u}{\mathrm{d}x}\right)^2 + \left(\frac{\mathrm{d}u}{\mathrm{d}y}\right)^2 \right] \mathrm{d}x \ \mathrm{d}y \ . \tag{7}$$

The velocities indexed by o, as v_o, refers to the growth rate of the projected crack length, L_o. This equation will be not considered in the calculations that will be followed here; it is just shown up in the text in order to form the body of the classic considerations, that cannot explain the phenomenon in question. A more general expression was included, however, the calculations do not require the explicit use of the expression of the kinetic energy, T_o, but general principles will just be used.

Mott in 1948 [19] proposed that all excess of the elastodynamic energy release rate, G_{oD}, above the energy necessary to create the fracture surfaces, $2\gamma_{\mathrm{eff}}$, is transformed into kinetic energy of crack growth. However, it is possible to generalize this proposition to the case of a material which shows a non-constant crack growth resistance, $R = 2\gamma_{\mathrm{eff}}$, that develops a ruggedness such as that of the kind $\mathrm{d}L/\mathrm{d}L_o$ expressed in Alves [17]. Rewriting (6):

$$G_{oD} = \frac{\mathrm{d}(F - U_{Lo})}{\mathrm{d}L_o} - \frac{\mathrm{d}T_o}{\mathrm{d}L_o} \, . \tag{8}$$

The first term on the right side of (8) corresponds to the usual elastic energy release rate, G_o given by (1), and so:

$$G_{oD} = G_o - \frac{\mathrm{d}T_o}{\mathrm{d}L_o} \, . \tag{9}$$

This equation shows the relationship between G_o and the elastodynamic energy release rate, G_{oD}. In analogous way to (2), the Irwin–Mott condition of dynamic crack growth is given by

$$G_{oD}(L_o, v_o) \geq \Gamma_0(L_o, v_o) \, , \tag{10}$$

where $\Gamma_0 = \Gamma_0(L_o, v_o)$ is the dynamic crack growth resistance, now depending on the crack growth length, L_o, and velocity, v_o.

In agreement with Griffith fracture criterion for the quasi-static case it is possible to write, in analogous way, the elastodynamic crack growth condition, as follows

$$G_{oD}(L_o, v_o) = \Gamma_0(L_o, v_o) \, , \tag{11}$$

where Γ_0 can be write as $\Gamma_0(L_o, v_o) = R_o - \mathrm{d}T_o(L_o, v_o)/\mathrm{d}L_o$ using (5) for crack resistance, R_o.

The expression (6) mentioned here will not be used in the same way, but completes the idea to be explained in the following section by using a variable separation in the non steady-state case, which is a quite knew procedure.

2.4 Advanced Dynamic Fracture Mechanics Considerations

The literature [13] points to the fact that experimental results confirm the continuum theory of dynamic brittle fracture for fast cracks, where expressions similar to 6 are used to explain the phenomena in focus. On the other hand, in classical dynamic fracture mechanics it is usually assumed that rupture is controlled by the expression [8]

$$G_{oD} \simeq G_o \left(1 - \frac{v_o}{c_R}\right) = 2\gamma_{\text{eff}} \, . \tag{12}$$

This expression equating G_{oD} with v_o is only true locally, i.e., in the immediate vicinity of the crack tip. For the configuration of plane stress applied to a medium of infinite extent, for example, a crack will constantly accelerate.

For this case, the quasi-static energy release rate, $G_o = \pi\sigma_0^2 L_o/2E_o$, in (12), is proportional to the fault crack length, L_o, so that as the rupture grows, L_o grows, and the velocity, v_o, tends to c_R in order to maintain $2\gamma_{\text{eff}}$ at a constant value. This formula fails because, accordingly to the well established fracture mechanics theory, as the velocity increases, $2\gamma_{\text{eff}}$ ceases to be a constant and depends on rupture speed. At velocities near 0.4 or 0.5 of c_R, rupture branches appear and (12) ceases to be applicable [3,20,21]. This explanation was well known to Kostrov, Slepyan, Knauss and Ravi–Chandar, Rice, etc and was rediscovered by physicists in the early 1990s [1,2,13,22,23]. In the dynamical case Slepyan [8,20] proposed that the elastodynamic energy release rate, $G_{oD}(L_o, v_o)$, can be written in terms of the elastic energy release rate, $G_o(L_o)$, for the stable case [16] in the following way:

$$G_{oD}(L_o, v_o) = G_o(L_o) g\left(\frac{v_o}{c_R}\right) . \tag{13}$$

where c_R is the Rayleigh velocity.

To write the functional dependence of this elastodynamic energy release rate, $G_{oD}(L_o, v_o)$, Slepyan [20] propose a Maximum Energy Dissipation Principle. In Alves [24] one has also demonstrated the need for correcting this principle in order to include the ruggedness of fracture surface [17]. For the Fineberg–Gross [1,2] experimental set up of a semi-infinite plane plate under Mode I loading, the elastodynamic crack growth condition, given by (11) yields:

$$G_{oD}(L_o, v_o) = \frac{2\gamma_{\mathrm{eff}}(\mathrm{d}L/\mathrm{d}L_o)}{1 - \frac{v_o}{c_R}(\mathrm{d}L/\mathrm{d}L_o)} = \Gamma_0(L_o, v_o) , \tag{14}$$

Equation (14) shows a great agreement with Gross experimental results [22] as was fitted by Alves [24].

Using the Maximum Energy Dissipation Principle of Slepyan [20], modified to include the ruggedness, $\mathrm{d}L/\mathrm{d}L_o$ of fracture surface, and the Irwin–Mott crack growth dynamic condition given by (10) in (14), including (13), one can assert that:

$$G_o(L_o) g\left(\frac{v_o}{c_R}\right) \geq \frac{2\gamma_{\mathrm{eff}}(\mathrm{d}L/\mathrm{d}L_o)}{1 - \frac{v_o}{c_R}(\mathrm{d}L/\mathrm{d}L_o)} . \tag{15}$$

Comparing (14) with (13) it can be concluded in agreement with Alves [17] that for the brittle materials case where $J \equiv G$, that:

$$G_o(L_o) \geq 2\gamma_{\mathrm{eff}} \frac{\mathrm{d}L}{\mathrm{d}L_o} , \tag{16}$$

and

$$g\left(\frac{v_o}{c_R}\right) \sim \frac{1}{1 - \frac{v_o}{c_R}} . \tag{17}$$

This is an equation that shows dependence on the particular experimental set up under use. The case studied here is of a plane plate under Mode I loading and boundary conditions of infinite body accordingly to Freund [25].

2.5 The Foundations of Non-Stationary Dynamic Fracture Mechanics

For, for the non-stationary case the dynamic energetic balance for the bi-dimensional case can be write written as

$$P = \dot{U} + \dot{T} + \phi \,, \tag{18}$$

where ϕ is the rate of work done by the traction on the surface S, $\dot{U}$ is the rate of increase of the strain energy, $\dot{T}$ is the kinetic energy in the region with R with area A and F is the energy flux into the crack-tip region.

$$P = \int_S T_i \dot{u}_i \; \mathrm{d}s \,, \tag{19}$$

$$U = \lim_{\Gamma^* \to 0} \int_R W(\epsilon_{ij}) \; \mathrm{d}A \,, \tag{20}$$

$$T = \lim_{\Gamma^* \to 0} \int_R \frac{1}{2} \rho \dot{u}_i \dot{u}_i \; \mathrm{d}A \,, \tag{21}$$

Since the loop Γ^* moves with the crack tip, the region R is time-dependent.

In this case the dynamic energy released rate, G_{oD} can be write as

$$\phi = G_{oD}(L_o, v_o, t) v_o = P - (\dot{U} + \dot{T}) \,, \tag{22}$$

For the purposes of our calculations we consider that the non-stationary energy released rate $G_{oD}(L_o, v_o, t)$ can be described by a function of kind:

$$G_{oD}(L_o, v_o, t) = G_{oD}(L_o, v_o) f(t) = G_o g\left(\frac{v_o}{c_R}\right) f(t) \,, \tag{23}$$

where $f(t \geq \tau) \to 1$

Considering that the Mott postulate [19] is valid not only at the beginning of the crack growth, at $t = 0$, but instantaneously at any time, t, during the whole crack growth process, i.e., while, as a crack grows with velocity $v_o(L_o(t))$. Thus the energy flux, $\phi_0(t)$, to the crack tip that is characterized by the velocity $v_o(L_o(t))$ is given by:

$$\phi_0(t) = G_{oD}[L_o(t), v_o(L_o(t)), t] \; v_o(L_o(t)) \,, \tag{24}$$

where one made use of (9). The coefficient of $v_o(L_o(t))$ is the elastodynamic energy release rate, $G_{oD}[L_o(t), v_o(L_o(t)), t]$, given by (23) [8]. Therefore

$$\phi_0(t) = G_o(L_o(t)) \; g(\frac{v_o(L_o(t))}{c_R}) \; f(t) \; v_o(L_o(t)) \,, \tag{25}$$

Based on (13) the energy flux, given in (25), can be written as

$$\begin{aligned} &G_{oD}[L_o(t), v_o(L_o(t)), t]\; v_o(L_o(t)) \\ &= G_o(L_o(t))\; g\left(\frac{v_o(L_o(t))}{c_R}\right) f(t)\; v_o(L_o(t)) \,. \end{aligned} \tag{26}$$

For the particular case of a semi-infinite plane plate under Mode I loading the energy flux to the crack tip is derived from (14) as

$$\phi_0(t) = \frac{2\gamma_{\text{eff}}(\mathrm{d}L/\mathrm{d}L_o)}{1 - \frac{v_o}{c_R}(\mathrm{d}L/\mathrm{d}L_o)} f(t) v_o(t) \,. \tag{27}$$

2.6 Advanced Considerations Based on Fractal Aspects of Fracture Surface for Dynamic Fracture Mechanics

In a fast crack growth experiment as that performed by Fineberg and Gross the oscillations produced in the crack growth velocity can make that it reaches velocities close to the values of Rayleigh waves speed, ($v_o \to c_R$), and when the correspondent energy injected into the crack tip is above this value, it is enough to create new paths for the crack generating branchings [1,2]. Starting from there, if to continue having an increase in the energy injected into the crack tip due to an indefinite loading stress, that is to say, $G(t) \to \infty$, the instability and branching process stays and it starts to happen in different scales, i.e., for each new crack made by branching, being obtained a self-affine geometric pattern (invariant by scale transformation). This way it is observed that the fractal nature of instability and of branching it is nothing else than a physical confirmation that the phenomenology of the process described above continues reproducing in scale, indefinitely, while there is an energy excess into the crack tip. It can also be understood, that the chaotic nature of the fracture possesses a kind of "phenomenological memory" that repeats in different scales being registered in the fractality of fracture surfaces generated in the dynamic crack growth.

Based in the Fineberg experimental results it can be assumed that the temporal dependence of rugged crack growth of crack length has a structured self-similar [8] or self-affine fractal behaviour [17, 20, 26, 27] in the time, such as, an appropriated function satisfying the relaxation process can be supposed as,

$$L[L_o(t+\tau)] = h[L(L_o(t))] \,, \tag{28}$$

where τ is a time lag with magnitude of the order of the viscoelastic relaxation time in the sample. This hypothesis means that the crack length generated in the dynamic fracture process elapses on itself.

For the crack growth velocity to be in agreement with (28) another appropriated self-affine equation can be written as

$$v_o[L_o(t+\tau)] = v_o[L_o(t)]\frac{\mathrm{d}h[L(t)]}{\mathrm{d}L_o(t)}\frac{\mathrm{d}L_o(t)}{\mathrm{d}L(t)} . \tag{29}$$

where $v_o[L_o(t)] = v[L(t)]\mathrm{d}L_o(t)/\mathrm{d}L(t)$.

This mathematical procedure, naturally, introduces in the equations of dynamic fracture the two necessary conditions to describe the instability process. To know: (i) a decoupling of the dynamical functions between the input and output of system which are the elastodynamic energy release rate, G_{oD}, and work of fracture, Γ_0, by means of the dynamical variables of length, L_o, and crack growth velocity, v_o, [28] as it is shown in (28) and (29) and (ii) the existence of at least two situations equally probable [20], unifying them into a single condition. This second condition will be focused latter on.

One observes that, while the input energy flux into the crack tip, $\phi_0(t)$, given by (25) depends on the the crack growth velocity function, $g(\frac{v_o}{c_R})$, the output energy flux, $\psi(t)$, must depend on the local ruggedness created instantaneously in the crack growth process. Therefore, the response in the form of fracture surface, designated by $\psi(t+\tau)$, as being the energy adsorbed to form the rugged crack (shown in Fig. 2), can be written as

$$\begin{aligned}\psi(t+\tau) &= \frac{\mathrm{d}U_\gamma[L(t),v(t)]}{\mathrm{d}L(t)}\frac{\mathrm{d}L(t)}{\mathrm{d}t}\frac{\mathrm{d}h[L(t)]}{\mathrm{d}L(t)} \\ &= \frac{\mathrm{d}U_\gamma[L(t),v(t)]}{\mathrm{d}L(t)}\frac{\mathrm{d}L(t+\tau)}{\mathrm{d}t} ,\end{aligned} \tag{30}$$

Since $\psi(t+\tau) = \psi_0(t+\tau)$, due to the energetic equivalence among the rugged and the projected crack path, then one can express (30) as

$$\psi_0(t+\tau) = \Gamma_0[L_o(t), v_o(L_o(t))]v_o[L_o(t+\tau)] , \tag{31}$$

where $L(t+\tau)$ is the ruggedness (i.e. actual) fracture surface length in the time, $t+\tau$, and $L_o(t+\tau)$ is its corresponding plane projected length on the direction of crack growth.

Substituting (14) into (31), one gets

$$\psi_0(t+\tau) = \frac{2\gamma_{\mathrm{eff}}(\mathrm{d}L/\mathrm{d}L_o(t))}{1-\frac{v_o[L_o(t)]}{c_R}(\mathrm{d}L/\mathrm{d}L_o(t))}v_o[L_o(t+\tau)] , \tag{32}$$

This equation describe the dissipation rate in the formation of the fracture surface formation in the time $t+\tau$.

2.7 Dynamic Fracture Model with a Time Delay

Sharon & Fineberg [13] admit that theory predicting the motion of a crack is governed by the balance between the energy flux into the crack tip, $\phi_0(t)$, and the dissipation rate, $\psi_0(t)$, here given by (31) including the time delay between them.

$$\phi_0(t) = \alpha\psi_0(t) \ , \tag{33}$$

where $0 < \alpha < 1$ is a conversion factor of elastodynamic energy into surface energy (work of fracture).

However it is necessary to take into account the relaxation process at the crack tip. Observe that (26) says that the energy balance between the instantaneous input flux and stationary flux at crack tip, in accordance with Irwin–Mott's (10), can be written in a general form as *Stationary Flux(time, t) = Input Flux(time, t)/Coupling function of time,* $f(t)$ or

$$\phi_{o(\text{stationary})}(t) = \frac{\phi_0(t)}{f(t)} \ , \tag{34}$$

where from (23), (24) and (25) we have:

$$\phi_{o(\text{stationary})}(t) = G_{oD}[L_o(t), v_o(L_o(t))]v_o[L_o(t)] \ , \tag{35}$$

depending on the projected crack length, $L_o(t)$, already formed in each instant, t.

Substituting (35) into (26) one gets:

$$\begin{aligned} &G_{oD}[L_o(t), v_o(L_o(t))]v_o[L_o(t)] \\ &= G_o(L_o)g\left(\frac{v_o(L_o(t))}{c_R}\right)v_o[L_o(t)] \ . \end{aligned} \tag{36}$$

The particular experimental apparatus in use influences the functional form of the kinetic energy, $T_o(L_o, v_o)$ developed by the cracks and this in turn affects the ruggedness developed instantly in the fracture. Therefore it should exist a relationship between the response in the form of a local ruggedness created instantaneously in the crack growth process and the of the crack growth velocity function, $g(\frac{v_o}{c_R})$, Thus, assuming a function $\frac{dh(L(t))}{dL_o(t)}\frac{dL_o}{dL} \sim 1/[g(\frac{v_o(L_o(t))}{c_R})]$ the energy flux, given by (26), can be written as

$$\begin{aligned} &G_{oD}[L_o(t), v_o(L_o(t))] \ v_o[L_o(t)]\frac{\mathrm{d}h[L(t)]}{\mathrm{d}L_o(t)}\frac{\mathrm{d}L_o}{\mathrm{d}L} \\ &= \alpha G_o[L_o(t)]v_o[L_o(t)] \ . \end{aligned} \tag{37}$$

Observe from (37) that the Mott postulate, enunciated before, becomes valid instantaneously during the whole crack grow process by means of a new insight in the energy flux balance with time delay. Therefore from the elastodynamic crack growth condition given in (11), and using (29) in (37), this equation can be expressed as

$$\Gamma_0[L_o(t), v_o(L_o(t))]v_o[L_o(t+\tau)] = \alpha G_o[L_o(t)]v_o[L_o(t)] \ . \tag{38}$$

In accordance with the self-affine hypothesis made in (28) and (29) the energy flux balance between the input and output with a time delay, τ, may be written as α. *Relaxed Flux(time past, t) = Output Flux(time present = time past plus relaxation time, $t+\tau$)*, or in accordance with Irwin–Mott's (10), (38) can be written in a general form as

$$\phi_{o(\text{relaxed})}(t) = \frac{1}{\alpha}\psi_0(t+\tau)\,. \tag{39}$$

Therefore at time, t, the flux of elastodynamic energy into the specimen to the crack tip, $\phi_0(t) = G_{oD}[L_o(t), v_o(L_o(t)), t]\ v_o[L_o(t)]$, will create in front of the crack tip, at time $t+\tau$, the conditions for the formation of a process zone (viscoelastic region of size $\sim l_o$) that will separate at time, $t+\tau$, forming a rugged crack having length $L(t+\tau)$, necessary for crack growth (e.g. breaking of chemical bonds, formation and nucleation of dislocations, etc.). This theoretical result resumes all the problematics found by Fineberg and Gross [1,2], already explained in the Sects. 2.1 and 2.2 of this chapter.

2.8 Chaotic Nature of Dynamic Fracture

The chaotic nature of fractures can be revealed by different forms depending of the particular experimental set up. For the case of a semi-infinite plane plate under Mode - I loading the function $g(v_o[L_o(t)]/c_R)$ determines a logistic map well known in the literature concerning chaos theory [29, 30]. Therefore, using (14) in (38) or (41) and (32) in (39) this becomes,

$$\begin{aligned}&\alpha G_o[L_o(t)]v_o[L_o(t)]\\ &= \frac{2\gamma_{\text{eff}}(\mathrm{d}L/\mathrm{d}L_o(t))}{1-\frac{v_o[L_o(t)]}{c_R}(\mathrm{d}L/\mathrm{d}L_o(t))}v_o[L_o(t+\tau)]\,. \end{aligned}\tag{40}$$

Rewriting (40) and multiplying the resulting by $\mathrm{d}L/\mathrm{d}L_o$ term it yields the following expression for the normalized crack growth velocity $v_o[L_o(t+\tau)]$ corresponding to projected surface

$$\begin{aligned}&\frac{v_o(L_o(t+\tau))}{c_R}\frac{dL}{dL_o(t)}\bigg|_{t+\tau}\\ &= \frac{\alpha G_o(L_o(t))}{2\gamma_{\text{eff}}(\frac{dL}{dL_o(t)})}\\ &\times\frac{v_o(L_o(t))}{c_R}\frac{dL}{dL_o(t)}[1-\frac{v_o(L_o(t))}{c_R}\frac{dL}{dL_o(t)}]\,, \end{aligned}\tag{41}$$

which has the form of the equation of the logistic map (Fig. 3) [29–31]. For convenience, instead of the delay in time, τ, (41) can be written, as a sequence of events, i.e.,

$$x_{ok+1} = \mu_0 x_{ok}(1-x_{ok})\,, \tag{42}$$

where x_{ok} corresponds to the normalized projected surface crack growth velocity and the coefficient, μ_0, to control parameter of (31), i.e.,

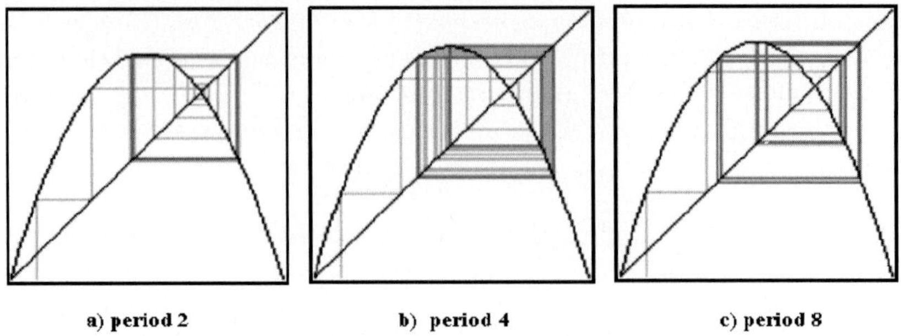

Fig. 3. Logistic map of the logistic equation $x_{ok+1} = \mu_0 x_{ok}(1 - x_{ok})$ related to the input energy flux, $\phi_0(t)$, and output flux of dissipated power, $P_0(t+\tau)$ in function of the normalized velocities $v_o(t)/c_R$ and $v_o(t+\tau)/c_R$, showing the cycles or periods of iteration

$$\mu_0 \equiv \frac{\alpha G_o(L_o)}{2\gamma_{\text{eff}}(\mathrm{d}L/\mathrm{d}L_o)} , \tag{43}$$

$$x_{ok} \equiv \frac{v_o[L_o(t)]}{c_R} \frac{\mathrm{d}L}{\mathrm{d}L_o(t)} , \tag{44}$$

and

$$x_{ok+1} \equiv \frac{v_o[L_o(t+\tau)]}{c_R} \frac{\mathrm{d}L}{\mathrm{d}L_o(t)} , \tag{45}$$

respectively.

In Fig. 4, the iterations of this equation are shown as a function of the control parameter, μ_0, [30] which can be identified with the crack growth characteristics.

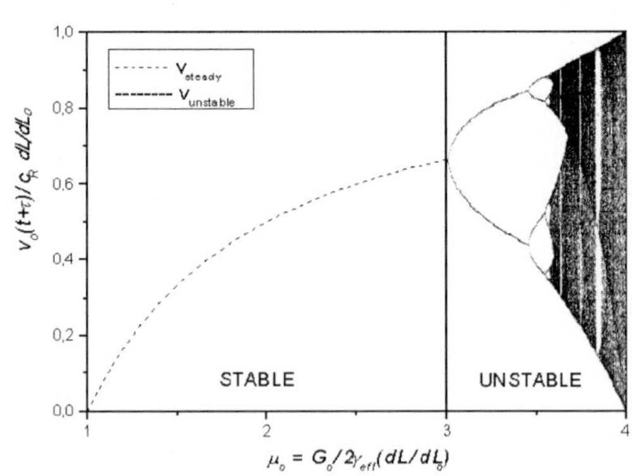

Fig. 4. Iterates of the logistic map of (48) as a function of μ_0 for $1.0 \leq \mu_0 \leq 4.0$. An transient of 200 points has been discarded in each case [30]. The control parameter, μ_0, is given by (43) and $v_o(t+\tau)/c_R$ is the normalized velocity as it was explained in the text

Observe that a normalized crack growth velocity to rugged surface, defined as $v(L(t)) = v_o[L_o(t)]\mathrm{d}L/\mathrm{d}L_o(t)$, can also be written in the logistic equation form, in analogous way to plane projected fracture surfaces, dropping all the zero subscripts in the (42), (43) and (44) in a such way that the new rugged crack growth velocity can be written as

$$\frac{v[L(t+\tau)]}{c_R} = \frac{\alpha G(L)}{2\gamma_{\text{eff}}} \frac{v[L(t)]}{c_R} \left(1 - \frac{v[L(t)]}{c_R}\right) , \tag{46}$$

where $G(L)$ is the elastic energy release rate to the rugged crack path, giving place to write $x_{k+1} = \mu x_k(1 - x_k)$. In analogous way it gets $x_k = v[L(t)]/c_R$ and $\mu = \alpha G(L)/2\gamma_{(\text{eff})}$. This means that the same conditions can be repeated at each new crack path created, in a indefinite branching process, originating a self-affine spatial fractal pattern formed by all cracks generated during the branched crack growth process.

2.9 The Instability Process Under the Sight of Time Delay

Therefore, the instability dynamic process on the fast crack growth and crack branching in brittle materials like the soda-lime glass and PMMA can be explained in the following way: The existence of a time delay, τ, between the energy flux injected into the crack tip, $\phi_{o(\text{inst})}$, (25), and the spent energy flux (or the dissipated power, ψ_0, (31)) to form the fracture surfaces, (32), (Fig. 3), produces from a critical velocity, (44), a "uncompassing" (Fig. 3) between the crack growth velocity, $v_o(t)$, and the rate of formation of the fracture surfaces, $v_o(t+\tau) = \mathrm{d}A_o(t+\tau)/\mathrm{d}t$. This "uncompassing" is responsible for an indetermination in the crack growth velocity, (Fig. 4), at crack tip that by its turn gives rise to a dynamic instability in the form of oscillations in the crack growth velocity. This instability produces a rugged fracture surface increasing the consumption of energy. This increase limits the crack growth velocity to a smaller value than the speed of Rayleigh waves in the material, ($v_o \leq c_R$), (46), producing a new delay in relation to fracture surfaces formation rate and increasing still more the "uncompassing" between the injected energy flux and the worn-out energy rate to form the fracture surfaces and so forth (Figs. 3 and 4).

3 Results

Equations (45) and (41) can be explored through the graphs of $\psi_0(t+\tau)$ and $\phi_0(t)$ versus $v_o(t)$ (Figs. 3 and 4) and of $v_o(t+\tau)/c_R$ versus μ_0 (Fig. 4). By means of this diagram one notices that combining the magnitudes, $\phi_0(t)$ and $\psi_0(t+\tau)$ it generates an instability whenever a temporal delay exists among them. Actually, (42) is the well-known logistic equation. However, Before analyzing Fineberg et al. experimental results [1] we start from the use of

Slepyan's criterion, given by Maximum Energy Dissipation Principle [20], and modify it in order to include the ruggedness term, $\mathrm{d}L/\mathrm{d}L_o$, of fracture surface [24]. From this result it was possible to obtain the energy balance between the input and output of the system, where in our notation it is given by (46). Observe that this expression can be written in the following way:

$$\begin{aligned}\alpha G_o[L_o(t)]\left(1-\frac{v_o[L_o(t)]}{c_R}\frac{\mathrm{d}L}{\mathrm{d}L_o(t)}\right)v_o[L_o(t)]\\ =2\gamma_{\mathrm{eff}}\frac{\mathrm{d}L}{\mathrm{d}L_o(t)}v_o[L_o(t+\tau)]\,. \qquad (47)\end{aligned}$$

Although its left side reminds the expression of the classical result proposed by Freund [8] where the elastodynamic energy release rate being, $G_{oD} = G_o[1 - v_o(t)/c_R]$, for a medium with infinite extension ((8.77) of Freund [8]). Therefore, a first result that must be observed in this paper is that the classical dynamic fracture mathematical formalism is reobtained from (41) when the relaxation process is almost negligible and the persistent creep phenomenon is not considered into a thin crack tip in short range effect described by punctual process zone. In this case it has that $v_o(t+\tau) = v_o(t)$ and therefore the classical comes out:

$$\frac{v_o(t)}{c_R}=\left(1-\frac{2\gamma_{\mathrm{eff}}}{G_o(L_o)}\frac{\mathrm{d}L}{\mathrm{d}L_o(t)}\right)\frac{\mathrm{d}L_o(t)}{\mathrm{d}L}\,. \qquad (48)$$

Whenever the elastic energy release rate, $G_o(L_o)$, increases linearly with the crack length, L_o, which in the absence of ruggedness, i.e. $\mathrm{d}L/\mathrm{d}L_o = 1$, (48) becomes,

$$v_o = c_R\left(1-\frac{L_{oc}}{L_o}\right)\,. \qquad (49)$$

This result shows the stable region of the fracture process formed by "fixed point of order one" of the logistic map (Fig. 4) where the conditions of the crack growth are considered to be slow because the viscoelastic properties of the material do not influence the process.

On the other side, the first graph (Fig. 3) shows the dependence of elastodynamic energy flux, $\phi_0(t)$, given by (43) and (44), that flows into the crack tip, and of the dissipated power spent to form the fracture surface, $\psi_0(t+\tau)$, given by (31) and (48), in function of the normalized crack growth velocities $v_o(t)/c_R$ and $v_o(t+\tau)/c_R$ respectively. This graph shows that the energy flux, $\phi_0(t)$, possesses a maximum value at $v_o/c_R = 0.5$ and a nonlinear dependence of dissipated power, ψ_0, on the normalized velocity $v_o(t+\tau)/c_R$ as it was proposed by the model (26).

The second graph (Fig. 4) shows the iterations of the logistic map, (41) and (48), of the normalized crack velocity, $v_o(t+\tau)/c_R$, (44) as a function of the control parameter, μ_0, given by (43). This control parameter, $\mu_0 = \frac{G_o}{2\gamma_{\mathrm{eff}}(dL/dL_o)}$, for a medium with infinite extension, is linearly proportional to the crack length, $L_o(t)$; then as the crack grows, the control parameter increases. Therefore this graph represents the dependence of the speed with the retardation, $v_o(t+\tau)$, in function of the projected crack length, L_o. For values of $1 \leq \mu_0 \leq 3$ the crack growth process is dynamically stable since at those speeds the viscoelastic process zone in front of the crack tip is too small to influence the process. However, the time delay that exists between the energy that flows into the crack tip, $\phi_0(t)$, and the dissipated power for crack growth, $\psi_0(t+\tau)$, causes instabilities that show up in the interval $3 \leq \mu_0 \leq 4$. Such instability begins only when the control parameter reaches the value, $\mu_0 = 3$. Exactly at this point, $\mu_0 = 3$, there is a flip bifurcation [29–31] and it begins the influence of the time delay (where $\tau \sim l_o/v_o$) into the process zone causing instabilities in the crack growth. As shown in Fig. 4 at $\mu_0 > 3$ the fixed points determined by (48) are not stable anymore. Considering that the relaxation with time delay is fixed it is seen that the length of the process zone oscillates in time becoming responsible for the oscillations in the crack growth velocity, $v_o(t)$. The velocity function, $v_o(t+\tau)$ in (41), which is related to the dissipated power, $\psi_0(t+\tau)$, to $\mu = 3$, has the value equal to $2/3c_R$. This value from (48) corresponds to a critical velocity

$$v_{o(\text{critic})} = 1/3c_R \,. \tag{50}$$

The main assumption of a time delay is expressed mathematically by formula (48) where the factor μ_0 of (43) is taken to be fixed. This means that the crack growth velocity at $t+\tau$ is proportional to the energy flux into the growing crack tip at time t. This relationship is very interesting since it leads to the oscillating type of instability for the steady motion (with $v_{o(\text{steady})} = c_R(1-1/\mu_0)\mathrm{d}L_o/\mathrm{d}L$ as the solution (48) if $3 \leq \mu_0 \leq 4$. By the way, if μ_0 increases (from $\mu_{oc} = 3$ to a limit value $\mu_{o\,\text{max}} = 4$), the minimum limit of the averaged speed, $v_{o\,\text{min}}$, is given approximately by

$$\frac{v_{o\,\text{min}}}{c_R} = \left(1-\frac{1}{3}\right)\left(\frac{\mathrm{d}L_o}{\mathrm{d}L}\right)_{\text{max}} = \frac{2}{3}\left(\frac{\mathrm{d}L_o}{\mathrm{d}L}\right)_{\text{min}} , \tag{51}$$

just depending on the maximum value of ruggedness, $\mathrm{d}L_{\text{min}}/\mathrm{d}L_o$, in this point. For a value of ruggedness equal to, $\mathrm{d}L/\mathrm{d}L_o = 3/2$ the limit of the averaged speed, v_o, is given by:

$$v_o = 0.44\,c_R \,, \tag{52}$$

and the maximum limit of the averaged speed, $v_{o\,\text{max}}$, is given approximately by

$$\frac{v_{o\,\text{max}}}{c_R} = \left(1-\frac{1}{4}\right)\left(\frac{\mathrm{d}L_o}{\mathrm{d}L}\right)_{\text{max}} = \frac{3}{4}\left(\frac{\mathrm{d}L_o}{\mathrm{d}L}\right)_{max} , \tag{53}$$

depending also on the maximum value of the ruggedness, $\mathrm{d}L_{\text{max}}/\mathrm{d}L_o$, at this point. Equally for a value to ruggedness value equal to $\mathrm{d}L/\mathrm{d}L_o = 3/2$ the limit of the averaged speed, v_o, is given by:

$$v_o = 0.5\, c_R \;, \tag{54}$$

from where the normalized crack velocity v_{ok}/c_R saturates at $c_R(\mathrm{d}L_o/\mathrm{d}L)_{\max}$, below of Rayleigh waves velocity, which is approximately the saturation velocity measured by Fineberg et al. [1, 27].

As can be seen in Fig. 3, $v_{0\,\max}$ and consequently G_{oD} has tendency to approach the maximum of the energy flux. After $\mu_0 \geq 4$ there is no sense to speak about the influence of μ_0 parameter over the crack growth because the G_{oD} reaches the maximum constant flux and v_o reaches a maximum constant velocity and after this the crack growth follows a new stationary state conserving the ruggedness already created to maintain the this maximum values. It can also be observed that (53) assumes an infinite body sample where the influence of external stress field it is negligible and after the maximum flux condition have been reached the dynamic energy release rate, G_{oD}, does not depend anymore on the sample length.

3.1 Comparison Between Theory and Experiment

Fineberg et al. [1, 2] observed that at the onset of instability the fracture surface changes from a featureless (on a scale larger than 1 μm) to a jagged structure which develops into coherent oscillations and coalesce downstream of the crack growth broadening to extend over the entire width of the sample. This change in morphology as the crack accelerates can be associated with the geometric increase in the number of flip bifurcations, or, with the number of non-trivial stable fixed points as can be seen in Fig. 4. To each stable fixed point corresponds a given velocity v_{ok+1}/c_R. As the crack grows the straight line crack velocity oscillates among the allowed velocities, looking for alternative paths of energy dissipation in excess at the crack tip by means the generation of a rugged surface that mathematically correspond to more energetic path. It is therefore reasonable to associate the issuing increase of ruggedness with the number of possible velocities. At $\mu_0 < 3$ there is only one possible velocity (one stable fixed point on the logistic map) and therefore the surface is smooth. As the crack propagates μ_0 (given by (43)) increases with the crack length, $L_o(t)$, and the number of allowed straight line crack velocities becomes larger and larger. The straight line crack velocity oscillates among the allowed values (projections of the actual velocity) resulting in morphologies of higher complexity, in agreement with the observations of Fineberg et al. [1, 2].

Peter Gumbsch [32–34], using molecular dynamics methods in your simulations, reported that the instability in dynamic crack growth sets at $G_o/2\gamma_{o(\mathrm{eff})} = 3$. This condition for initiate initiating instabilities, considering $\alpha \mathrm{d}L/\mathrm{d}L_o \simeq 1.0$, is equivalent to $\mu_0 = 3.0$, as is given by (43). Therefore this result is in reasonable agreement with the result shown by the chaotic model presented in this paper.

Taking for the energy conversion factor $\alpha = 1.0$ one obtains from (48) the value of 0.33 for the normalized crack velocity at which instabilities should occur. However, Fineberg and co-workers observed that beyond a critical value of 0.34 the normalized velocities, v_o/c_R, start to show instabilities [1,2]. Therefore this theoretical value of $0.33\,c_R$ is in excellent agreement with the value of experimental result of 0.34 [1, 2], measured by Fineberg–Gross et al., at the moment the instability starts. Gross [22] also observed that a visible rough structure (branching) appears near from $0.42\,c_R$ in the form of a parabolic crack branching.

4 Discussion

The experiments performed by Fineberg and co-workers [1, 2] provide evidence for instability in the brittle fracture of isotropic materials. On the other hand, theories based on conventional concepts such as energy balance [16] and quasi-static configurational forces at crack tips show no indication of strong oscillatory or branching instabilities [35]. Yoffe [36] by analyzing the stresses in the neighborhood of a crack tip growing at high velocity hinted about the emergence of instabilities but the analysis is not a truly dynamic theory of forces and accelerations of fractured surfaces.

The basic property of dynamic fracture mechanics is that the processes near the fault tip occur at near wave velocities, and for this reason the crack tip is independent of the details of loading. This is a famous theorem proved independently by Kostrov and Eshelby in 1964 and 1969, respectively, for the antiplane case and by Kostrov and Nikitin in 1970 for general loading. Observe that (17) refers to a particular experimental set up. Therefore, in accord to the functional dependence of this equation for the $g(v_o/c_R)$ term or depending on of the particular form of the experiments other kinds of logistic maps can be obtained, since that the same procedure of calculations accomplished until now can be done. Equation (46) developed in this paper is equivalent to (12), and the same improvements (finite size of the sample, influence of boundary, etc.) proposed to (12) [8] can be incorporated into (43) and (46) without consequences on the results presented in this paper. This is corroborated by experimental evidence [1, 2, 13] showing that the onset of instabilities is independent of the size of the sample and/or of the geometrical set up of the experiment (see also [3]).

5 Summary and Conclusions

This chapter presents arguments in favour of chaotic behaviour of rupture. The arguments are general and based on energy conservation principle which are totally valid in fracture mechanics. The central hypothesis is the energy flux through the crack tip is converted there into fracture energy with a time delay, τ, due to the development of a viscoelastic process zone in front of the crack tip. It is tacitly assumed in this paper that such a delay exists and it has a well defined time scale, τ, being a characteristic property of the material. Its magnitude is of the order of the viscoelastic relaxation time of the sample material under local fracture conditions. A key assumption of the theory is that the onset of instability observed in the velocity of dynamic crack growth is due to the time delay, which yields (41) and (46). This time delay factor, τ, in (28) implies the possibility to derive an equation for the crack growth the velocity in the form of a logistic map equation.

Before concluding this chapter it is necessary to notice that the hypothesis of linear energy transfer as given by (46) and (45) is an oversimplified approach. The energy release rate, G_o, is linearly dependent on the crack length whereas the crack resistance, R_o, rises in a non-linear form [10, 12]. Based on (28), (29) and (41), this property will be used in a forthcoming paper in which it is shown that the energy dissipation can also be written in the form of a logistic map having as consequence crack branching and other phenomena so far not explained by the classical fracture theory.

The purpose of this chapter is to show that contrary to what has been thought previously, the most familiar models in fracture mechanics are intrinsically incomplete. Therefore, it was used an as simple as possible case of dynamic crack growth of a semi-infinite body with plane strain condition, and well established concepts and results, to derive an expression for the crack velocity in the well known form of a logistic equation and map. From this map conclusions regarding instabilities of crack growth are drawn and compared with the experimental results obtained by Fineberg and co-workers [1]. This work shows that other logistic maps can be built, accordingly to the particularity of the experiment and accordingly to the expression of its the kinetic energy. This article presents a new picture for dealing with fracture dynamics, making use of logistic maps as a new method for predicting the possible velocities that a crack can reach and from there to try to reproduce its geometric fractal behaviour, ruggedness, etc. Each particular material and each particular experimental testing condition will determine the type of map and the type of crack as well. The logistic map built has an interpretation which enables to understand even more complex situations for the phenomenon under study.

From the above results it is concluded that the instabilities involved in dynamic fracture are consequences of the mathematical structure of chaos that underlies such phenomena. It was possible to write the straight line crack velocity in the form of a logistic map explaining the onset of instabilities observed by Fineberg et al. [1]. This achievement brings into fracture mechanics all the mathematical structure developed for complex systems. This theoretical approach provides a single and concise tool to determine among others properties the conditions under which crack growth becomes dynamically unstable and branching takes place as will be shown in a forthcoming paper.

The literature usually shade that there are fractals in the quasi-static fracture of surfaces. Then undoubtedly, in fracture dynamics there will be chaotic behaviour in the formation of the same ones. Therefore, if the model proposed in this chapter is not the final answer for the subject, at least it is an initial step, it lifts and it opens a new proposal for studying fracture dynamics. Therefore, we want to say that experimental research is needed to illuminate the theoretical evidences more closely.

Acknowledgments

This research work was in part supported financially by CNPq, FAPESP, CAPES and one of the authors, Lucas Máximo Alves, thanks the Brazilian program PICDT/CAPES and PROPESQ-UEPG (for concession of a scholarship)and to Profs. Drs. Bernhard Joachim Mokross and Leonid Slepyan for helpful discussions and Prof. Benjamin de Melo Carvalho for concession of a computer obtained through the Project (Ref. Proc. n. 46.8808/00-0) supported by CNPq-BRASIL to accomplish the calculations, typing and printing the final version of this work.

References

1. J. Fineberg, S.P. Gross, M. Marder, H.L. Swinney: Phys. Rev. Lett. **67**, 457 (1991)
2. J. Fineberg, S.P. Gross, M. Marder, H.L. Swinney: Phys. Rev. B **45**, 5146 (1992)
3. B.V. Kostrov and L. V. Nikitin: Archiwum Mechaniki Stosowanej **22**, 749 (1970)
4. R.M Christensen: *Theory of Viscoelasticity: An Introduction* (Academic Press, New York 1982)
5. J.R. Willis: J. Mech. Phys. Solids **15**, 151 (1967)
6. J. R. Walton: J. App. Mech. **54**, 635 (1987)
7. G. Golenievski: Int. J. of Fracture **33**, 39 (1988)
8. L.B. Freund: *Dynamic Fracture Mechanics* (Cambridge University Press, Cambridge 1990)
9. B. Lawn: *Fracture of Brittle Solids* (Cambridge University Press, Cambridge 1995)
10. H.L. Ewalds, R.J.H. Wanhill: *Fracture Mechanics* (Edward Arnold Publishers 1993)
11. M.F. Kanninen, C. H. Popelar: *Advanced Fracture Mechanics* (CRC Press, Oxford 1985) pp 437
12. T.L. Anderson: *Fracture Mechanics, Fundamentals and Applications*, 2nd edn (CRC Press, Oxford 1995) pp 215–218

13. E. Sharon, J. Fineberg: Nature **397**, 333 (1999)
14. P.D. Washabaugh and W.G. Knauss: Int. J. of Fracture **65**, 97 (1994)
15. L.L. Mishnaevsky Jr.: Int. J. of Fracture **79**, 341 (1996)
16. A.A. Griffith: Phil. Trans. R. Soc. London **A221**, 163 (1920)
17. L.M. Alves, Rosana Vilarim da Silva, B.J. Mokross: Physica A **295**, 144 (2001)
18. H. Gao: J. Mech. Phys. Solids **41**, 457 (1993); J. Mech. Phys. Solids **44** 1453 (1996)
19. N.F. Mott: Engineering **165**, 16 (1948)
20. L.I. Slepyan: J. Mech. Phys. Solids **41**, 1019 (1993)
21. W.G. Knauss, K. Ravi-Chandar: Int. J. of Fracture **27**, 127 (1985)
22. S.P. Gross: Dynamics of fast fracture. PhD Dissertation, Faculty of the Graduate School of the University of Texas at Austin, Texas August (1995)
23. M.P. Marder, S.P. Gross: J. Mech. Phys. Solids **43**, 1 (1995)
24. L.M. Alves: *Fractal geometry concerned with stable and dynamic fracture mechanics*, accepted for publication in Theoretical and Applied Fracture Mechanics
25. L.B. Freund: Journal of Elasticity **2**, 341 (1972)
26. E. Sharon, S.P. Gross, J. Fineberg: Phys. Rev. Lett. **74**, 5096 (1995)
27. J.F. Boudet; S. Ciliberto, V. Steinberg: J. Phys. II France **6**, 1493 (1996)
28. J.G. Williams, A. Ivankovic: Int. J. of Fracture **51**, 319 (1991)
29. N. Fiedler-Ferrara, C.P.C. do Prado: *Caos, Uma Introdução*, (Ed. Edgard Blűcher Ltda Brazil 1994)
30. S. N. Rasband: *Chaotic Dynamics of Non-Linear Systems* (John Wiley, New York 1990) p 23
31. C. Beck and F. Schlögl: *Thermodynamics of Chaotic Systems* (Cambridge University Press, Cambridge 1993)
32. P. Gumbsch: Journal of Materials Research **10**, 2897 (1995)
33. P. Gumbsch, S.J. Zhou and B.L. Holian: Phys. Rev. B **55**, 3445 (1997)
34. P. Gumbsch, In: *Computer Simulation in Materials Science*, ed by H.O. Kirchner, L. Kubin and V. Pontikis (Kluwer Academic Publishers, Dordrecht 1996) pp 227–244
35. B. Cotterell, J.R. Rice: Int. J. Fract. Mech. **16**, 155 (1980)
36. E. Yoffe: Phil. Mag. **42**, 739 (1951)
37. E. Sharon, S. P. Gross, J. Fineberg: Phys. Rev. Lett. **76**, 2117 (1996)
38. H. Tan, W. Yang: J. Appl. Phys. **78**, 7026 (1995)
39. R. Mohan, A.J. Markworth, R.W. Rollins: Modeling Simul. Mater. Sci. Eng. **2**, 659 (1994)

Nonlinear Dynamics and Fractal Avalanches in a Pile of Rice

Rinke J. Wijngaarden[1], Kinga A. Lőrincz[2], and Christof M. Aegerter[3]

[1] Division of Physics and Astronomy, Faculty of Sciences, Vrije Universiteit, De Boelelaan 1081, 1081HV Amsterdam, The Netherlands
rw@nat.vu.nl

[2] Division of Physics and Astronomy, Faculty of Sciences, Vrije Universiteit, De Boelelaan 1081, 1081HV Amsterdam, The Netherlands
ka.lorincz@few.vu.nl

[3] Division of Physics and Astronomy, Faculty of Sciences, Vrije Universiteit, De Boelelaan 1081, 1081HV Amsterdam, The Netherlands
present address:
Fachbereich Physik, Universität Konstanz, Universitätstrasse 10, P.O. Box 5560, D-78457 Konstanz, Germany
christof.aegerter@uni-konstanz.de

1 Introduction

The result of a measurement x e.g. of the length of Dutch men or the weight of rice grains is usually Gaussian distributed with a well-defined average $\bar{x}$ and width σ. Quite to the contrary, around 1944 Gutenberg and Richter [1] discovered that the size of earthquakes follow a power law distribution: an earthquake of energy E has a probability to occur given by

$$P(E) \sim E^{-\tau} \tag{1}$$

where τ is close to 1. Such power law behaviour is quite special since an average and width cannot be defined. A particularly nasty property is that the probability decays only as a power law, while in the Gaussian case it decays faster than exponentially. As a result there is a non-negligible chance for a huge earthquake.

Many systems in nature display dynamics dominated by similar punctuated behaviour, which we call here generically 'avalanches'. Other examples are: snow-avalanches [2], forest fires, rain fall [3], stock-market indices [4] and the extinction of species in biology [5]. In all these cases, due to a power law distribution function, there is a finite chance for very big, catastrophic events.

In 1988, Bak, Tang and Wiesenfeld [6] introduced the Self-Organized Criticality (SOC) model; it was made accessible for a broad audience by the book "How Nature Works" [7]. In fact, apart from the power law behaviour, there are now many more criteria [8] for SOC behaviour, which enable a more stringent test.

Although there has been a very significant amount of numerical simulations on SOC systems, there are only very few controlled experimental investigations. Up to now experimental work was much hampered by the difficulty to determine whether a particular system is SOC or not. In fact, for sand piles both SOC and non-SOC behaviour has been reported [9] (based on power law scaling of avalanches only).

We study experimentally a 3-dimensional pile of rice with a 1×1 m^2 floor area. Using the more stringent criteria, we discovered [10] that rice behaves as SOC: not only is the avalanche distribution a power law, but also finite size scaling (FSS) is obeyed: the maximum avalanche size scales in a particular manner with the size of the system, as predicted by SOC-theory. We determine the exponent (power) of the power law distribution function as well as the fractal dimensions of the avalanche cluster (avalanche exponents). On the other hand, we also determine the growth and roughness exponents (surface exponents) for the rough surface that remains after many avalanches. SOC theory does not fix the exact values of the individual exponents, but does predict scaling relations e.g. between the avalanche and surface exponents. We find that as far as we tested these relations, they are obeyed for our experimental data.

In addition, SOC theory predicts that a system which is not yet 'critical' moves in a particular manner toward the SOC state described by the so-called gap-equation. This approach is characterized by another exponent, the value of which, interestingly, is given through another scaling relation containing only values for the SOC state [11]. Also this relation is obeyed for our pile.

After introducing our experimental set-up, we review these ideas and corresponding experimental work in more detail below. We conclude by some (experimentally tested) ideas on how to prevent big catastrophes.

2 Experiment: A Big Rice Pile

For the experiments we use long grained rice with dimensions of typically $2 \times 2 \times 7$ mm^3, similar to rice A of [12]. Our experimental set-up is shown in Fig. 1. It consists of a rice pile of 1×1 m^2 floor area on the top of which rice rains down from a linear source. Uniformity within 5% in the distribution along this line is guaranteed by the use of a distributor board, see Fig. 2.

By the distributor board the stream of rice is continuously split such that at the bottom we end up with a row of 64 uniformly distributed sub-streams. A plastic flap at the bottom is used for further spreading and to slow down the rice before it impinges onto the surface of the pile.

To rain down at a uniform rate, the rice is fed to the distribution board from a mechanically stirred funnel, from which the rice emerges at an approximately constant rate of typically 1500 grains per image (taken at 30 s interval), distributed over the whole width of the distributor board.

To monitor the shape of the pile as function of time, a set of coloured (red, green and blue) lines (see Fig. 3) is projected onto the pile using an overhead projector. Photographs are taken from a different viewpoint with

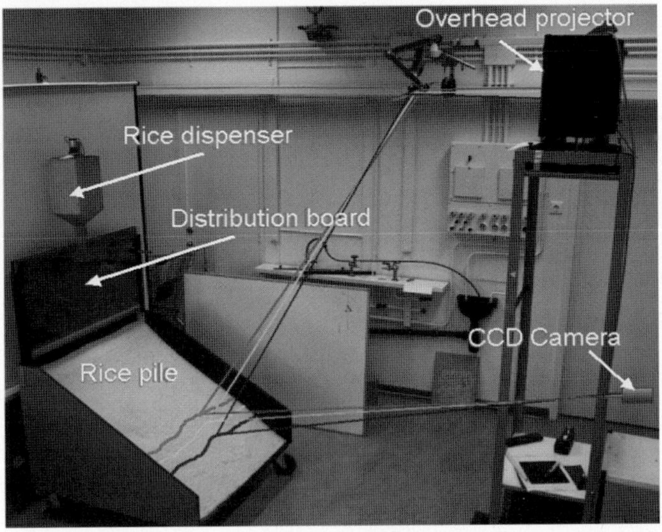

Fig. 1. The rice set-up with $1 \times 1\ \mathrm{m}^2$ floor area. Rice rains down from the distributor board onto the top of the pile. The 3-dimensional shape of the pile is reconstructed from the shape and position of a set of coloured lines projected onto the pile

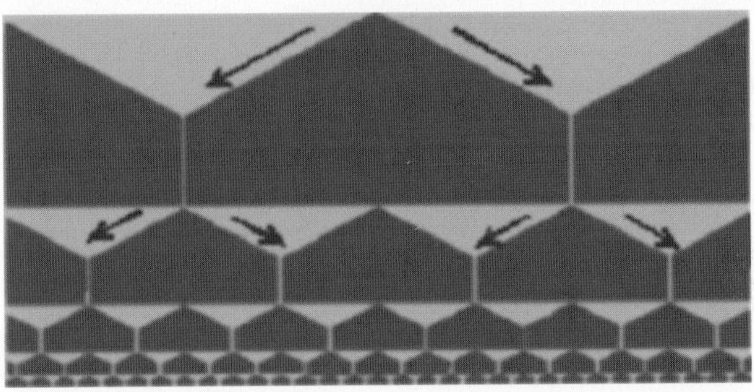

Fig. 2. The binary distributor board used to create a uniform line source

respect to the projector, such that the shape of the pile can be reconstructed using the geometry of stereoscopy.

Typically an experimental run lasts 4 hr, with a picture being taken every 30 s. The pictures are taken with a digital camera with a resolution of 2048 × 1536 pixels. For each picture the centre of gravity of each line is detected (see Fig. 3, lower panel), from which the 3-dimensional surface is reconstructed. From the 3-dimensional surfaces, the roughening properties (see Sect. 3 below) and the avalanche properties (see Sect. 4 below) are calculated.

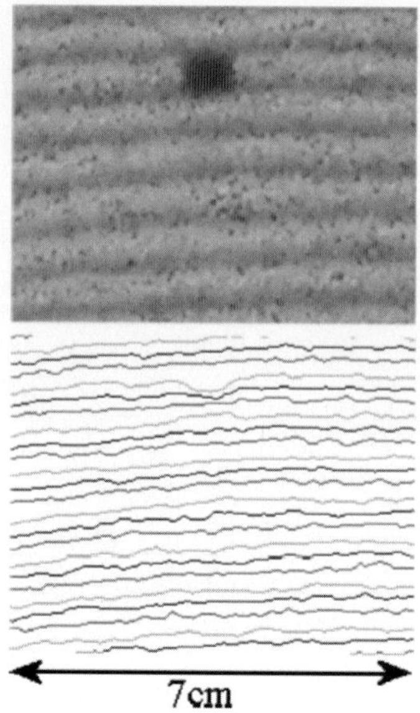

Fig. 3. (Top panel) detail of the set of coloured lines projected onto the pile. (Bottom panel) the centre of gravity of the lines as recognized by the software

3 The Rough Surface of the Pile

The surface of the rice pile is not completely smooth, but shows surface fluctuations, which are mainly due to the occurrence of avalanches that remove material from a more elevated part of the pile and deposit it at a lower position. Since the avalanches have all kinds of sizes, they lead to surface fluctuations of all kinds of size.

A common way to analyze such 'rough' surfaces is to consider the spatial and temporal dependences of the root-mean-square (RMS) 'width' of the surface. For this we first fit a plane $\bar{h}(x,y,t)$ to the surface $h(x,y,t)$ of the pile. The RMS deviations $w(L,t)$ of the surface with respect to this plane are then easily calculated from

$$w(L,t) = \left(\frac{1}{L^2} \sum_{x,y=1}^{L} \left[h(x,y,t) - \bar{h}(x,y,t) \right]^2 \right)^{\frac{1}{2}} . \quad (2)$$

It is well known [13], that initially this width grows as a power law of time $w(L,t) \sim t^{\beta}$ where β is called the growth exponent, while at later times, when lateral correlations span the whole size of the system L, this width grows as a power law of system size $w(L,t) \sim L^{\alpha}$ where α is called the roughness exponent. Hence the exponents α and β can be obtained from the slope in a log-log plot of w vs. L and t respectively, see Fig. 4. In fact, a

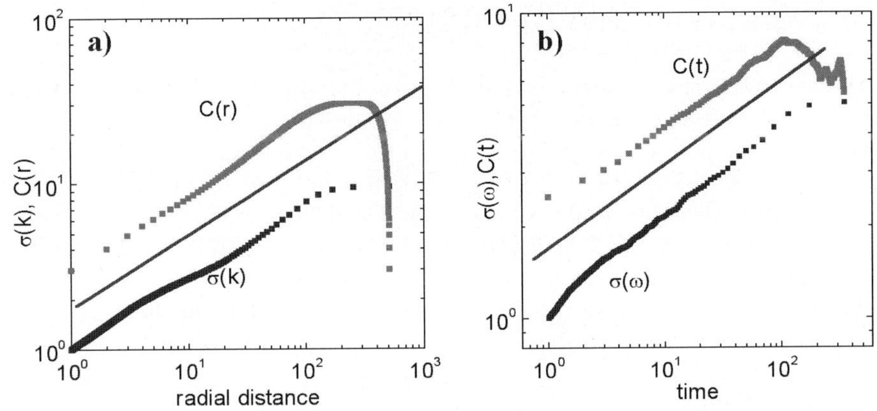

Fig. 4. (**a**) Determination of the roughness exponent from the spatiotemporal correlation function $C(L,0)$ and the distribution function $\sigma(k)$ as defined in the text. The line indicates the average slope, which defines the value for the roughness exponent obtained. (**b**) Determination of the growth exponent from the spatiotemporal correlation function $C(0,t)$ and the distribution function $\sigma(\omega)$ as defined in the text. The line indicates the average slope, which defines the value for the growth exponent obtained

more accurate way to determine these exponents was used for this figure. It is based on the correlation function

$$C(L,t) = \left(\langle [h(x+\xi, y+\eta, t+\tau)\ h(\xi,\eta,\tau)] \rangle_{L,\tau} \right)^{\frac{1}{2}} \tag{3}$$

where the $\langle \cdot \rangle_\tau$ indicates averaging over all $\tau > 0$, and where $\langle \cdot \rangle_L$ indicates averaging over all points (ξ,η) and all (x,y) at a radius L from the origin. The behaviour of this correlation function is similar to that of the width i.e. $C(0,t) \sim t^\beta$ and $C(L,0) \sim L^\alpha$. Alternatively [14], one may also start with the radially averaged power spectrum $S(k)$ of the surface defined by

$$S(k) = \left| \hat{h}(k_x, k_y) \right|^2 \tag{4}$$

where $\hat{h}$ denotes the 2-dimensional Fourier transform of the surface and $k^2 = k_x^2 + k_y^2$. The corresponding spatial distribution function $\sigma(k)$ behaves in the same manner [14] as the correlation function $C(x,t)$ where

$$\sigma^2(k) = \int_0^k S(\kappa)\,\kappa\,\mathrm{d}\kappa\,. \tag{5}$$

Similar relations are defined for the temporal distribution function $\sigma(\omega)$. In Fig. 4 we show this scaling behaviour.

The resulting exponents are $\alpha = 0.42(3)$ for the roughness exponent and $\beta = 0.28(3)$ for the growth exponent. Before discussing these values, we shall now turn our attention to the avalanche behaviour.

4 Avalanches on the Rice Pile

A typical avalanche is shown in Fig. 5. The size and shape of the avalanches can be determined from the height difference of the surface between two consecutive images. In this manner we determine the size of the avalanches (as a volume, which can be expressed in the number of rice grains participating) and also the shape of the avalanches. In particular, we determine the fractal dimension D of the avalanche cluster and also the fractal dimension d_B of the area on which the avalanche happened: this area is a projection of the avalanche shape on the surface of the pile. Both fractal dimensions are determined using box counting [15].

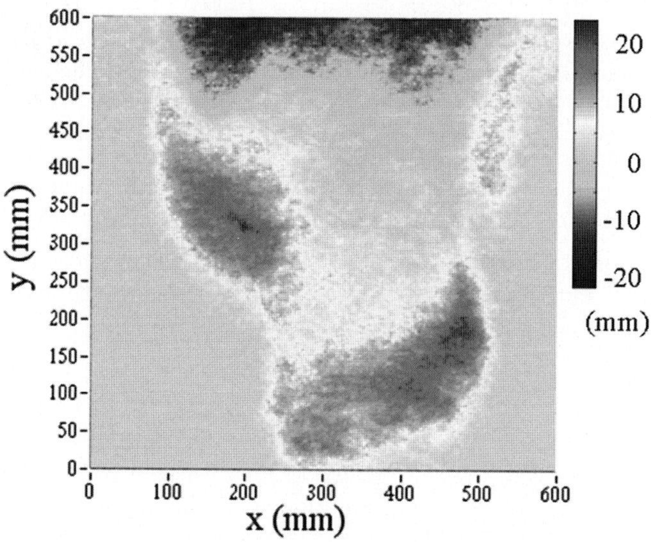

Fig. 5. Typical avalanche on the rice pile. The picture shows the avalanche as the difference in height between two successive images

A plot of avalanche size as a function of time, presented in Fig. 6, clearly shows the punctuated behaviour.

The avalanches were determined in the central part of the pile with a size of 600×600 mm^2. In Fig. 7a we plot the size distribution for the avalanches in this area, but also for subsets of this area.

We observe power law behaviour with a size distribution exponent $\tau = 1.21\,(2)$ (corresponding to the straight line), but clear deviations from this behaviour occur above a certain size, which is related to the size of the area of observation. Interestingly, SOC-theory makes a precise prediction for this deviation: it occurs because the avalanche 'feels' the size of the system L, i.e. because the linear size of the avalanche becomes comparable to L. Clearly

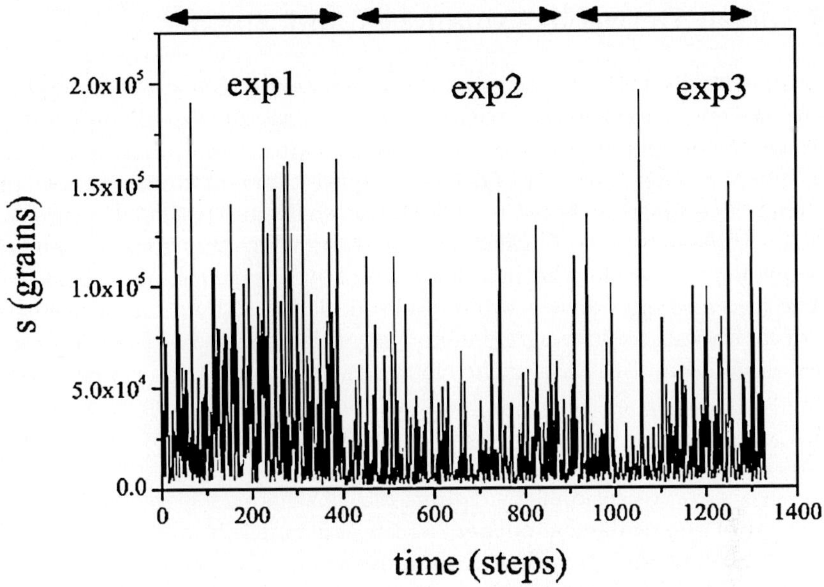

Fig. 6. Size of the avalanches as a function of time in the steady state, i.e. after the system has been left running for a long time. Results for three experiments are shown. Note the punctuated behaviour

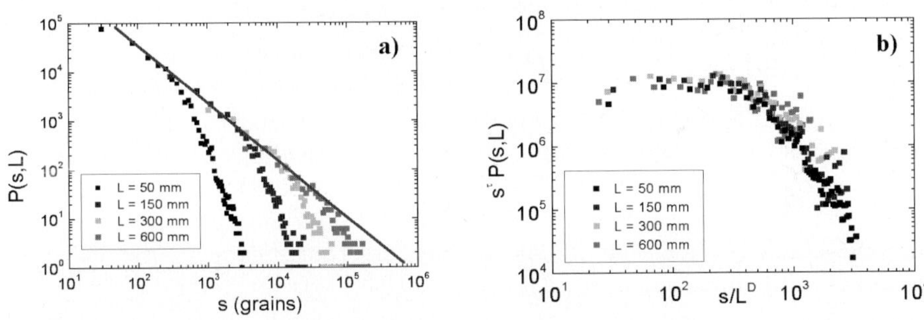

Fig. 7. (**a**) Size distribution of the avalanches. $P(s, L)$ is the number of avalanches of size s that were observed while monitoring a window of size $L \times L$ (**b**) Finite size scaling of the size distribution of the avalanches. Note the nice data-collapse: the deviation from power law behaviour starts from a size $s \sim L^D$ where D is the fractal dimension of the avalanche volume

this happens for $s \approx L^D$ and hence the deviation should scale with s/L^D. This is verified in Fig. 7b , where, in addition, the vertical axis was multiplied by s^τ to take out the power law. Clearly a very nice data-collapse is observed confirming the "finite size scaling"-prediction of SOC theory. This data collapse yields an accurate value for the exponents: $D = 1.99\,(2)$ and $\tau = 1.21\,(2)$.

5 Relation Between Avalanches and Surface

An avalanche disturbs and changes the surface on which it occurs and intuitively one might expect that there exists a relation between the properties of an avalanche and the surface that remains after the avalanche has taken place. On the other hand, avalanches are quite different from granular piles, and hence a comparison is not trivial. Below we discuss two relations between the statistical properties of avalanches and the surface that they leave behind. Subsequently, we verify whether these are obeyed in our experiments.

The first and most simple scaling relation is found by calculating the volume of a (fractal) avalanche. By definition, this volume is L^D since D is the fractal dimension of the avalanche cluster (which can be determined directly in our experiment). On the other hand, the volume should also be equal to the fractal surface times the fractal height. The fractal surface area is the projection of the avalanche cluster on an average (flat) plane trough the pile surface. The area of this projection is by definition L^{d_B} where d_B is the surface fractal dimension of the avalanche cluster, which can be easily obtained in our experiment. The height or thickness of an avalanche is obtained by subtracting the heights (in the direction perpendicular to the average plane, mentioned above) of the piles before and after the avalanche. Since the avalanche modifies the pile only locally, this fractal thickness scales as the surface roughness and is proportional to L^α. Combining these ingredients yields $L^D \sim L^{d_B} L^\alpha$ from which we obtain the scaling relation

$$D = d_B + \alpha \ . \tag{6}$$

The second scaling relation follows from the fact that the deviation from power law behaviour seen in Fig. 7 occurs because avalanches above a certain size 'feel' that the pile is finite. We shall now make this more explicit.

As stipulated above, the time evolution of an initially flat surface subject to roughening is such that at short times its root-mean-squared width w increases with time as $w \sim t^\beta$. After some time, however, lateral correlations extend over the whole area of the pile and w does not increase anymore: it is limited to a pile-size dependent value $w \sim L^\alpha$, where L is the linear size of the pile and α is the roughness exponent. At the cross-over time $t_\times$, both relations hold from which

$$t_\times \sim L^{\alpha/\beta} \ . \tag{7}$$

On the other hand, since this cross-over is due to correlations that start to span the whole system, $t_\times$ is also the moment when the first avalanche occurs that spans the whole pile; by definition the size of this avalanche is $s_\times$. Since we seed the pile with a constant rate, $t_\times$ also is proportional to the amount of mass M we must add to the pile to obtain such a pile-spanning avalanche. However, before we create an avalanche of size $s_\times$, many smaller avalanches have occurred, adding to the amount of material that we have to add before a pile-spanning avalanche takes place. Thus the total mass M is

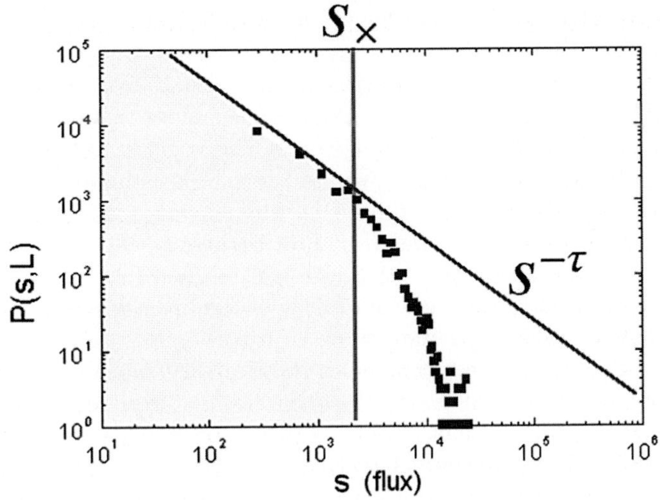

Fig. 8. The scaling relation $D(2-\tau) = \alpha/\beta$ is derived by comparing the time needed for correlations to span the whole pile and the time needed to create avalanches that span the whole pile (see text)

equal to the integral of the size distribution function up to $s_\times$, i.e. the shaded area in Fig. 8. This gives

$$t_\times \sim M = \int_0^{s_\times} sP(s)\,\mathrm{d}s \sim \int_0^{s_\times} s\, s^{-\tau}\mathrm{d}s \sim s_\times^{2-\tau} . \tag{8}$$

In the finite size scaling analysis we found that $s \sim L^D$, hence we obtain

$$t_\times \sim s_\times^{2-\tau} \sim L^{D(2-\tau)} . \tag{9}$$

Combination with (7) yields the exponent relation

$$D(2-\tau) = \frac{\alpha}{\beta} . \tag{10}$$

Equations (6) and (10), which were previously derived by Paczuski et al. [8], offer an interesting possibility: one can calculate the roughness and growth exponents, α and β, from the *avalanche* properties only. In the table below we compare the values [10] from such an analysis with those obtained above from a direct roughness analysis of the surface of the pile:

	α	β
from roughness analysis	0.42 (3)	0.28 (3)
from avalanche analysis	0.41 (3)	0.26 (2)

Clearly, an excellent agreement is found, which supports the underlying assumption i.e. that SOC theory yields a valid description of the avalanche behaviour in our pile.

It is interesting to note that the values we find for the roughening of our rice pile are very close to those observed for Kardar-Parisi-Zhang [16] (KPZ) systems. For a 2-dimensional interface it was found from simulations that for KPZ systems $\alpha = 0.39$ and $\beta = 0.24$, rather close to the values given in the table above. In addition, if we consider the roughening of contour lines (lines of equal height) on our pile, we find for these 1-dimensional interfaces $\alpha = 0.48\,(3)$ and $\beta = 0.33\,(3)$, while it can be shown [13] rigorously that for 1-dimensional KPZ $\alpha = 1/2$ and $\beta = 1/3$. Strangely enough, our system is very unlike a KPZ system: the latter is governed by a Langevin (differential) equation, which describes the development of the interface in a deterministic and local fashion (although the effects of disorder are very important and must be included in the description). To the contrary, our system seems much more random and non-local due to the occurrence of avalanches. Nevertheless the most likely manner of change of our pile is by increase or decrease of its height, which is most easy in a direction that is parallel to the local normal to the surface. Exactly this is also the main ingredient in the derivation of the KPZ equation: growth always proceeds along the local normal. Possibly, the KPZ model extends to non-Langevin systems obeying this same growth rule.

6 Avalanches as Spatiotemporal Fractals

In this section we discuss the spatiotemporal structure of the avalanches. It is quite natural under our condition of constant seeding of the pile, to consider all avalanches together as belonging to one superavalanche, which is punctuated in time. In fact, using similar theoretical ideas as above, one can make a testable prediction for the power spectrum of this superavalanche.

To derive a scaling relation for the avalanche power spectrum exponent, let us consider the development of an avalanche, see Fig. 9. Shown are five successive stages of the development of an avalanche cluster, which is shown full-grown at the top. At the bottom of the figure, the substrate area of this avalanche is shown. At a certain moment during the avalanche a site is changed (indicated in the middle three pictures by a red colour) or not (blue). If we consider a single site, then the behaviour as a function of time is punctuated and a time-line through such a site (indicated by the vertical line) is a fractal. We define the length of this time-line as T^θ where T is the total time needed to create the full-grown avalanche and θ is the fractal dimension of the time-line (to be calculated below). If the linear size of the full-grown avalanche cluster is L (note the change in definition of L), then (according to similar arguments as given above concerning $t_\times$) the time needed to create such correlated cluster is $T \sim L^{\alpha/\beta}$. Hence the length of the time-line is $T^\theta \sim L^{\theta\alpha/\beta}$. The full-grown cluster is created during the time T and has a volume that is easily calculated from its substrate area L^{d_B} and its height, corresponding to the number of times T^θ that activity occurred during its

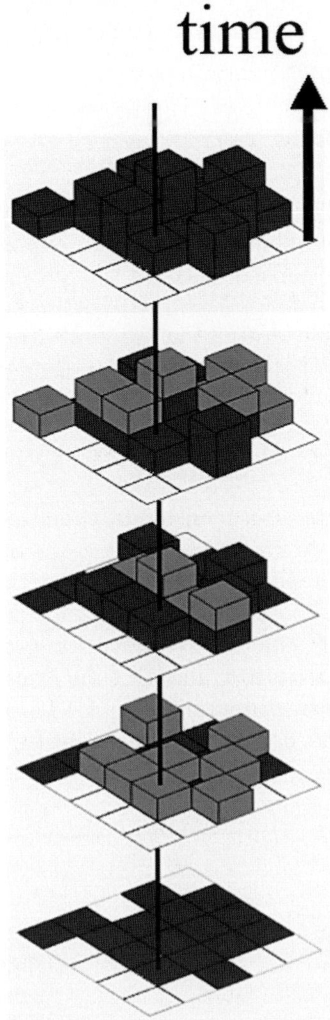

Fig. 9. Schematic representation of the spatiotemporal structure of an avalanche. Shown are five successive stages of development of an avalanche cluster, shown in full-grown state at the top. At the bottom the substrate area of the full-grown avalanche is shown. As time progresses there is activity (grey) or not (black) at a particular site. A time line running through one site is indicated by the vertical line. For further explanation see text

growth. From this we obtain for the volume of the full-grown avalanche cluster $V = L^{d_B} \times L^{\theta\alpha/\beta}$. On the other hand, this size is also L^D, since D is the fractal dimension of the full-grown avalanche. Combining yields the scaling relation

$$D = d_B + \theta \frac{\alpha}{\beta} . \tag{11}$$

Solving this for θ yields

$$\theta = (D - d_B) \frac{\beta}{\alpha} = \beta \tag{12}$$

where we have used (6). Hence the fractal dimension of the time line through a single site has the same value as the growth exponent.

We now determine the consequence for the power spectrum. Consider first the activity-function $A(t)$ for a certain timeline, such as in Fig. 9. $A(t)$ is one at each moment of activity and zero at all other times. On the average the total number of ones of A during a time t, which we call $n(t)$ is given by

$$n(t) = \int_0^t A(\tau)\, \mathrm{d}\tau \sim t^{\beta} \tag{13}$$

since the number of points on a length t of time-line with fractal dimension β is given by t^{β}.

By definition, the temporal correlation function for the avalanche is $C(t) = \langle A(\tau) A(\tau + t) \rangle_{\tau}$ (note that this $C(t)$ for activity during avalanches is different from the $C(t)$ calculated above for rough surfaces), where the average is over all starting times τ and all time lines. In this context, $C(t)$ is also called [8] the all-return probability $P_{\mathrm{all}}(t)$ i.e. the probability for a site that is active at $\tau = 0$ to become active again at $\tau = t$. Hence [8] on the average $n(t+1) = n(t) + C(t)$. Thus

$$C(t) \sim \frac{dn(t)}{dt} \sim t^{\beta - 1} . \tag{14}$$

And for the power spectrum [17]:

$$S(f) \sim f^{-\beta} . \tag{15}$$

An experimental power spectrum is shown in Fig. 10. The red line indicates power law behaviour with a negative power of 0.27 (3). According to (15) one expects a value equal to the growth exponent $\beta = 0.28\,(3)$ as determined above. Clearly we find a good agreement supporting again the applicability of SOC theory.

7 How to Prevent Avalanches

It is rather worrying that despite vigorous efforts to prevent snow avalanches by controlled explosions, accidents still occur. In fact, the same applies to forest fires (notably in the National Parks in the US), where 'controlled' burning sometimes gets out-of-hand and starts large fires. In fact, it is a

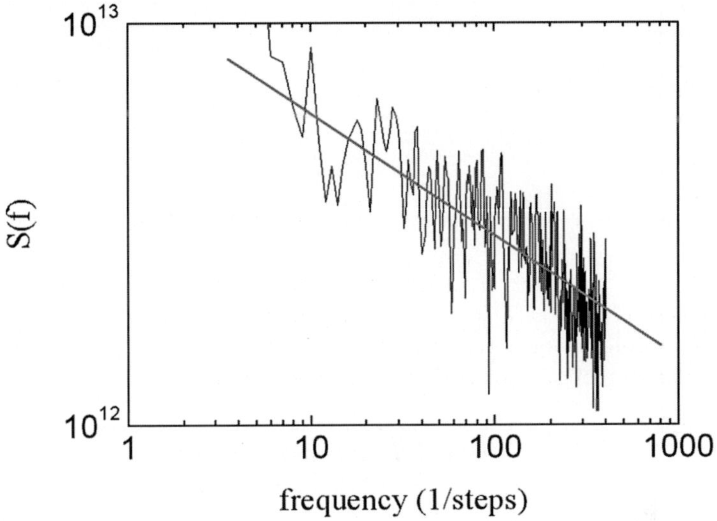

Fig. 10. Power spectrum of the avalanches of Fig. 6. The red line indicates power law behaviour with a negative power of 0.27 (3), very close to the value of the growth exponent, 0.28 (3), in good agreement with SOC theory

property of SOC that 'small' disturbances may lead to big avalanches, making control very difficult.

In Fig. 11 we replot the data of a comparative study [2] of snow avalanches in two regions of the US, one where controlled explosions were used to try to prevent the occurrence of large avalanches and another, where no explosions were made. Clearly, the power law distributions are very similar, with nearly equal slope. The disturbed region shows even slightly worse behaviour: the sizes are slightly larger and the decay is slightly slower, making large avalanches relatively more likely.

To investigate another method for the prevention of devastating avalanches, we study the *transient* state. This is the state of the system before it reaches criticality. The rice pile was prepared manually to be at an angle much smaller than the critical angle (we refer to this state hereafter as 'flat'), after which the measurement was immediately started. A detailed analysis of the multi-scaling properties of the temporal correlation function was made [18], but here we shall concentrate on the avalanche properties.

The punctuated behaviour of the avalanches is shown in Fig. 12, where we see that initially large avalanches do not occur. Indeed averaging the avalanche sizes over 100 consecutive time-steps shows a linear increase with time, see Fig. 13, while the system is moving from the prepared 'flat' state towards steady-state SOC behaviour.

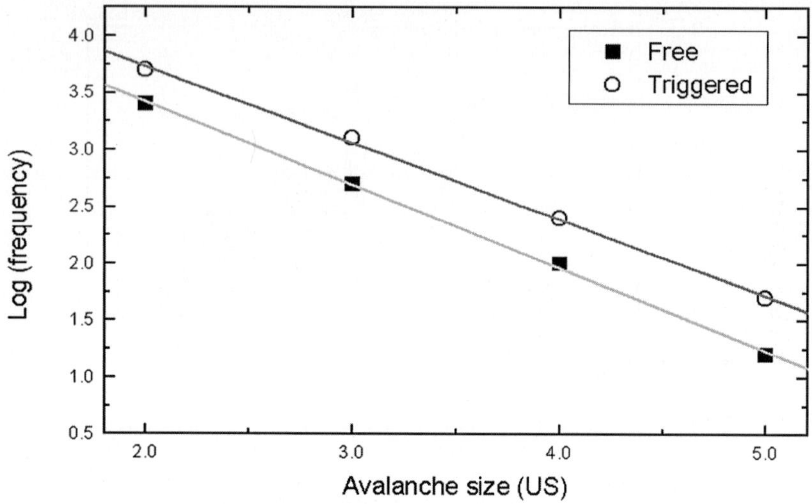

Fig. 11. Replot of the data of Birkeland and Landry [2], showing that the avalanche size distribution for snow avalanches is slightly 'worse' for the region where attempts were made to prevent large avalanches by controlled explosions

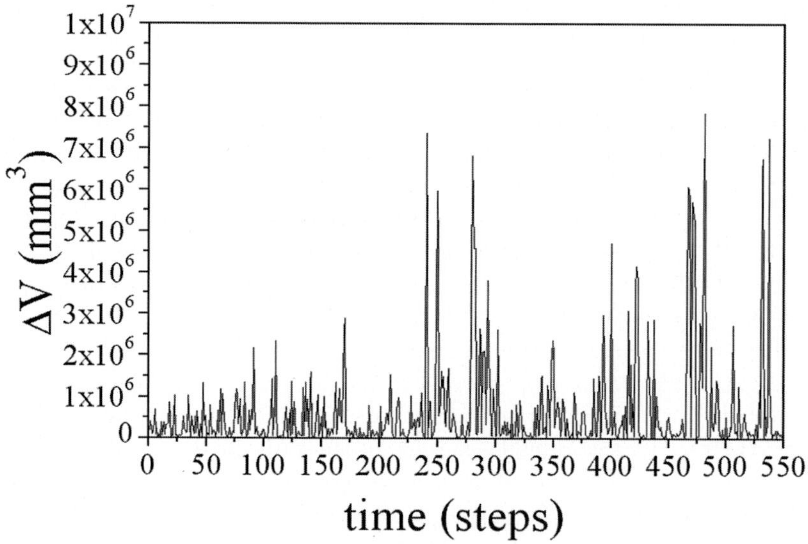

Fig. 12. Avalanche volume versus time immediately after preparing the pile manually in a rather flat state. Clearly, in the beginning large avalanches do not occur

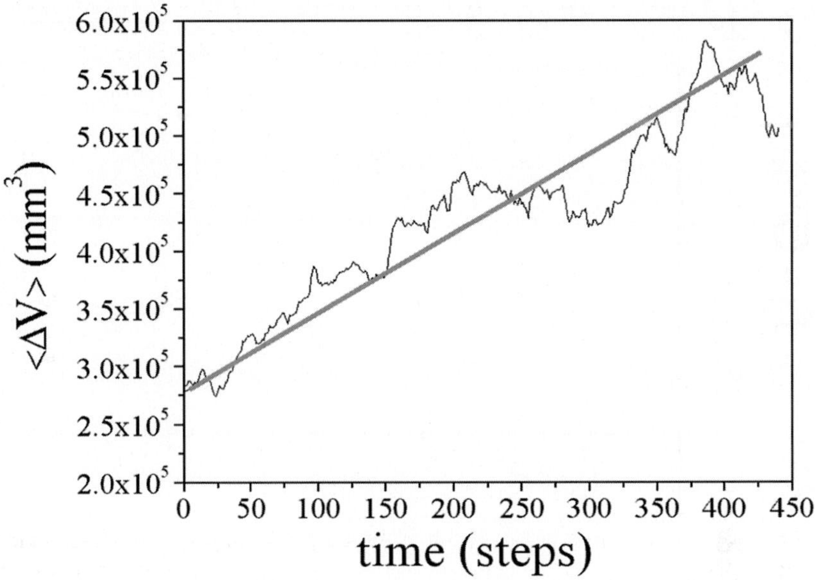

Fig. 13. Size of the avalanches of Fig. 12 averaged over 100 time steps. Clearly the average avalanche size increases linearly with time while the system is developing towards the SOC steady state

Interestingly, SOC theory [8] prescribes how the system should approach the SOC state. We call the critical slope of the rice f_c. This is the maximum slope that a rice pile can have without starting to slide. The value $f_c = 0.92\,(1)$ was determined experimentally [11] (i) by slowly tilting a small box with rice until it started to slide and (ii) from the maximum slope observed anywhere in our large pile during many experiments.

We call the maximum slope observed anywhere in the pile at a certain instant in time G. SOC theory describes how G approaches f_c after starting from a 'flat' state. In particular, it predicts that the average avalanche size $\langle \Delta V \rangle$ diverges as the SOC state is approached according to $\langle \Delta V \rangle = (f_c - G)^{-\gamma}$. Combining this with the experimental observation that $\langle \Delta V \rangle \sim t$ yields

$$f_c - G \sim t^{-1/\gamma} \,. \tag{16}$$

Indeed, from the experimental data we find a reasonable agreement to this behaviour, see Fig. 14, except for the very beginning of the experiment.

In fact, according to SOC theory [8,11], the observed power $1/\gamma$ is given by

$$\frac{1}{\gamma} = \frac{1 + d_B/D - \tau}{2 - \tau} \,. \tag{17}$$

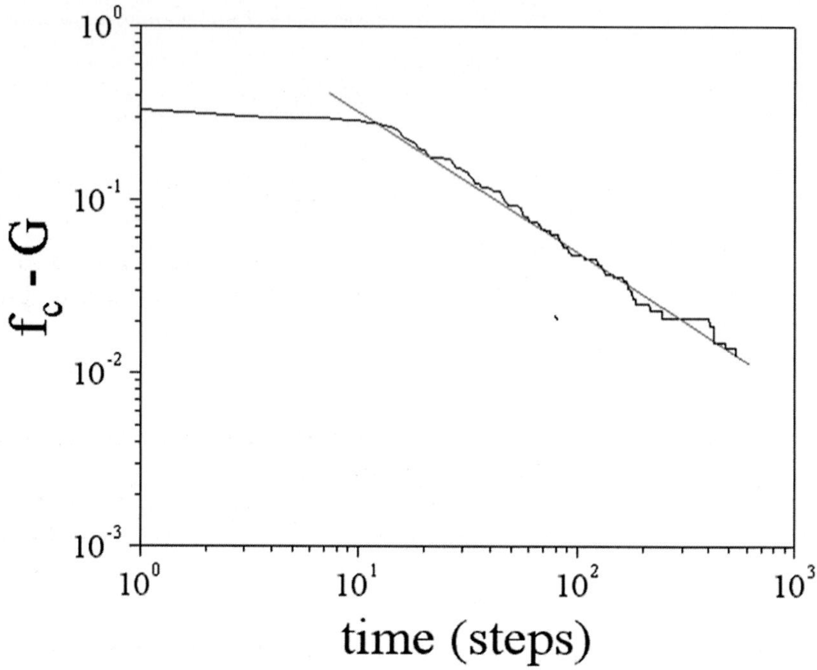

Fig. 14. Difference between actual maximal slope anywhere in the pile G and the critical slope f_c of rice as a function of time. The red line indicates power law behaviour with slope 0.8 (1)

This is an interesting relation, because it connects the behaviour in the transient non-SOC state as given by γ to exponents (D, d_B and τ) of the steady SOC state. Experimentally, we find $1/\gamma = 0.8\,(1)$, while using the values obtained above for D, d_B and τ, one would expect $1/\gamma = 0.74\,(2)$, which is a nice agreement.

In addition to the change in average avalanche size, also the avalanche size distribution exponent τ changes during the transient state, see Fig. 15. Initially τ is larger, indicating a faster decrease of the distribution function with increasing avalanche size and hence a smaller chance for large avalanches. This makes the transient state safer. An alternative to firing explosives in (near-)critical snow masses is to disturb the snow at an early stage. Thus the snow may be kept from developing SOC behaviour. In addition, due to the larger τ in the transient, even if triggering provokes an avalanche, the chances that it is very big are significantly reduced. Similar arguments hold for controlled burning in the prevention of forest fires.

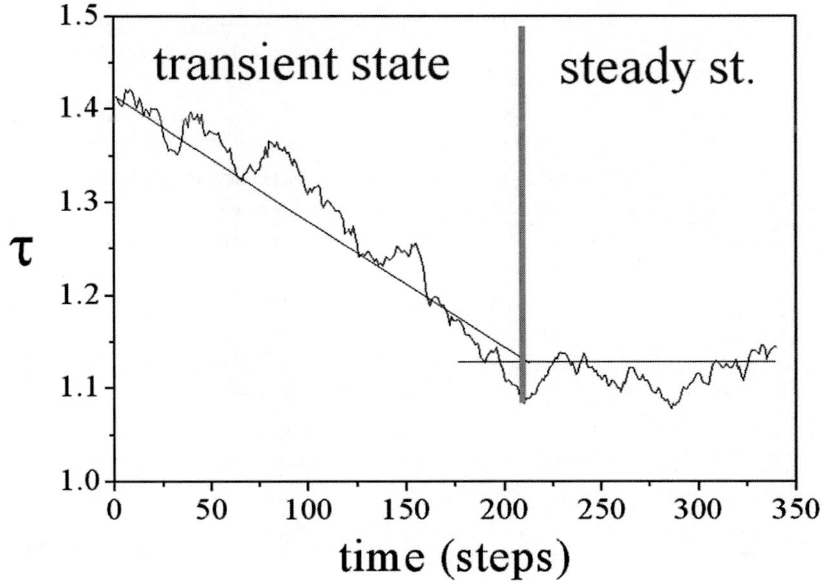

Fig. 15. Change of the avalanche size distribution exponent τ as a function of time in the transient state. Clearly at early times the distribution function is steeper, making large avalanches less likely

8 Conclusions

Self-organized criticality is a class of models meant to describe punctuated behaviour in naturally occurring phenomena such as earthquakes, avalanches and extinctions of species in biology. These wildly different phenomena have characteristic properties, such as the size of events, that all have power law distribution functions. The exponents of these power laws may vary from system to system, however, SOC theory [8] gives explicit relations between the exponents. In fact, many of the exponents can be expressed in terms of only three of them: D, d_B and τ.

We have experimentally investigated the statistical properties of avalanches on a 3-dimensional pile of rice and find that avalanche sizes are indeed power law distributed. In addition, we find that the starting point for deviations from power law behaviour scales as $s_\times \sim L^D$, as predicted by SOC theory. Using this and direct imaging of the avalanches, we have directly measured D, d_B and τ. From the properties of the surface of the pile, we independently determined its roughness and growth exponents α and β. We have thus verified that within experimental accuracy, the scaling relations $D = d_B + \alpha$ and $D\,(2 - \tau) = \alpha/\beta$ are obeyed.

In addition, we have verified that the value of the exponent of the power spectrum of the avalanches is equal to β as predicted by another scaling relation.

It is striking that, while we find SOC behaviour in our rice pile and are thus able to verify the exponent relations proposed by Paczuski et al. [8] for our system, for sand the occurrence of SOC is highly debated [9]. It is conceivable that a certain minimum amount of disorder is necessary for SOC behaviour to occur. This idea is supported by the experiments of Altshuler et al. on steel balls [19] and by some of us on superconductors [20], and may also explain why in the Oslo experiments [12], long grained rice (like ours) behaved as SOC while more rounded rice did not.

In experiments that started from a manually prepared 'flat' pile, we investigated the approach from this 'flat' state towards the SOC state. Again, within experimental accuracy, the scaling relation (17) is corroborated by our experiment. In the approach towards SOC, the avalanches are smaller and their distribution is steeper, making the chance for large avalanches much smaller than in the SOC state. It seems that the best strategy to avoid huge avalanches is to stay away from the SOC state. That is indeed possible by continuously disturbing the pile.

References

1. B. Gutenberg, C.F. Richter: *Seismicity of the Earth and Associated Phenomena* (Princeton University Press, Princeton 1949); B. Gutenberg, C.F. Richter: Bull. Seis. Soc. Am. **34**, 185 (1944) and B. Gutenberg, C.F. Richter: Ann. Geofis. **9**, 1 (1956)
2. K.W. Birkeland, C.C. Landry: Geophys. Res. Lett. **29**, 1554 (2002)
3. C.M. Aegerter: Physica A **319**, 1 (2003)
4. X. Gabaix, P. Gopikrishnan, V. Plerou, H.E. Stanley: Nature **423**, 267 (2003)
5. P. Bak, K. Sneppen: Phys. Rev. Lett: **71**, 4083 (1993)
6. P. Bak, C. Tang, K. Wiesenfeld: Phys. Rev. A **38**, 364 (1988)
7. P. Bak: *How Nature Works* (Copernicus, New York 1966)
8. M. Paczuski, S. Maslov, P. Bak: Phys. Rev. E **53**, 414 (1996)
9. H.M. Jaeger, S.R. Nagel, R.P. Behringer: Rev. Mod. Phys. **68**, 1259 (1996); S.R. Nagel: ibid. **64**, 321 (1992) and references therein
10. C.M. Aegerter, R.Günther, R.J. Wijngaarden: Phys. Rev. E **67**, 051306 (2003)
11. C.M. Aegerter, K.A. Lőrincz, M.S. Welling, R.J. Wijngaarden: Phys. Rev. Lett. **92**, 058702 (2004)
12. V. Frette, K. Christensen, A. Malthe-Sørenssen, J. Feder, T. Jøssang, P. Meakin: Nature **379**, 49 (1996)
13. A.L. Barabasi and H.E. Stanley: *Fractal Concepts in Surface Growth* (Cambridge University Press, Cambridge, 1995)
14. J. Schmittbuhl, J.-P. Vilotte, and S. Roux: Phys. Rev. E **51** (1995) 131; M. Siegert: ibid. **53**, 3209 (1996); J.M. Lopez, M.A.Rodriguez, R. Cuerno: ibid. **56**, 3993 (1997)
15. B.B. Mandelbrot: *The Fractal Geometry of Nature* (Freeman, New York 1983)

16. M. Kardar, G. Parisi, Y.C. Zhang: Phys. Rev. Lett. **56**, 889 (1986)
17. H.J. Jensen: *Self Organized Criticality* (Cambridge University Press, Cambridge 1998)
18. C.M. Aegerter, K.A. Lőrincz, R.J. Wijngaarden: Europhys. Lett. **67**, 342 (2004)
19. E. Altshuler, O. Ramos, C. Martinez, L.E. Flores, C. Noda: Phys. Rev. Lett. **86**, 5490 (2001)
20. M.S. Welling, C.M. Aegerter, R.J. Wijngaarden: cond-mat/0410369.

Part V

Miscellaneous

A Recent Appreciation of the Singular Dynamics at the Edge of Chaos

E. Mayoral and A. Robledo

Instituto de Física, Universidad Nacional Autónoma de México, Apartado Postal 20-364, México 01000 D.F., Mexico
robledo@fisica.unam.mx

Summary. We study the dynamics of iterates at the transition to chaos in the logistic map and find that it is constituted by an infinite family of Mori's q-phase transitions. Starting from Feigenbaum's σ function for the diameters ratio, we determine the atypical weak sensitivity to initial conditions ξ_t associated to each q-phase transition and find that it obeys the form suggested by the Tsallis statistics. The specific values of the variable q at which the q-phase transitions take place are identified with the specific values for the Tsallis entropic index q in the corresponding ξ_t. We also describe the bifurcation gap induced by external noise and show that its properties exhibit the characteristic elements of glassy dynamics close to vitrification in supercooled liquids, e.g. two-step relaxation, aging and a relationship between relaxation time and entropy.

1 Introduction

The logistic equation was introduced in 1845 by the Belgian mathematician and sociologist Pierre-François Verhulst to model the growth of populations limited by finite resources [1]. The discrete time variable version of Verhulst's growth law, the logistic map, has become a foundation stone for the theory of nonlinear dynamics. The logistic map is the archetypal example of how the use as starting point of simple nonlinear discrete maps have often led to significant developments in the theory of nonlinear dynamical systems [2]. The discovery of the universal properties associated to the renowned period-doubling and intermittency routes to chaos displayed by the logistic map, similar to those of conventional critical phenomena in statistical physics, triggered, about three decades ago, an upsurge of activity in the field and now both routes, as well as many other remarkable features displayed by the logistic map, are well understood.

Here we might argue that at the present time the logistic map is again becoming a prototypical model system. This time for the assessment of the validity and understanding of the reasons for applicability of the nonextensive generalization of the Boltzmann–Gibbs (BG) statistical mechanics [3,4]. This is a formalism assumed to be appropriate for circumstances where the system is out of the range of validity of the canonical BG theory. And these circumstances are believed in some cases to be a breakdown in the chain of increasing randomness from non-ergodicity to completely developed phase-space mixing.

The logistic map contains infinite families of critical attractors at which the ergodic and mixing properties breakdown. These are the tangent bifurcations and the accumulation point(s) of the pitchfork bifurcations, the so-called onset of chaos [2]. At each of the map critical attractors the Lyapunov coefficient λ_1 vanishes, and the sensitivity to initial conditions ξ_t for large iteration time t ceases to obey exponential behaviour, exhibiting instead power-law or faster than exponential behaviour. The pitchfork bifurcations are also critical attractors at which the negative Lyapunov coefficient of periodic orbits goes to zero. There are other attractors at which the Lyapunov coefficient diverges to minus infinity, where there is faster than exponential convergence of orbits. These are the superstable attractors located between successive pitchfork bifurcations and their accumulation point is also the onset of chaos.

Here we review briefly specific and rigorous results on the dynamics associated to the critical attractors of the logistic map, or of its generalization to nonlinearity of order $z > 1$, $f_\mu(x) = 1 - \mu |x|^z$, $-1 \leq x \leq 1$, $0 \leq \mu \leq 2$. (The phase space variable is x, the control parameter is μ and the conventional logistic map corresponds to $z = 2$). Our results relate to the anomalous sensitivity to initial conditions at the onset of chaos, the associated spectrum of Tsallis q-Lyapunov coefficients, and the relationship of these with the Mori q-phase transitions [5], one of which was originally observed numerically for the Feigenbaum attractor [5, 6]. In particular, we identify the Mori singularities in the Lyapunov spectra with the appearance of special values for the Tsallis entropic index q. As the properties of the logistic map are very familiar and well understood it is of interest to see how previous knowledge fits in with the new perspective.

Tsallis suggested [7] that for critical attractors ξ_t (defined as $\xi_t(x_0) \equiv \lim_{\Delta x_0 \to 0}(\Delta x_t / \Delta x_0)$ where Δx_0 is the initial separation between two orbits and Δx_t that at time t), has the form

$$\xi_t(x_0) = \exp_q[\lambda_q(x_0)\, t] \equiv [1 - (q-1)\lambda_q(x_0)\, t]^{-1/(q-1)} , \qquad (1)$$

that yields the customary exponential ξ_t with λ_1 when $q \to 1$. In (1) q is the entropic index and λ_q is the q-generalized Lyapunov coefficient; $\exp_q(x) \equiv [1-(q-1)x]^{-1/(q-1)}$ is the q-exponential function. Tsallis also suggested [7] that the Pesin identity $K_1 = \lambda_1$ (where the rate of entropy production K_1 is given by $K_1 t = S_1(t) - S_1(0)$, t large, and $S_1 = -\sum_i p_i \ln p_i$) would be generalized to $K_q = \lambda_q$, where the q-generalized rate of entropy production K_q is defined via $K_q t = S_q(t) - S_q(0)$, t large, and where

$$S_q \equiv \sum_i p_i \ln_q \left(\frac{1}{p_i}\right) = \frac{1 - \sum_i^W p_i^q}{q-1} \qquad (2)$$

is the Tsallis entropy; $\ln_q y \equiv (y^{1-q} - 1)/(1-q)$ is the inverse of $\exp_q(y)$.

To check on the above, we have recently analysed [8–11] both the pitchfork and tangent bifurcations and the onset of chaos of the logistic map and found that indeed the Tsallis suggestions hold for these critical attractors, though with some qualifications. For the case of the tangent bifurcation it is important to neglect the feedback mechanism into the neighbourhood of the tangency to avoid the crossover out of the q-exponential form for ξ_t. As we explain below, for the onset of chaos there is a multiplicity of q-indices, appearing in pairs q_j, and $Q_j = 2 - q_j$, $j = 0, 1, \ldots$, and a spectrum of q-Lyapunov coefficients $\lambda_{q_j}^{(k,l)}$ and $\lambda_{Q_j}^{(k,l)}$ for each q_j and Q_j, respectively. (The superindex (k, l) refers to starting and final trajectory positions). The dynamics of the attractor confers the q-indexes a decreasing order of importance. Retaining only the dominant indices q_0 and Q_0 yields a quite reasonable description of the dynamics and considering the next few leading indexes provides a satisfactorily accurate account.

Although in brief, we also describe our finding [12] that the dynamics at the noise-perturbed edge of chaos in logistic maps is analogous to that observed in supercooled liquids close to vitrification. The three major features of glassy dynamics in structural glass formers, two-step relaxation, aging, and a relationship between relaxation time and configurational entropy, are displayed by orbits with vanishing Lyapunov coefficient. The known properties in control-parameter space of the noise-induced bifurcation gap play a central role in determining the characteristics of dynamical relaxation at the chaos threshold.

2 Critical Attractors in the Logistic Map

For our purposes it is convenient to recall some essentials of logistic map properties. The accumulation point of the period doublings and also of the chaotic band splittings is the Feigenbaum attractor that marks the threshold between periodic and chaotic orbits, at $\mu_\infty(z)$, with $\mu_\infty = 1.40115\ldots$ when $z = 2$. The locations of period doublings, at $\mu = \mu_n < \mu_\infty$, and band splittings, at $\mu = \widehat{\mu}_n > \mu_\infty$, obey for large n power laws of the form $\mu_n - \mu_\infty \sim \delta(z)^{-n}$ and $\mu_\infty - \widehat{\mu}_n \sim \delta(z)^{-n}$, with $\delta = 0.46692\ldots$ when $z = 2$, which is one of the two Feigenbaum's universal constants. For our use below we also recall the sequence of parameter values $\overline{\mu}_n$ employed to define the diameters d_n of the bifurcation forks that form the period-doubling cascade sequence. At $\mu = \overline{\mu}_n$ the map displays a 'superstable' periodic orbit of length 2^n that contains the point $x = 0$. For large n the distances to $x = 0$, of the iterate positions in such 2^n-cycle that are closest to $x = 0$, $d_n \equiv f_{\overline{\mu}_n}^{(2^{n-1})}(0)$, have constant ratios $d_n/d_{n+1} = -\alpha(z)$; $\alpha = 2.50290\ldots$ when $z = 2$, which is the second of the Feigenbaum's constants. A set of diameters with scaling properties similar to those of d_n can also be defined for the band splitting sequence [2]. Other diameters in the 2^n-supercycles are defined as the distance of the mth element

x_m to its nearest neighbour $f_{\overline{\mu}_n}^{(2^{n-1})}(x_m)$. That is

$$d_{n,m} \equiv f_{\overline{\mu}_n}^{(m+2^{n-1})}(0) - f_{\overline{\mu}_n}^{(m)}(0), \; m = 0, 1, 2, \ldots \; , \tag{3}$$

with $d_{n,0} = d_n$. Feigenbaum [13] constructed the auxiliary function

$$\sigma_n(m) = \frac{d_{n+1,m}}{d_{n,m}} \tag{4}$$

to quantify the rate of change of the diameters and showed that it has finite (jump) discontinuities at all rationals, as can be seen by considering the variable $y = m/2^{n+1}$ with n large (and therefore omitting the subindex n). One obtains [2,13] $\sigma(0) = -1/\alpha$, but $\sigma(0^+) = 1/\alpha^z$, and through the antiperiodic property $\sigma(y + 1/2) = -\sigma(y)$, also $\sigma(1/2) = 1/\alpha$, but $\sigma(1/2 + 0^+) = -1/\alpha^z$. Other discontinuities in $\sigma(y)$ appear at $y = 1/4, 1/8$, etc. As these decrease rapidly in most cases it is only necessary to consider the first few.

An important factor of our work is that the sensitivity to initial conditions $\xi_t(x_0)$ can be evaluated for trajectories within the Feigenbaum attractor via consideration of the discontinuities of $\sigma_n(m)$. If the initial separation is chosen to be a diameter $\Delta x_0 = d_{n,m}$ and the final time t is chosen to have the form $t = 2^n - 1$, then $\Delta x_t = d_{n,m+2^n-1}$, and $\xi_t(x_0)$ can be written [14] as

$$\xi_t(m) \simeq \left| \frac{\sigma_n(m-1)}{\sigma_n(m)} \right|^n, \; t = 2^n - 1, \; n \text{ large} \; . \tag{5}$$

For the clarity of the presentation, we shall only use absolute values of positions so that the dynamics of iterates do not carry information on the self-similar properties of "left" and "right" symbolic dynamic sequences [2]. This choice does not affect the results on the sensitivity to initial conditions.

3 Mori's *q*-Phase Transitions in the Logistic Map

During the late 1980's Mori and coworkers developed a comprehensive thermodynamic formalism to characterize drastic changes at bifurcations and at other singular phenomena in low dimensional maps [5]. The formalism was also adapted to the study of critical chaotic attractors and was illustrated by considering the specific case of the onset of chaos in the logistic map [5,6,15]. For critical attractors the scheme involves the evaluation of fluctuations of the generalized finite-time Lyapunov coefficient

$$\lambda(t, x_0) = \frac{1}{\ln t} \sum_{i=0}^{t-1} \ln \left| \frac{\mathrm{d} f_{\mu_\infty}(x_i)}{\mathrm{d} x_i} \right|, \quad t \gg 1. \tag{6}$$

Notice the replacement of the customary t by $\ln t$ above, as the ordinary Lyapunov coefficient

$$\lambda_1(x_0) = \lim_{t\to\infty} \frac{1}{t} \sum_{i=0}^{t-1} \ln \left| \frac{\mathrm{d}f_{\mu_\infty}(x_i)}{\mathrm{d}x_i} \right| \tag{7}$$

vanishes at μ_∞, $t \to \infty$.

The probability density distribution for the values of λ, $P(\lambda,t)$, is written in the form

$$P(\lambda,t) = t^{-\psi(\lambda)} P(0,t), \quad t \gg 1 \,, \tag{8}$$

where $\psi(\lambda)$ is a concave spectrum of the fluctuations of λ with minimum $\psi(0) = 0$ and is obtained as the Legendre transform of the 'free energy' function $\phi(q)$, defined as

$$\phi(q) \equiv -\lim_{t\to\infty} \frac{1}{\ln t} \ln Z(t,q) \,, \tag{9}$$

where $Z(t,q)$ is the dynamic partition function

$$Z(t,q) \equiv \int \mathrm{d}\lambda \; P(\lambda,t) \; t^{-(q-1)\lambda} \,. \tag{10}$$

The 'coarse-grained' function of generalized Lyapunov coefficients $\lambda(q)$ is given by $\lambda(q) \equiv \mathrm{d}\phi(q)/\mathrm{d}q$ and the variance $v(q)$ of $P(\lambda,t)$ by $v(q) \equiv \mathrm{d}\lambda(q)/\mathrm{d}q$ [5, 6]. Notice the special weight $t^{-(q-1)\lambda}$ in the partition function $Z(t,q)$ as this shapes the quantities derived from it. These functions are the dynamic counterparts of the Renyi dimensions D_q and the spectrum $f(\widetilde{\alpha})$ that characterize the geometric structure of the attractor.

A "q-phase" transition is indicated by a section of linear slope $m_c = 1-q_c$ in the spectrum (free energy) $\psi(\lambda)$, a discontinuity at q_c in the Lyapunov function (order parameter) $\lambda(q)$, and a divergence at q_c in the variance (susceptibility) $v(q)$. For the onset of chaos at $\mu_\infty (z = 2)$ a single q-phase transition was numerically determined [5, 6, 15] and found to exhibit a value close to $m_c = -(1-q_c) \simeq -0.7$; arguments were provided for this value to be $m_c = -(1-q_c) = -\ln 2/\ln\alpha = -0.7555\ldots$ Our analysis described below shows that the older results give a broad picture of the dynamics at the Feigenbaum attractor and that actually an infinite family of q-phase transitions of decreasing weights take place at μ_∞.

4 Tsallis Dynamics at the Edge of Chaos

By taking as initial condition $x_0 = 0$ we found that the resulting orbit consists of trajectories made of intertwined power laws that asymptotically reproduce the entire period-doubling cascade that occurs for $\mu < \mu_\infty$ [9, 11]. This orbit captures the properties of the so-called 'superstable' orbits at $\overline{\mu}_n < \mu_\infty$, $n = 1, 2, \ldots$ [2] (see Fig. 1), and can be used as reference to read all other orbits within the attractor. At μ_∞ the Lyapunov coefficient λ_1 vanishes and

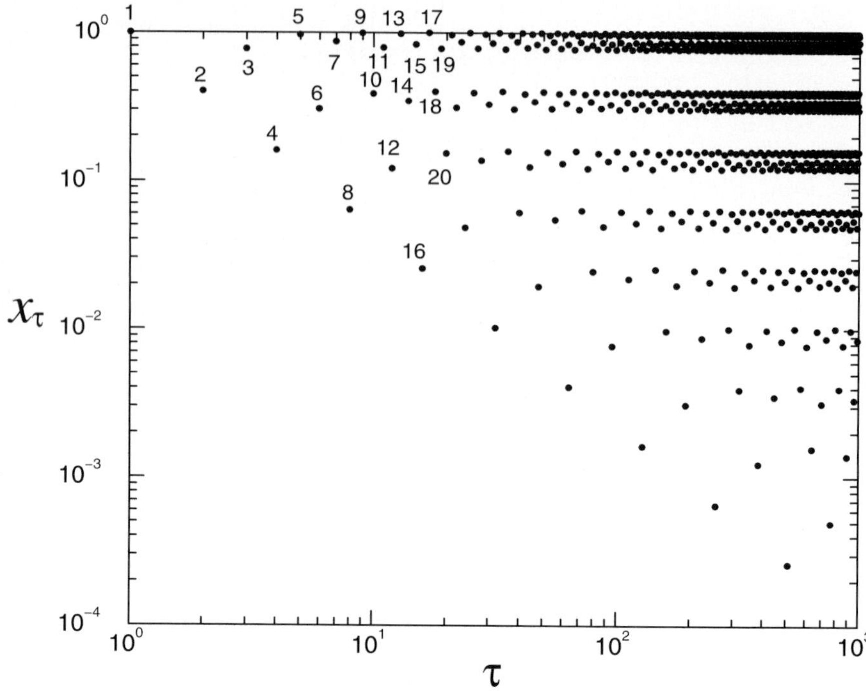

Fig. 1. Absolute values of positions in logarithmic scales of iterations τ for a trajectory at μ_∞ with initial condition $x_0 = 0$. The numbers correspond to iteration times. The power-law decay of the time subsequences can be clearly appreciated

in its place there appears a spectrum of q-Lyapunov coefficients $\lambda_q^{(k,l)}$. This spectrum was originally studied in [6] and [15] and our recent interest has been to study its properties in more detail to examine their relationship with the Tsallis statistics. Recent analytical results about the q-Lyapunov coefficients and the q-generalized Pesin identity are given in [9] and [11].

Now, consider a first approximation to the function $\sigma_n(m)$ for n large, applicable to general non-linearity $z > 1$. This is to assume that half of the diameters scale as α (as in the most sparse region of the attractor) while the other half scale as $\alpha_0 = \alpha^z$ (as in the most crowded region of the attractor). This approximation captures the effect on ξ_t of the most dominant trajectories within the attractor. With these two scaling factors $\sigma_n(m)$ becomes the periodic step function

$$\frac{1}{\sigma_n(m)} = \begin{cases} \alpha_0 = \alpha^z, & 0 < m \leq 2^{n-1}, \\ \alpha, & 2^{n-1} < m \leq 2 \cdot 2^{n-1}, \\ -\alpha_0 = -\alpha^z, & 2 \cdot 2^{n-1} < m \leq 3 \cdot 2^{n-1}, \\ -\alpha\ , & 3 \cdot 2^{n-1} < m \leq 4 \cdot 2^{n-1}, \ldots \end{cases} \tag{11}$$

Use of (11) into (5) for the sensitivity $\xi_t(m)$ yields the result

$$\xi_t(m) = \begin{cases} \alpha^{-(z-1)n}, & m = (2k+1)2^{n-1}, \\ \alpha^{(z-1)n}, & m = (2k+2)2^{n-1}, \end{cases} \tag{12}$$

where $k = 0, 1, \ldots$ and where the final observation time is of the form $t = l2^n - 1$, $l = 1, 2, \ldots$. With the introduction of the total time variable $\tau \equiv m + 1 + t$, (12) can be rewritten in terms of the q-exponential functions

$$\xi_\tau(m) = \begin{cases} [1 + (1-q_0)\lambda_{q_0}^{(k,l)}\tau]^{1/(1-q_0)}, & m = (2k+2)2^{n-1}, \\ [1 + (1-Q_0)\lambda_{Q_0}^{(k,l)}\tau]^{1/(1-Q_0)}, & m = (2k+1)2^{n-1}, \end{cases} \tag{13}$$

where

$$q_0 = 1 - \frac{\ln 2}{\ln \alpha_0/\alpha} = 1 - \frac{\ln 2}{(z-1)\ln \alpha}, \tag{14}$$

$$\lambda_{q_0}^{(k,l)} = \frac{(z-1)\ln \alpha}{(k+2l+1)\ln 2}, \tag{15}$$

$$Q_0 = 1 + \frac{\ln 2}{\ln \alpha_0/\alpha} = 1 + \frac{\ln 2}{(z-1)\ln \alpha}, \tag{16}$$

$$\lambda_{Q_0}^{(k,l)} = -\frac{2(z-1)\ln \alpha}{(2k+4l+1)\ln 2}. \tag{17}$$

Notice that $Q_0 = 2 - q_0$. For $z = 2$ one obtains $Q_0 \simeq 1.7555$ and $q_0 \simeq 0.2445$, this latter value agrees with that obtained in several earlier studies [7,9,11,16]. The two scales considered describe correctly only trajectories that start at the most sparse region of the multifractal ($x \simeq 0$) and terminate at its most crowded region ($x \simeq 1$), or the inverse. (This is why we obtain the two conjugate values q and $Q = 2 - q$, as the inverse of the q-exponential function satisfies $\exp_q(y) = 1/\exp_{2-q}(-y)$). The vertical lines in Fig. 2a represent the ranges of values obtained when $z = 2$ for $\lambda_{q_0}^{(k,l)}$ and $\lambda_{Q_0}^{(k,l)}$. See [14] for more details.

We consider the next discontinuities of importance in $\sigma_n(m)$. Independently of the number of discontinuities taken into account one obtains q-exponential forms for ξ_t. The value of $\sigma(y = m/2^{n+1})$ at $y = 1/4$ measures $1/\alpha_1$, with $\alpha_1 \simeq 5.4588$ for $z = 2$, and it is associated to one 'midway' region between the most crowded and most sparse regions of the attractor (the other 'midway' region being associated to $\sigma(3/4)$). With three scaling factors, α, α_0 and α_1, we have three values for the q index, q_0, q_1 and q_2 (together with the conjugate values $Q_0 = 2 - q_0$, $Q_1 = 2 - q_1$ and $Q_2 = 2 - q_2$ for the inverse trajectories). To each value of q there is a set of q-Lyapunov coefficients running from a maximum $\lambda_{q_i,\max}$ to zero (or a minimum $\lambda_{Q_i,\min}$ to zero). The results when $z = 2$ for the ranges of values obtained for the q-Lyapunov coefficients are shown as the vertical lines in Figs. 2(a), 3(a) and 4(a). Similar results are obtained for the case of four discontinuities in $\sigma_n(m)$, etc.

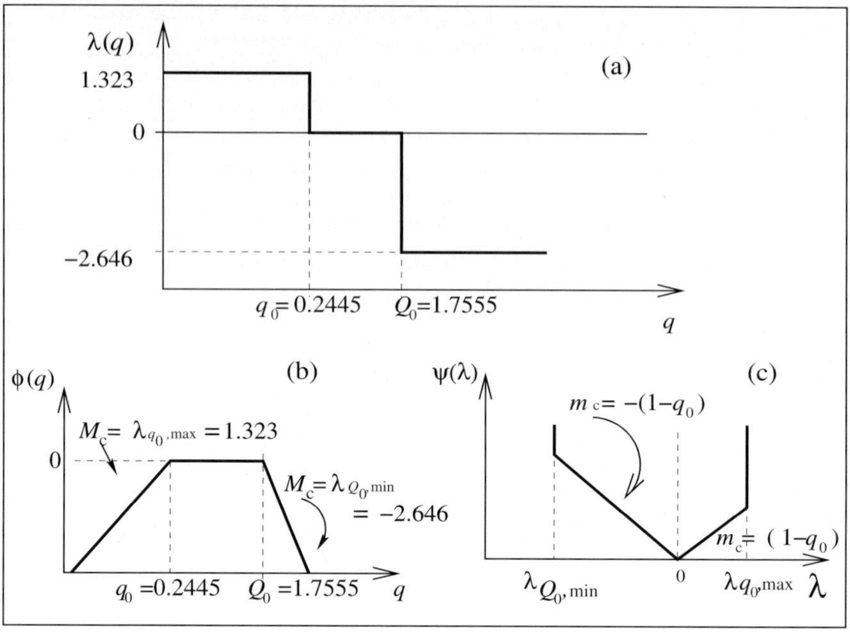

Fig. 2. q-phase transitions with index values q_0 and $Q_0 = 2 - q_0$ obtained from the main discontinuity in $\sigma_n(m)$. See text for details

5 A Family of q-Phase Transitions at the Edge of Chaos

As a function of the variable $-\infty < q < \infty$ the q-Lyapunov coefficients obtained in the previous section are functions with two steps with the jumps located at $q = q_i = 1 - \ln 2/\ln \alpha_i(z)/\alpha(z)$ and $q = Q_i = 2 - q_i$. Immediate contact can be established with the formalism developed by Mori and coworkers and the q phase transition obtained in [6]. Each step function for $\lambda(q)$ can be integrated to obtain the spectrum $\phi(q)$ ($\lambda(q) \equiv \mathrm{d}\phi/\mathrm{d}q$) and from this its Legendre transform $\psi(\lambda)$ ($\equiv \phi - (1-q)\lambda$). We illustrate this first with $\sigma_n(m)$ obtained with two scale factors, as in (11). We show numerical values for the case $z = 2$. From (14) to (17) we obtain

$$\lambda(q) = \begin{cases} \lambda_{q_0,\max}, & q \le q_0 = 1 - \ln 2/(z-1)\ln \alpha \simeq 0.2445 \,, \\ 0, & q_0 < q < Q_0 \,, \\ \lambda_{Q_0,\min}, & q \ge Q_0 = 2 - q_0 \simeq 1.7555 \,, \end{cases} \tag{18}$$

where $\lambda_{q_0,\max} = \ln \alpha/\ln 2 \simeq 1.323$ and $\lambda_{Q_0,\min} = -2\ln \alpha/\ln 2 \simeq -2.646$.

The free energy functions $\phi(q)$ and $\psi(\lambda)$ that correspond to (18) are given by

$$\phi(q) = \begin{cases} \lambda_{q_0,\max}(q - q_0), & q \le q_0 \,, \\ 0, & q_0 < q < Q_0 \,, \\ \lambda_{Q_0,\min}(q - Q_0), & q \ge Q_0 \,, \end{cases}$$

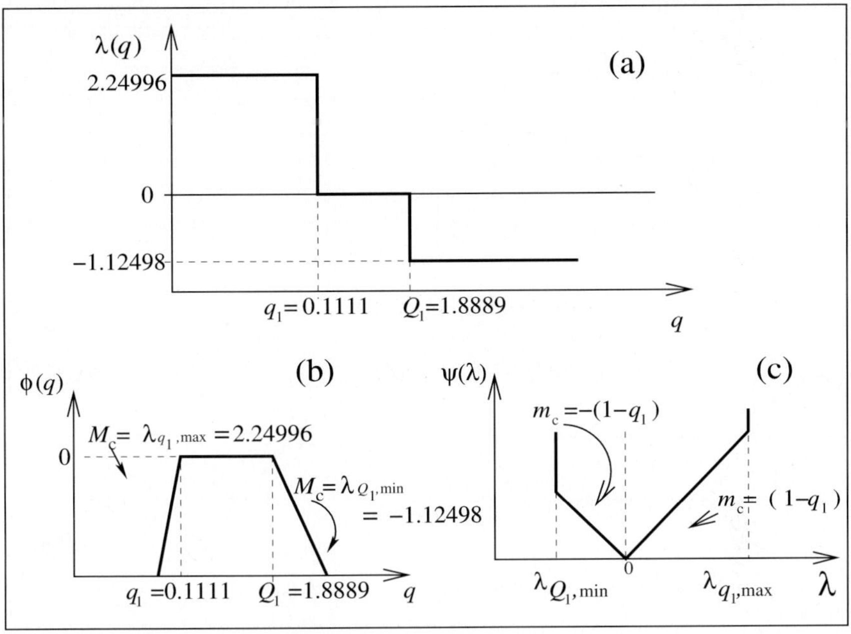

Fig. 3. q-phase transitions with index values q_1 and $Q_1 = 2 - q_1$ obtained from the main two discontinuities in $\sigma_n(m)$. See text for details

and

$$\psi(\lambda) = \begin{cases} (1 - Q_0)\lambda, & \lambda_{Q_0,\min} < \lambda < 0 \, , \\ (1 - q_0)\lambda, & 0 < \lambda < \lambda_{q_0,\max} \, . \end{cases}$$

We show these functions in Fig. 2. The constant slopes of $\psi(\lambda)$ represent the q-phase transitions associated to trajectories linking two regions of the attractor, that in this case are its most crowded and most sparse, and their values $1 - q_0$ and $q_0 - 1$ correspond to those obtained for the q-exponential ξ_t (13). The slope $q_0 - 1 \simeq -0.7555$ coincides with that originally detected by Mori and colleagues [5].

When we consider also the next discontinuity of importance in $\sigma_n(m)$, at $\sigma(1/4) = 1/\alpha_1$, we obtain a set of two q-phase transitions for each of the three values of the q index, q_0, q_1 and q_2. We show in Figs. 2, 3 and 4 the functions $\lambda(q)$, $\phi(q)$, and $\psi(\lambda)$ obtained for these three cases. The parameter values for the q-phase transitions at $1 - q_0$ and $q_0 - 1$ appear again, but now we have also two other sets at $1 - q_1$ and $q_1 - 1$, and at $1 - q_2$ and $q_2 - 1$, that correspond, respectively, to orbits that link a 'midway' region of the attractor with the most sparse region, and with the most crowded region of the attractor.

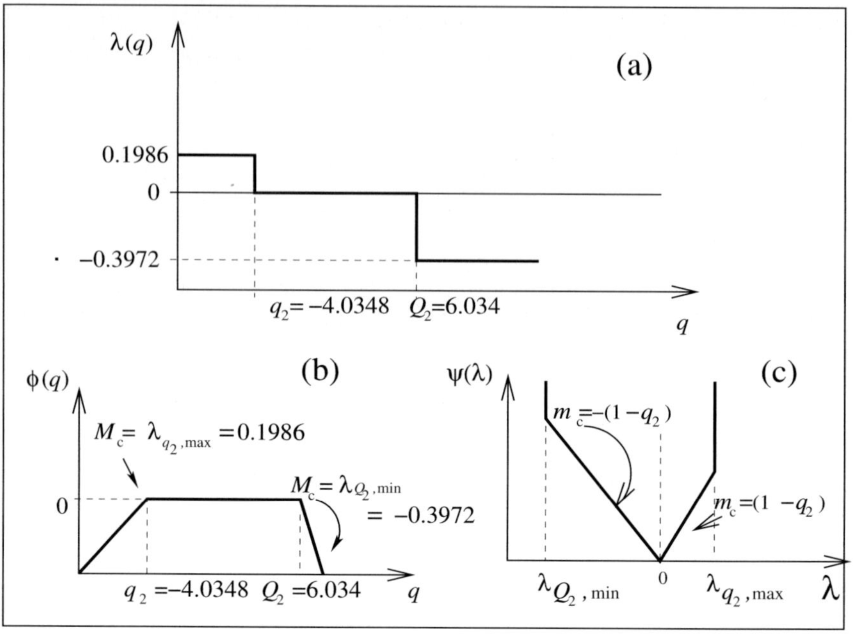

Fig. 4. q-phase transitions with index values q_2 and $Q_2 = 2 - q_2$ obtained from the main two discontinuities in $\sigma_n(m)$. See text for details

6 Noisy Dynamics at the Edge of Chaos

Consider now the logistic map $z = 2$ in the presence of additive noise

$$x_{t+1} = f_\mu(x_t) = 1 - \mu x_t^2 + \chi_t \varepsilon, \ -1 \leq x_t \leq 1, 0 \leq \mu \leq 2 \, , \tag{19}$$

where χ_t is the random variable with average $\langle \chi_t \chi_{t'} \rangle = \delta_{tt'}$, and ε measures the noise intensity [2,17]. Except for a set of zero measure, all the trajectories with $\mu_\infty(\varepsilon = 0)$ and initial condition $-1 \leq x_0 \leq 1$ fall into the attractor with fractal dimension $d_f = 0.5338\ldots$. These trajectories represent nonergodic states, since as $t \to \infty$ only a Cantor set of positions is accessible out of the total phase space. For $\varepsilon > 0$ the noise fluctuations erase the fine features of the periodic attractors as these widen into bands similar to those in the chaotic attractors, yet there remains a definite transition to chaos at $\mu_\infty(\varepsilon)$ where the Lyapunov exponent λ_1 changes sign. The period doubling of bands ends at a finite value $2^{N(\varepsilon)}$ as the edge of chaos transition is approached and then decreases at the other side of the transition. This effect displays scaling features and is referred to as the bifurcation gap [2, 17]. When $\varepsilon > 0$ the trajectories visit sequentially a set of 2^n disjoint bands or segments leading to a cycle, but the behaviour inside each band is completely chaotic. These trajectories represent ergodic states as the accessible positions have a fractal dimension equal to the dimension of phase space. Thus the removal of the noise $\varepsilon \to 0$ at μ_∞ leads to an ergodic to nonergodic transition in the map.

In the absence of noise ($\varepsilon = 0$) the diameter positions $x_{2^n} = d_n = \alpha^{-n}$ visited at times $\tau = 2^n$ by the trajectory starting at $x_0 = 1$ is given by the Feigenbaum fixed-point map solution $g(x)$,

$$x_\tau = \left| g^{(\tau)}(x_0) \right| = \tau^{-1/(1-q_0)} \left| g(\tau^{1/(1-q_0)} x_0) \right| , \tag{20}$$

that in turn is obtained from the $n \to \infty$ convergence of the 2^nth map composition to $(-\alpha)^{-n} g(\alpha^n x)$ with $\alpha = 2^{1/(1-q_0)}$. When $x_0 = 0$ one obtains in general [9]

$$x_\tau = \left| g^{(2l+1)}(0) g^{(2^n-1)}(0) \right| = \left| g^{(2l+1)}(0) \right| \alpha^{-n}, \tau = (2l+1)2^n, \; l, n = 0, 1, \ldots \tag{21}$$

When the noise is turned on (ε always small) the 2^nth map composition converges instead to

$$(-\alpha)^{-n} [g(\alpha^n x) + \chi \varepsilon \kappa^n G_\Lambda(\alpha^n x)] , \tag{22}$$

where κ a constant whose numerically determined [18, 19] value $\kappa \simeq 6.619$ is well approximated by $\nu = 2\sqrt{2}\alpha(1 + 1/\alpha^2)^{-1/2}$, the ratio of the intensity of successive subharmonics in the map power spectrum [2, 19]. The connection between κ and the ε-independent ν stems from the necessary coincidence of two ratios, that of noise levels causing band-merging transitions for successive 2^n and 2^{n+1} periods and that of spectral peaks at the corresponding parameter values μ_n and μ_{n+1} [2, 19]. Following the same procedure as above we see that the orbits x_τ at $\mu_\infty(\varepsilon)$ satisfy, in place of (20), the relation

$$x_\tau = \tau^{-1/(1-q_0)} \left| g(\tau^{1/(1-q_0)} x) + \chi \varepsilon \tau^{1/(1-r)} G_\Lambda(\tau^{1/(1-q_0)} x) \right| , \tag{23}$$

where $G_\Lambda(x)$ is the first-order perturbation eigenfunction, and where $r = 1 - \ln 2 / \ln \kappa \simeq 0.6332$. So that use of $x_0 = 0$ yields

$$x_\tau = \tau^{-1/(1-q_0)} \left| 1 + \chi \varepsilon \tau^{1/(1-r)} \right| \tag{24}$$

or

$$x_t = \exp_{2-q_0}(-\lambda_{q_0} t) \left[1 + \chi \varepsilon \exp_r(\lambda_r t) \right] \tag{25}$$

where $t = \tau - 1$, $\lambda_{q_0} = \ln \alpha / \ln 2$ ($\lambda_{q_0,\max}$ of the previous section) and $\lambda_r = \ln \kappa / \ln 2$.

At each noise level ε there is a 'crossover' or 'relaxation' time $t_x = \tau_x - 1$ when the fluctuations start destroying the detailed structure imprinted by the attractor on the orbits with $x_0 = 0$. This time is given by $\tau_x = \varepsilon^{r-1}$, the time when the fluctuation term in the perturbation expression for x_τ becomes ε-independent, i.e.

$$x_{\tau_x} = \tau_x^{-1/(1-q_0)} \left|1 + \chi\right| \; . \qquad (26)$$

Thus, there are two regimes for time evolution at $\mu_\infty(\varepsilon)$. When $\tau < \tau_x$ the fluctuations are smaller than the distances between adjacent subsequence positions of the noiseless orbit at $\mu_\infty(0)$, and the iterate positions in the presence of noise fall within small non overlapping bands each around the $\varepsilon = 0$ position for that τ. In this regime the dynamics follows in effect the same subsequence pattern as in the noiseless case. When $\tau \sim \tau_x$ the width of the fluctuation-generated band visited at time $\tau_x = 2^N$ matches the distance between two consecutive diameters, $d_N - d_{N+1}$ where $N \sim -\ln\varepsilon/\ln\kappa$, and this signals a cut-off in the advance through the position subsequences. At longer times $\tau > \tau_x$ the orbits are unable to stick to the fine period-doubling structure of the attractor. In this 2nd regime the iterate follows an increasingly chaotic trajectory as bands merge progressively. This is the dynamical image – observed along the *time evolution* for the orbits *of a single state* $\mu_\infty(\varepsilon)$ – of the static bifurcation gap first described in the map space of position x and control parameter μ [17, 18].

7 Analogy with Glassy Dynamics

We recall the main dynamical properties displayed by supercooled liquids on approach to glass formation. One is the growth of a plateau and for that reason a two-step process of relaxation, as presented by the time evolution of correlations e.g. the intermediate scattering function F_k [20]. This consists of a primary power-law decay in time t (so-called 'β' relaxation) that leads into the plateau, the duration t_x of which diverges also as a power law of the difference $T - T_g$ as the temperature T decreases to a critical value T_g. After t_x there is a secondary power-law decay (so-called 'α' relaxation) that leads to a conventional equilibrium state [20]. A second characteristic dynamical property of glasses is the loss of time translation invariance, a characteristic known as aging [21]. The time decay of relaxation functions and correlations display a scaling dependence on the ratio t/t_w where t_w is a waiting time. A third distinctive property is that the experimentally observed relaxation behaviour of supercooled liquids is described, via regular heat capacity assumptions [20], by the so-called Adam–Gibbs equation,

$$t_x = A \exp(B/TS_c) \; , \qquad (27)$$

where the relaxation time t_x can be identified with the viscosity, and the configurational entropy S_c is related to the number of minima of the fluid's potential energy surface (and A and B are constants).

Returning to the map, phase space is sampled at noise level ε by orbits that visit points within the set of 2^N bands of widths $\sim \varepsilon$, and this takes place in time in the same way that period doubling and band merging proceeds in the presence of a bifurcation gap when the control parameter is run through the interval $0 \leq \mu \leq 2$. That is, the trajectories starting at $x_0 = 0$ duplicate the number of visited bands at times $\tau = 2^n$, $n = 1, \ldots, N$, the bifurcation gap is reached at $\tau_x = 2^N$, after which the orbits fall within bands that merge by pairs at times $\tau = 2^{N+n}$, $n = 1, \ldots, N$. The sensitivity to initial conditions grows as $\xi_t = \exp_{q_0}(\lambda_{q_0} t)$ ($q_0 = 1 - \ln 2/\ln \alpha < 1$) for $t < t_x$, but for $t > t_x$ the fluctuations dominate and ξ_t grows exponentially as the trajectory has become chaotic and so one anticipates an exponential ξ_t (or $q = 1$). We interpret this behaviour to be the dynamical system analogue of the 'α' relaxation in supercooled fluids. The plateau duration $t_x \to \infty$ as $\varepsilon \to 0$. Additionally, trajectories with initial conditions x_0 not belonging to the attractor exhibit an initial relaxation stretch towards the plateau as the orbit falls into the attractor. This appears as the analog of the 'β' relaxation in supercooled liquids. See [12].

Next, we determine the entropy of the orbits starting at $x_0 = 0$ as they enter the bifurcation gap at $t_x(\varepsilon)$ when the maximum number 2^N of bands allowed by the fluctuations is reached. The entropy $S_c(\mu_\infty, 2^N)$ associated to the state at $\mu_\infty(\varepsilon)$ has the form

$$S_c(\mu_\infty, 2^N) = 2^N \varepsilon s \ , \tag{28}$$

since each of the 2^N bands contributes with an entropy εs where

$$s = -\int_{-1}^{1} p(\chi) \ln p(\chi) \mathrm{d}\chi \ , \tag{29}$$

and where $p(\chi)$ is the distribution for the noise random variable. In terms of t_x, given that $2^N = 1 + t_x$ and $\varepsilon = (1 + t_x)^{-1/(1-r)}$, one has

$$S_c(\mu_\infty, t_x)/s = (1 + t_x)^{-r/(1-r)} \tag{30}$$

or, conversely,

$$t_x = (s/S_c)^{(1-r)/r}. \tag{31}$$

Since $t_x \simeq \varepsilon^{r-1}$, $r - 1 \simeq -0.3668$ and $(1 - r)/r \simeq 0.5792$ then $t_x \to \infty$ and $S_c \to 0$ as $\varepsilon \to 0$, i.e. the relaxation time diverges as the 'landscape' entropy vanishes. We interpret this relationship between t_x and the entropy S_c to be the dynamical system analog of the Adam–Gibbs formula for a supercooled liquid. See [12]. Notice that (31) is a power law in S_c^{-1} while for structural glasses it is an exponential in S_c^{-1} [20].

Last, we examine the aging scaling property of the trajectories x_t at $\mu_\infty(\varepsilon)$. The case $\varepsilon = 0$ is more readily appraised because this property is actually built into the position subsequences $x_\tau = \left|g^{(\tau)}(0)\right|$, $\tau = (2l + 1)2^n$, $l, n = 0, 1, \ldots$. These subsequences are relevant for the description of trajectories that are 'detained' at a given attractor position for a waiting period of time t_w and then 'released' to the normal iterative procedure. We chose the holding positions to be any of those along the top band shown in Fig. 1 for a waiting time $t_w = 2l + 1$, $l = 0, 1, \ldots$. Notice that, as shown in Fig. 1, for the $x_0 = 0$ orbit these positions are visited at odd iteration times. The lower-bound positions for these trajectories are given by those of the subsequences at times $(2l+1)2^n$ (see Fig. 1). Writing τ as $\tau = t_w + t$ we have that $t/t_w = 2^n - 1$ and

$$x_{t+t_w} = g^{(t_w)}(0)g^{(t/t_w)}(0) \tag{32}$$

or

$$x_{t+t_w} = g^{(t_w)}(0)\exp_{q_0}(-\lambda_{q_0}t/t_w)\ . \tag{33}$$

This property is gradually modified when noise is turned on. The presence of a bifurcation gap limits its range of validity to total times $t_w + t < t_x(\varepsilon)$ and so progressively disappears as ε is increased. See [12].

8 Concluding Remarks

We have re-examined the dynamical properties at the onset of chaos in the logistic map and obtained further understanding about their nature. We exhibited links between original developments, such as Feigenbaum's σ function, Mori's q-phase transitions and the noise-induced bifurcation gap, with more recent advances, such as q-exponential sensitivity to initial conditions [9], q-generalized Pesin identity [11], and dynamics of glass formation [12]. The dynamics at the edge of chaos is anomalous because it is an incipient chaotic attractor with vanishing ordinary Lyapunov coefficient λ_1. Chaotic orbits possess a time irreversible property that stems from mixing in phase space and loss of memory, but orbits within critical attractors are non-mixing and have no loss of memory. As a classic illustration of the latter case the attractor at the onset of chaos presents dynamical properties with self-similar structure that result in a set of power laws for the sensitivity to initial conditions. We determined exact analytical expressions for ξ_t.

Our most striking finding is that the dynamics at the onset of chaos is constituted by an infinite family of Mori's q-phase transitions, each associated to orbits that have common starting and finishing positions located at specific regions of the attractor. Each of these transitions is related to a discontinuity in the σ function of 'diameter ratios', and this in turn implies a q-exponential ξ_t and a spectrum of q-Lyapunov coefficients for each set of orbits. The transitions come in pairs with specific conjugate indexes q and $Q = 2 - q$, as these correspond to switching starting and finishing orbital positions. Since the amplitude of the discontinuities in σ diminishes rapidly, in practical terms there is only need of evaluation for the first few of them. The dominant discontinuity is associated to the most crowded and sparse regions of the attractor and this alone provides a very reasonable description, as found in earlier studies [7,9,11,16]. Thus, the special values for the Tsallis entropic index q in ξ_t are equal to the special values of the variable q in the formalism of Mori and colleagues at which the q-phase transitions take place.

As described, the dynamics of noise-perturbed logistic maps at the chaos threshold presents the characteristic features of glassy dynamics observed in supercooled liquids. In particular our results are [12]: (i) The two-step relaxation that takes place when $\varepsilon \to 0$ is obtained in terms of the bifurcation gap properties, specifically, the plateau duration t_x is given by a power law in the noise amplitude ε. (ii) The map analogue of the Adam–Gibbs law is given also as a power-law relation between $t_x(\varepsilon)$ and the entropy $S_c(\varepsilon)$ associated to the noise widening of chaotic bands. (iii) The trajectories at $\mu_\infty(\varepsilon \to 0)$ are shown to obey a scaling property, characteristic of aging in glassy dynamics, of the form $x_{t+t_w} = h(t_w)h(\, t/t_w)$ where t_w is a waiting time.

The limit of vanishing noise amplitude $\varepsilon \to 0$ (the counterpart of the limit $T - T_g \to 0$ in the supercooled liquid) brings about loss of ergodicity. This nonergodic state with $\lambda_1 = 0$ corresponds to the limiting state, $\varepsilon \to 0$, $t_x \to \infty$, for a family of small ε states with glassy properties, which are expressed for $t < t_x$ via the q-exponentials of the Tsallis formalism. It has been suggested on several occasions [22, 23] that the setting in which nonextensive statistics appears to come out is linked to the prevalence of nonuniform convergence, such as that involving the thermodynamic $N \to \infty$ and the infinitely large time $t \to \infty$ limits. Here a similar situation happens, that is, if $\varepsilon \to 0$ is taken before $t \to \infty$ a nonergodic orbit restrained to the Feigenbaum attractor and with fully-developed glassy properties is obtained, whereas if $t \to \infty$ is taken before $\varepsilon \to 0$ a chaotic orbit with $q = 1$ would be observed.

Acknowledgement

AR is grateful for the support provided by the Verhulst 200 organizers. We also acknowledge support from CONACyT and DGAPA-UNAM, Mexican agencies.

References

1. P.-F. Verhulst: *Recherches mathématiques sur la loi d'accroissement de la population,* Nouv. Mem. de l'Acad. Roy. des Sciences et Belles-Lettres de Bruxelles XVIII.8, 1-38, 1845
2. See, for example, H.G. Schuster: *Deterministic Chaos. An Introduction*, 2nd Revised Edition (VCH Publishers, Weinheim 1988)
3. C. Tsallis: J. Stat. Phys. **52**, 479 (1988)
4. For recent reviews see, *Nonextensive Entropy – Interdisciplinary Applications*, ed by M. Gell-Mann, C. Tsallis (Oxford University Press, New York 2004). See http://tsallis.cat.cbpf.br/biblio.htm for full bibliography
5. H. Mori, H. Hata, T. Horita, T. Kobayashi: Prog. Theor. Phys. Suppl. **99**, 1 (1989)
6. T. Horita, H. Hata, H. Mori, K. Tomita: Prog. Theor. Phys. **82**, 897 (1989)

7. C. Tsallis, A.R. Plastino, W.-M. Zheng: Chaos, Solitons and Fractals **8**, 885 (1997)
8. F. Baldovin, A. Robledo: Europhys. Lett. **60**, 518 (2002)
9. F. Baldovin, A. Robledo: Phys. Rev. E **66**, 045104(R) (2002)
10. A. Robledo: Physica D **193**, 153 (2004)
11. F. Baldovin, A. Robledo: Phys. Rev. E **69**, 045202(R) (2004)
12. A. Robledo: Phys. Letters A **328**, 467 (2004)
13. M.J. Feigenbaum: Commun. Math. Phys. **77**, 65 (1980); Physica 7D, 16 (1983)
14. E. Mayoral-Villa, A. Robledo: submitted
15. G. Anania, A. Politi: Europhys. Lett. **7**, 119 (1988)
16. M.L. Lyra, C. Tsallis: Phys. Rev. Lett. **80**, 53 (1998)
17. J.P. Crutchfield, J.D. Farmer, B.A. Huberman: Phys. Rep. **92**, 45 (1982)
18. J. Crutchfield, M. Nauenberg, J. Rudnick: Phys. Rev. Lett. **46**, 933 (1981)
19. B. Shraiman, C.E. Wayne, P.C. Martin: Phys. Rev. Lett. **46**, 935 (1981)
20. For a recent review see, P.G. De Benedetti, F.H. Stillinger, Nature **410**, 267 (2001)
21. See, for example, J.P. Bouchaud, L.F. Cugliandolo, J. Kurchan, M. Mezard. In *Spin Glasses and Random Fields*, ed by A.P. Young (World Scientific, Singapore 1998)
22. C. Tsallis. In *Nonextensive Statistical Mechanics and Its Applications*, ed by S. Abe, Y. Okamoto, Lecture Notes in Physics 560, 3 (Springer, Berlin 2001)
23. C. Tsallis, A. Rapisarda, V. Latora, F. Baldovin. In *Dynamics and Thermodynamics of Systems with Long-Range Interactions*, ed by S. Ruffo, E. Arimondo, M. Wilkens, Lecture Notes in Physics 602, 140 (Springer, Berlin 2002)

Quantum Chaos Versus Classical Chaos: Why is Quantum Chaos Weaker?

H. Kröger, J.-F. Laprise, G. Melkonyan, and R. Zomorrodi

Département de Physique, Université Laval, Québec, Québec G1K 7P4, Canada
hkroger@phy.ulaval.ca

1 Introduction

Chaotic systems occur in nature everywhere. Examples are: the turbulent flow of gases and liquids, e.g. when pouring milk into a cup of coffee, in the bath tub, in hurricanes, or in the formation of galaxies. Chaos occurs in biology in the dynamics of populations and species. It occurs in the rate of heart beats, in attacks of epilepsy, and in the brain. It occurs also in chemical reactions. Most of the dynamics occurring in nature is chaotic. The occurrence of non-chaotic systems in nature is more of an exception. The mathematical description of chaos has reached a mature state over the last decade. The tools being used are classical phase space, Lyapunov exponents, Poincaré sections, Kolmogorov–Sinai entropy and others.

There is also chaos in the quantum systems [1–3]. People have investigated in quantum physics the analogues of classically chaotic systems. For example, a stadium-shaped billiard is classically chaotic. The corresponding quantum system is a system of ultra-cold atoms bouncing against walls of stadium shape, being created by interaction of the atoms with laser beams. Quantum chaos has been found to play an important role in dynamical tunnelling [4, 5]. The fingerprints of chaos in quantum systems and the mathematical tools of its description are quite different from those used in classical chaos. The necessity for a different treatment is due to the nature of quantum mechanics: There is no proper phase space in quantum systems – Heisenberg's uncertainty principle forbids that a point in phase space (uncertainty zero in position and momentum) exists. Heisenberg's uncertainty relation is a direct consequence of quantum mechanical fluctuations. A common approach to describe quantum chaos is random matrix theory and the use of energy level spacing distributions [3]. A fundamental conjecture by Bohigas, Giannoni and Schmit [6] postulates that the energy level spacing distributions possesses a dominant part – which depends on the particular system – and a subleading universal part – independent of the particular system. The universal part gives a level spacing distribution of Wignerian type, if the corresponding classical system is fully chaotic. Such level spacing distribution can be generated also by random matrices of a certain symmetry class. The conjecture has not been rigorously proven yet, but has been verified and found to be valid in almost all cases.

This approach is successful and very popular. However, it has some shortcomings: First, the conjecture by Bohigas et al. holds strictly only in the case of a fully chaotic system, while in nature most systems are only partly chaotic, i.e. so-called mixed systems where its classical counterpart has coexistence of regular and chaotic phase space. In such cases the quantum system yields a level spacing distribution, which is neither Wignerian nor Poissonian (where the latter corresponds to a completely regular system). There is no mathematical prediction of the functional form of such distribution. However, a number of interpolations between the Poisson and Wigner distribution have been proposed [7–11]. Second, we may ask: How about the comparison of the classical with the quantum system? And what is the quantitative degree of chaos? How can we answer this when the instruments used to measure chaos are quite different for both systems?

Starting from this perspective and having in mind the goal to compare classical with quantum chaos, one may try the following strategy: Find a uniform description of chaotic phenomena, valid for both, classical and quantum systems. In more detail: Starting from nonlinear dynamics and phase space in classical systems, one may look for a suitable quantum analogue phase space. Starting from random matrix theory and energy level spacing distributions of quantum systems, one may seek a random matrix description and a level spacing distribution of suitable dynamical objects in classical physics. In the following we shall discuss some progress recently made in this direction. Using some of those results, we shall compare for a particular system the chaotic behaviour of the quantum system with the classical system. The numerical analysis shows that the quantum system is globally less chaotic than the classical system. We believe that such finding is not limited to the particular system. In particular, we want to understand the underlying reason for such behaviour.

2 Cases Where Quantum Chaos Was Found to Be Weaker

First, Casetti et al. [12] considered the N-component Φ^4 theory in the presence of an external field and in the limit of large N. They used mean field theory and observed a strong suppression of chaos in the quantum system, due to quantum corrections causing the system to move away from a hyperbolic fixed point responsible for classical chaos. Second, Matinyan and Müller [13] studied another field theoretic model which is classically chaotic, namely massless scalar electrodynamics. They investigated the corresponding quantum field theory using effective field theory and loop expansion. They noticed that quantum corrections increase the threshold for chaos due to a modification of the ground state of the system. A third example is the kicked rotor which is a classically chaotic system in 1-D. Schwengelbeck and

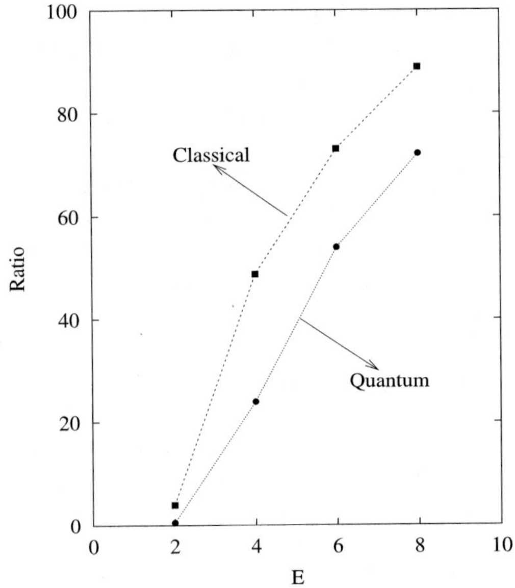

Fig. 1. Volume of chaotic phase space over total phase space versus energy. Comparison of results from classical action (*dotted line*) with quantum action (*full line*)

Faisal [14] considered the corresponding quantum system using Bohm's interpretation of quantum mechanics to introduce trajectories and a quantum equivalent phase space. They found that the Kolmogorov–Sinai entropy goes to zero in the quantum system, i.e. it is non-chaotic. This approach has also been applied to study chaos in anisotropic harmonic oscillators [15], coupled anharmonic oscillators [16] and the hydrogen atom in an external electromagnetic field [17]. Finally, Caron et al. [18] considered the anharmonic oscillator in 2-D, which is classically a mixed chaotic system. Using the concept of the quantum action functional, a quantum analogue phase space has been constructed. As a result, the phase space portrait of the quantum system was found to be slightly but globally less chaotic for all energies (see Fig. 1). Because the quantum action has been constructed from the classical action by taking into account quantum fluctuations [19–25], hence the softening of chaos in the quantum system must be due to quantum fluctuations. We suspect that such softening effect may not solely show up in chaos. Indeed, looking at a double-well potential $V(x) = \lambda\left(x^2 - \frac{1}{8\lambda}\right)^2$ in the context of tunnelling, we observe that quantum effects cause the quantum potential to be much "weaker" than the classical potential (see Fig. 2), i.e. the potential wells are less pronounced and the potential barrier is much lower for the quantum potential (note that the quantum potential has a triple-well shape). The shape of the potential for tunnelling translates into the shape of the instantons. Thus it comes as no surprise that the instanton of the quantum

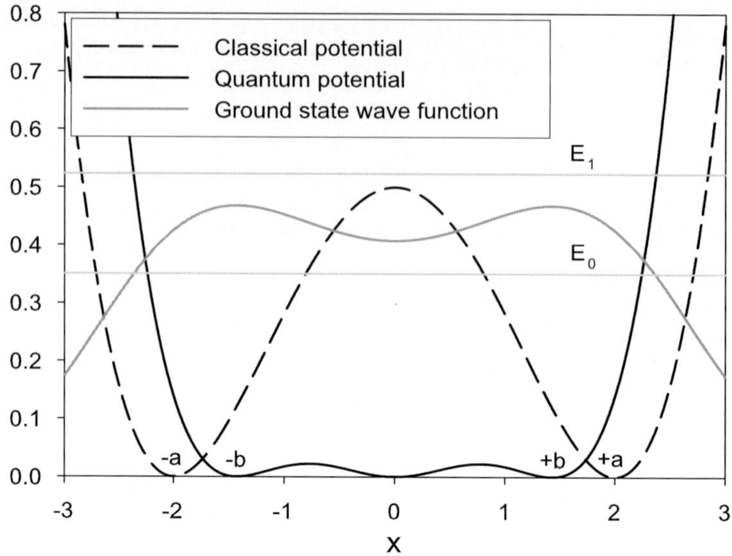

Fig. 2. Comparison of classical potential and quantum potential (multiplied by a scale factor 1.7331) for quartic coupling $\lambda = 1/32$. Also shown is the ground state wave function and two lowest energy levels

action (actually a double-instanton) is softer than the classical instanton (see Fig. 3).

3 Uniform Description of Chaos

Let us recall that the conjecture by Bohigas et al. is about random matrix theory and energy level spacing distributions to describe chaos in quantum systems, while chaos in classical systems is conventionally described in terms of phase space. That is, the tools to describe chaos in classical and quantum physics are different. For the purpose to better understand the physical content of this conjecture or eventually to find a mathematical proof, it is highly desirable to use the same language, respectively, tools in quantum physics as in classical physics. This can be viewed in two ways: Either one adopts the point of view that chaos should be analysed in terms of phase space. Then a uniform description is achieved by use of classical phase space in classical physics and the quantum analogue phase space in quantum systems. How about the point of view that chaos should be analysed in terms of random matrix theory and level spacing distributions? In the following we propose how to construct such a level spacing distribution, which will play the same role in classical physics as the energy level spacing distribution in the quantum system.

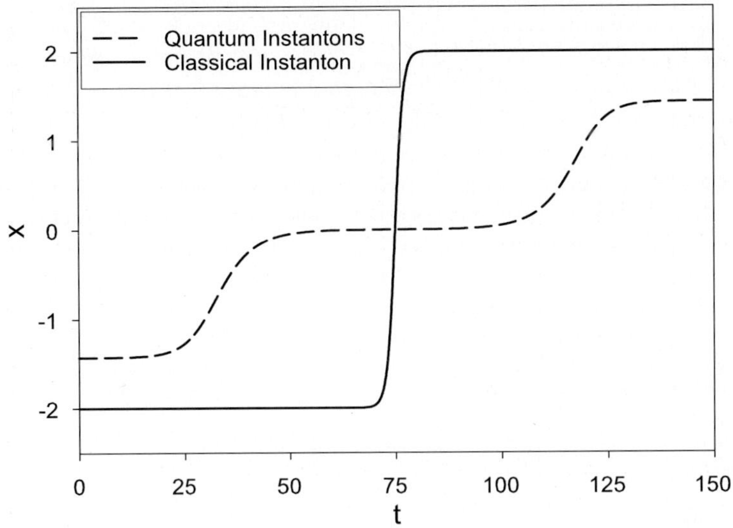

Fig. 3. Comparison of classical instanton with instanton from the quantum action

One may immediately object that the energy level spacing distribution in quantum physics is due to the fact that one considers a system of bound states (with a "confining" potential), which due to the rules of quantum mechanics gives a discrete spectrum. The discreteness of the spectrum is a quantum effect, i.e., a physical effect. Having in mind to construct a counterpart in classical physics, one may object that classical physics is continuous. There is no discreteness inherent in classical physics. If any discreteness occurs it will be due to some (mathematical) approximation. Hence, how can a level spacing distribution derived from a classical function be discrete and physical? First we want to propose a function and show how to construct a meaningful level spacing distribution. Afterwards, we shall try to answer the last question.

We would like to emphasize two properties: First, the level spacing distribution of the quantum system – in the case of a classically fully chaotic system – corresponds to a Wigner distribution. The type of Wigner distribution is determined by the symmetry of the Hamiltonian. E.g., there is the Gaussian orthogonal ensemble (GOE), the unitary ensemble and the symplectic ensemble. Thus, we need a function which carries the same symmetries. A function convenient for this purpose is the classical action. Another property to be emphasized is locality. Let us recall that the level spacing distribution, e.g. the GOE distribution, is invariant under orthogonal transformations. Because an orthogonal transformation maps any orthogonal basis of states onto another orthogonal basis, the level spacing distribution is essentially independent of the particular choice of the basis. That means, we may choose a basis that is quasi-local, i.e., built from square integrable functions which are identically zero everywhere except in a small interval where they are non zero and constant. Those box functions are almost local. The advantage of locality is the fact that symmetries of the Hamiltonian often have to do with transformations in position space, and locality facilitates the analysis of such symmetries. Why do we need to bother about symmetries? Because the energy level spacing distribution is meaningful only (and gives the Wigner distribution for a chaotic system, respectively a Poissonian distribution for an integrable system) if the energy levels have all the same quantum numbers (except for the quantum number of energy). For example, for the spectrum of the hydrogen atom, one should take the bound states all with the same angular momentum quantum numbers, e.g., $l = m = 0$.

3.1 Action Matrix

We consider a system defined by a Lagrange function $L(q, \dot{q}, t)$. Let S denote the corresponding action,

$$S[q(t)] = \int_0^T \mathrm{d}t \; L(q, \dot{q}, t) \; . \tag{1}$$

Let q_{traj} denote the trajectory, i.e. the solution of the Euler–Lagrange equation of motion. Such trajectory is a function, which makes the action functional stationary. Each trajectory is specified by indicating the initial and final boundary points, i.e., $q(t = 0) = q_{\text{in}}$ and $q(t = T) = q_{\text{fi}}$. We assume that those boundary points are taken from some finite set of nodes, $q_{\text{in}}, q_{\text{fi}} \in \{q_1, \ldots, q_N\}$. Thus with each pair of boundary points, (q_k, q_l) we associate a trajectory $q_{kl}^{\text{traj}}(t)$. Then we introduce an action matrix Σ, where the matrix element Σ_{kl} corresponds to the value of the action S evaluated along the trajectory $q_{kl}^{\text{traj}}(t)$,

$$\Sigma_{kl} = S[q_{kl}^{\text{traj}}(t)] \; . \tag{2}$$

All matrix elements Σ_{kl} are real. They are also symmetric, $\Sigma_{kl} = \Sigma_{lk}$. Thus the action matrix Σ is a Hermitian $N \times N$ matrix. Consequently, Σ has a discrete spectrum of action eigenvalues, which are all real,

$$\sigma(\Sigma) = \{\sigma_1, \ldots, \sigma_N\} \; . \tag{3}$$

3.2 Symmetry

For the purpose to compute a level spacing distribution from the action eigenvalues, one must first address the issue of symmetry. Let us consider for example the harmonic oscillator in 2-D. We may take the coordinates q_k to be located on the nodes of a regular grid (reaching from $-\Lambda$ to $+\Lambda$ on the x and y axis), with a spacing $\Delta x = \Delta y = a = const.$ For example, the action of the harmonic oscillator, with the classical trajectory going from $\boldsymbol{x}_a$ to $\boldsymbol{x}_b$ in time T is given by

$$\Sigma_{a,b} = \frac{m\omega}{2\sin(\omega T)} \left[(x_a^2 + x_b^2)\cos(\omega T) - 2\boldsymbol{x}_a \cdot \boldsymbol{x}_b \right] . \tag{4}$$

This function is invariant under rotations. When choosing the coordinates q_k to be located on the regular grid, the continuous symmetry of rotation will become a discrete symmetry of finite rotations (a group). We have to find the irreducible representations of such group and sort the action eigenvalues according to those irreducible representations (this can be done by inspecting the properties of the corresponding eigenvector). Then one has to select a particular representation and retain a subset of eigenvalues in that representation. The action level spacing distribution can then be obtained from the action eigenvalues in such subset.

This procedure is feasible. However, it has two disadvantages. First, finding the irreducible representations and classifying the eigenvectors accordingly is laborious. More importantly, the fact that one has to work with a subset of eigenvalues only and not the whole ensemble of eigenvalues means a drastic reduction of the size of the statistical ensemble. In other words, the statistics of the resulting level spacing distribution will deteriorate.

For those reasons it would be highly desirable to avoid the above strategy. This is indeed possible by using the following trick. One can camouflage the symmetry by choosing the coordinates off the nodes of the regular grid. That means, for example to define new coordinates as follows,

$$q_k^{\text{deform}} = q_k + \epsilon_k , \tag{5}$$

where ϵ_k denotes a randomly chosen small deformation (in angle and length). In this way the nodes are irregularly distributed. Consequently, the discrete symmetry of the action matrix Σ_{kl} disappears when replacing it by the "deformed" action matrix,

$$\begin{aligned} \Sigma_{kl}^{\text{deform}} = & \text{ action evaluated along the classical trajectory} \\ & \text{from node } q_k^{\text{deform}} \text{ to node } q_l^{\text{deform}} . \end{aligned} \tag{6}$$

In doing so we avoid a laborious symmetry analysis and secondly shall have a better statistics!

3.3 Action Level Spacing Distribution

Now we want to construct a level spacing distribution of action levels. We proceed in analogy to random matrix theory and the method of constructing an energy level spacing distribution. For an overview on how to compute level spacing distributions see [26]. One has to separate the dominant system dependent part from the subleading universal part which describes the properly normalized fluctuations. Because we are here interested only in the fluctuation part, we suppress the leading part. One should note that this means to discard all physical information which depends on the particular system.

For example thermodynamical functions can not be computed from the subleading fluctuating part. The strategy to obtain the subleading part is called unfolding. One constructs a fit to the original spectrum and multiplies the spectrum such that on average the mean spacing distribution becomes unity. Also the integrated level spacing distribution will be normalized to unity.

We have applied this to simple integrable systems in 1-D. For integrable systems one would expect a Poissonian distribution for the action level spacing distribution. Preliminary results are compatible with a Poissonian distribution. The following remarks are in order. The first numerical results for integrable systems have to be repeated with precision and analysed carefully. Second, one wants to see what happens in chaotic systems. Possibly such strategy applied to a fully chaotic system will result in a Wignerian action level spacing distribution. Finding an answer will be computationally much more involved, simply because the action functions for the integrable systems considered above are analytically known, while for a chaotic system (i.e. non-integrable) this needs to be calculated numerically. Moreover, the numerical precision required needs to be sufficient to resolve small fluctuations. In the statistical sense, one is interested in a sample of large size. But that means that after unfolding the fluctuations will become small and hence require a high numerical precision for its resolution. Presently, numerical studies of such question are under way.

Let us get back to the question posed at the beginning of Sect. 3: How can a level spacing distribution derived from a classical function be discrete and physical? In our opinion, the answer lies in the fact that the level spacing distribution is universal, that is, it does not depend, for example, on the parameters of the discrete grid. Different grids give the same result. This has been verified numerically. It also should not depend on the deformation ϵ_k (as long it is not too close to zero and the discrete symmetry is restored). Also this has been verified numerically and found to be satisfied. Thus one could in principle go with the volume of the lattice $V = (2\Lambda)^D$ to infinity and with the lattice spacing $\Delta x = \Delta y$ to zero. The result should not change, but one would have reached the continuum limit. The discreteness would then disappear.

4 Renormalisation Flow of Parameters of the Quantum Action

In Sect. 2 we have seen examples for the observation that quantum chaos seems to be weaker than chaos in the corresponding classical system. Of course it would be interesting to explore a much wider class of systems in order to see if such observation holds more generally. Here we want to pick one of the above examples, namely the chaotic anharmonic oscillator in 2-D and try to understand why quantum chaos is weaker than classical chaos. The classical action is given

$$S = \int_0^T \mathrm{d}t \, \frac{1}{2m}(\dot{x}^2 + \dot{y}^2) - V(x,y) \, ,$$
$$V = \frac{m\omega^2}{2}(x^2 + y^2) + \lambda x^2 y^2 = v_0 + v_2(x^2 + y^2) + v_{22} x^2 y^2 \, . \quad (7)$$

For $\lambda = 0$ the system is reduced to the standard harmonic oscillator, which is integrable. The chaoticity is introduced and controlled by the parameter λ. Thus small λ causes mild chaos, while large λ makes the system strongly chaotic. The quantum action has been postulated to be of the functional form like the classical action, i.e. the kinetic term of the quantum action may differ in the value of the mass, and the potential of the quantum action should also be local, depend only on coordinates, but may have a different functional form. Here let us consider an Ansatz of the following form

$$\tilde{S} = \int_0^T \mathrm{d}t \, \frac{1}{2\tilde{m}}(\dot{x}^2 + \dot{y}^2) - \tilde{V}(x,y) \, ,$$
$$\tilde{V} = \tilde{v}_0 + \tilde{v}_2(x^2 + y^2) + \tilde{v}_{22} x^2 y^2 + \text{higher order polynomials} \, . \quad (8)$$

As a quantitative measure of the strength of chaos we take the strength of the parameters of the action, in particular, the parameter λ. The parameters of the quantum action can be interpreted as a "renormalisation effect" of the parameters of the classical action. The calculation of those parameters has to be done numerically, following the definition of the quantum action to be a functional which fits the transition amplitudes [19]. However, in a certain limit, the quantum action is known to be an exact representation of the transition amplitudes and moreover the action is related via differential equations to the classical action [22]. This limiting case is using imaginary time and let time go to infinity (Feynman–Kac limit). Because we want to obtain an analytical result, we shall use perturbation theory. This means that we consider the regime of small $\lambda = v_{22}$.

In order to simplify the matter, let us start by considering the system in 1-D. Thus we have the potential

$$V(x) = \frac{1}{2} m\omega^2 x^2 + \lambda x^4 \equiv v_2 x^2 + v_4 x^4 \, . \quad (9)$$

We assume λ to be small,

$$\frac{\lambda \Lambda_{\mathrm{sc}}}{v_2} \ll 1 \, , \quad (10)$$

where Λ_{sc} introduces a physical length scale, e.g. the analogue of the Bohr radius. According to [23] the following relation between the classical and the quantum potential holds,

$$2m(V(x) - E_{\mathrm{gr}}) = 2\tilde{m}(\tilde{V}(x) - \tilde{v}_0) - \frac{\hbar}{2} \frac{\frac{d}{dx} 2\tilde{m}(\tilde{V}(x) - \tilde{v}_0)}{\sqrt{2\tilde{m}(\tilde{V}(x) - \tilde{v}_0)}} \operatorname{sgn}(x) \, . \quad (11)$$

We define the functions

$$W(x) = 2m(V(x) - E_{\mathrm{gr}}) \; , \tag{12}$$

$$U(x) = 2\tilde{m}(\tilde{V}(x) - \tilde{v}_0) \; , \tag{13}$$

where E_{gr} denotes the ground state energy of the lowest eigenstate of the quantum mechanical system (obtained from the Schrödinger equation). Due to (9), the function $W(x)$ must have the form

$$W(x) = w_0 + w_2 x^2 + w_4 x^4 \; . \tag{14}$$

This function is symmetric with respect to parity. Then the function $U(x)$ representing the quantum potential, will be parity symmetric too. We make an Ansatz of the form

$$U(x) = u_0 + u_2 x^2 + u_4 x^4 + u_6 x^6 + \cdots \tag{15}$$

The assumption that the expansion parameter λ is small is now expressed by

$$w_4 \equiv w_4^{(0)} \epsilon \; , \quad \epsilon \ll 1 \tag{16}$$

Now using (14) and (15) in combination with (11), we obtain

$$w_0 + w_2 x^2 + w_4 x^4 = u_2 x^2 + u_4 x^4 + u_6 x^6 + \cdots - \frac{\hbar}{2} \frac{2u_2 + 4u_4 x^2 + 6u_6 x^4 + \cdots}{\sqrt{u_2 + u_4 x^2 + u_6 x^4 + \cdots}} \quad \text{for } x > 0 \; . \tag{17}$$

The smallness of w_4 implies that the terms of fourth order and higher in x occurring in the function $U(x)$ are small compared to the second order, i.e.

$$u_4 x^2 + u_6 x^4 + \cdots \ll u_2 \; . \tag{18}$$

Now doing a Taylor expansion in the small terms $u_4 x^2 + u_6 x^4 + \cdots$ allows to express the r.h.s. of (17) as a polynomial in x. Then comparing terms in x order by order, we find the following relations

$$\begin{aligned} w_0 &= -\frac{\hbar}{4\sqrt{u_2}^3} 4u_2^2 \; , \\ w_2 &= u_2 - \frac{\hbar}{4\sqrt{u_2}^3} 6u_2 u_4 \; , \\ w_4 &= u_4 - \frac{\hbar}{4\sqrt{u_2}^3} (10u_2 u_6 - 4u_4^2) \; . \end{aligned} \tag{19}$$

Now we try to find the parameters u_2 and u_4 as solution of those equations. The first equation gives

$$u_2 = (w_0/\hbar)^2 \; . \tag{20}$$

On the other hand we have, due to (13),

$$w_0 = -2mE_{\rm gr} \ . \tag{21}$$

Due to the anharmonic perturbation the ground state energy is different from the ground state energy of the harmonic oscillator. However, because of the smallness of the perturbation, we can express the ground state energy $E_{\rm gr}$ using perturbation theory as a power series in ϵ,

$$E_{\rm gr} = E^{(0)} + \epsilon E^{(1)} + \epsilon^2 E^{(2)} + \cdots \ , \tag{22}$$

where $E^{(0)} = E^{\rm osc}_{\rm gr}$. Thus from (21) and (22) we obtain

$$\sqrt{u_2} = m\omega + \epsilon \frac{2mE^{(1)}}{\hbar} + O(\epsilon^2) \ , \tag{23}$$

or

$$\begin{aligned} u_2 &= m^2\omega^2 + \epsilon \frac{4m^2E^{(1)}}{\hbar} + O(\epsilon^2) \\ &= w_2[1 + \frac{2E^{(1)}}{E^{(0)}}\epsilon + O(\epsilon^2)] \ . \end{aligned} \tag{24}$$

Next let us consider (19b). We obtain

$$u_4 = -\frac{2}{3\hbar}\sqrt{u_2}(w_2 - u_2) \ . \tag{25}$$

Recalling $w_2 = m^2\omega^2$ and (24), we find

$$u_4 = -\frac{2}{3}\frac{m^3\omega^3}{\hbar}\frac{E^{(1)}}{\hbar\omega}\epsilon + O(\epsilon^2) \ . \tag{26}$$

In (24), (26) we have expressed the parameters of the quantum action in terms of the parameters of the classical action. However, it remains to compute the energy $E^{(1)}$. Again we use stationary perturbation theory. The Hamiltonian is given by, taking into account (14), (16),

$$\begin{aligned} H &= H^{(0)} + \epsilon H^{(1)} \\ H^{(0)} &= \frac{p^2}{2m} + \frac{1}{2}m\omega^2x^2 \\ H^{(1)} &= \frac{w_4^{(0)}}{2m}x^4 \ . \end{aligned} \tag{27}$$

To first order of perturbation theory in ϵ the energy $E^{(1)}$ is given by

$$E^{(1)} = \langle \psi^{\rm osc}_{\rm gr} | \frac{w_4^{(0)}}{2m} x^4 | \psi^{\rm osc}_{\rm gr} \rangle \ . \tag{28}$$

which yields the result

$$E^{(1)} = \frac{3w_4^{(0)}\hbar^2}{8m^3\omega^2} \ . \tag{29}$$

Substituting this result into (24), (26) we finally obtain

$$u_2 = w_2 \left[1 + (\frac{3\hbar}{2m^3\omega^3}) w_4 + O(\epsilon^2) \right] \ , \tag{30}$$

and

$$u_4 = -\frac{1}{4} w_4 + O(\epsilon^2) \ . \tag{31}$$

5 Interpretation

Let us see what happens when we keep the classical parameters fixed, except for w_4, i.e., we keep $w_4^{(0)}$ fixed and vary ϵ. Note that $w_2 > 0$ and $w_4 > 0$. We also have $w_4^{(0)} > 0$ and $\epsilon > 0$. Now we want to study what happens when $\epsilon \to 0$. Equation (30) yields

$$\begin{aligned} &u_2 > w_2 \\ &u_2 \longrightarrow_{\epsilon \to 0} w_2 \ . \end{aligned} \tag{32}$$

Likewise, (31) yields

$$\begin{aligned} &u_4 < w_4 \\ &u_4 \longrightarrow_{\epsilon \to 0} w_4 \longrightarrow_{\epsilon \to 0} 0 \ . \end{aligned} \tag{33}$$

In other words, in the limit $\epsilon \to 0$ the classical potential approaches the potential of the harmonic oscillator. The potential of the quantum action asymptotically approaches the classical potential, hence also the harmonic oscillator potential. That is, the renormalisation group flow of the parameters $u_2(\epsilon), u_4(\epsilon)$ goes to a Gaussian fixed point. Second, for any value of ϵ the value the quadratic term of the potential is larger for the quantum potential than for the classical potential. Third, for any value of ϵ the value the quartic term of the potential is smaller for the quantum potential than for the classical potential. Recall that the quadratic term is the term, which, if it would stand alone, would make the system integrable. On the other hand, the quartic term is the term which drives the system away from integrability (and introduces chaos in 2-D). Thus we find that quantum fluctuations, which are the cause for the differences $\Delta_2 = w_2 - u_2$ and $\Delta_4 = w_4 - u_4$ to be nonzero, have the tendency to drive the quantum system closer to the regime of integrability.

Now the above perturbative calculations were performed in 1-D. Chaos in time-independent Hamilton systems exists only for $D \geq 2$. A similar, but more tedious calculation can be performed in 2-D. It confirms the above result that quantum fluctuations drive the system closer to the regime of integrability and away from the regime of chaos.

Of course such perturbative calculations are meaningful only in a neighbourhood of the Gaussian fixed point. It would be desirable to extend the calculations to a larger regime. However going to higher order of perturbation theory would make those calculation much more tedious. Nevertheless the perturbative result gives some insight into the dynamical consequences of quantum fluctuations.

References

1. H. Friedrich, D. Wintgen: Phys. Repts. **183**, 37 (1989)
2. O. Bohigas, S. Tomsovic, D. Ullmo: Phys. Repts. **223**, 43 (1993)
3. T. Guhr, A. Müller-Gröling, H.A. Weidenmüller: Phys. Repts. **299**, 189 (1998)
4. W.K. Hensinger, H. Häffner, A. Browaeys, N.R. Heckenberg, K. Helmerson, C. McKenzie, G.J. Wilburn, W.D. Phillips, S.L. Roiston, H. Rubinsztein-Dunlop, B. Upcroft: Nature **412**, 52 (2001)
5. D.A. Steck, W.H. Oskay, M.G. Raizen, Science **293**, 274 (2001)
6. O. Bohigas, M.J. Giannoni, C. Schmit: Phys. Rev. Lett. **52**, 1 (1984)
7. I.C. Percival: J. Phys. B **6**, L229-232 (1973)
8. T.A. Brody, J. Flores, J.B. French, P.A. Mello, A. Pandey, S.S.M. Wong: Rev. Mod. Phys. **53**, 385 (1981)
9. F.M. Izrailev: Phys. Rep. **196**, 299 (1990)
10. G. Lenz, F. Haake: Phys. Rev. Lett. **67**, 1 (1991)
11. M.V. Berry, M. Robnik: J. Phys. A **17**, 2413 (1984)
12. L. Casetti, R. Gatto, M. Modugno: Phys. Rev. E **57**, 1223 (1998)
13. S.G. Matinyan, B. Müller: Phys. Rev. Lett. **78**, 2515 (1997)
14. U. Schwengelbeck, F.H.M. Faisal: Phys. Lett. A **199**, 281 (1995)
15. R.H. Parmenter, R.W. Valentine: Phys. Lett. A **201**, 1 (1995)
16. M.H. Partovi: Phys. Rev. Lett. **89**, 144101 (2002)
17. G. Iacomelli, M. Pettini: Phys. Lett. A **212**, 29 (1996)
18. L.A. Caron, D. Huard, H. Kröger, G. Melkonyan, K.J.M. Moriarty, L.P. Nadeau: J. Phys. A: Math. Gen. **37**, 1 (2004)
19. H. Jirari, H. Kröger, X.Q. Luo, K.J.M. Moriarty, S.G. Rubin: Phys. Rev. Lett. **86**, 187 (2001)
20. H. Jirari, H. Kröger, X.Q. Luo, K.J.M. Moriarty, S.G. Rubin: Phys. Lett. A **281**, 1 (2001)
21. L.A. Caron, H. Jirari, H. Kröger, X.Q. Luo, G. Melkonyan, K.J.M. Moriarty: Phys. Lett. A **288**, 145 (2001)
22. H. Kröger: Phys. Rev. A **65**, 052118(2002)
23. H. Jirari, H. Kröger, X.Q. Luo, G. Melkonyan, K.J.M. Moriarty: Phys. Lett. A **303**, 299 (2002)
24. D. Huard, H. Kröger, G. Melkonyan, K.J.M. Moriarty, L.P. Nadeau: Phys. Rev. A **68**, 034101 (2003)
25. L.A. Caron, D. Huard, H. Kröger, G. Melkonyan, K.J.M. Moriarty, L.P. Nadeau: Phys. Lett. A **322**, 60 (2004)
26. F. Haake: *Quantum Signatures of Chaos* (Springer, Berlin 2001)

On the Prediction of Chaos in the Restricted Three-Body Problem

Houman Safaai[1] and Mohammad Hasan Ghaffari Saadat[2]

[1] Department of Mechanical Engineering, Amirkabir University of Technology, Tehran, Iran,
Houman1359@yahoo.com

[2] Department of Mechanical Engineering, Amirkabir University of Technology, Tehran, Iran,
Ghaffari@aut.ac.ir

In this chapter it is shown that the elementary tools of Riemannian differential geometry can be successfully used to explain the origin of Hamiltonian chaos beyond the usual picture of homoclinic intersections. The stability of dynamics is related to the curvature of the configuration space manifold, this is possible because the natural motion of standard Hamiltonian systems (i.e. with a quadratic kinetic energy term) can be seen as geodesics of the configuration space manifold equipped with the standard Jacobi metric. The stability properties of such manifold can be investigated through the Jacobi equation for geodesic spread and a geometric indicator of chaos similar to the largest Lyapunov exponent can be used. The case of a general two degree of freedom system is considered. The method is applied to the restricted three-body problem and the results of the Jacobi equation is compared with the detailed qualitative information provided by the approximate Lyapunov exponents. Complete agreement is found. The configuration space manifold associated with the Hamiltonian studied here is everywhere of positive curvature. The fluctuations of curvature of manifold along the geodesics yield parametric instability of the trajectories and thus chaos. Chaotic flows of physics have nothing to do with Anosov flows defined on negative curvature manifolds.

1 Introduction

The differential geometric approach of Hamiltonian chaos has recently been proposed and successfully applied to the study of chaos in Hamiltonian systems [1–3]. This method provides a geometric explanation of Hamiltonian chaos and an effective method to quantify it [4]. The starting point of this method is that trajectories of a standard Newtonian system can be viewed as the geodesics of a Riemannian manifold endowed with a suitable metric. The instability properties of geodesics, and therefore of corresponding dynamics are shown to be related to the curvature properties of the underlying manifold through the Jacobi equation for the evolution of geodesic separation.

By combining analytic computations with numerical simulations, we have a powerful tool for the study of chaos in Hamiltonian systems.

In the present work, we verify that the geometry of the configuration space appears to enfold all the information of a nonlinear Hamiltonian system for what concerns regular and chaotic dynamics.
We show that the Riemannian approach is effective for systems with two degrees of freedom. For such a system, we can solve the Jacobi equation and we show that the solution of Jacobi equation brings all those detailed information about order and chaos that can be obtained by analyzing Lyapunov exponent at different energies.

In Sect. 2, we recall the basic definitions and concepts of the geometric formulation of Newtonian mechanics, including the stability criteria and the Jacobi equation for a generic two-dimensional system. In Sect. 3, we apply the theory to the restricted three-body problem. We show that geometry contains all the information about order and chaos and is in complete agreement with the results of the largest Lyapunov exponent.

2 The Relationship Between Mechanics and Riemannian Geometry

We aim at considering Newtonian mechanics from a geometrical point of view. More precisely, the trajectories of a Hamiltonian flow are regarded as geodesics of a Riemannian manifold, equipped with a suitable metric, so that weak or strong chaotic instability can be related with geometrical properties of the underlying manifold.

The natural geometrical setting of Hamiltonian dynamics is within the framework of symplectic or, more generally, Poisson geometry. However, since we are mostly interested in the study of Hamiltonian systems that have a standard kinetic part, also Riemannian geometry can be used. The advantage of using a Riemannian manifold relies on the possibility of measuring the distances between two points, thus the separation between two different trajectories. It is well known how Newtonian mechanics can be rephrased in the Riemannian geometrical language.

Different choices are possible for the ambient space (configuration space, configuration space-time, phase space); therefore, different metrics can be used. Besides, for more general problems, when the Riemannian description is not possible, still a geometrical approach in the same spirit can be developed with the aid of Finsler spaces.

In the following, as the first step to use the geometry, we use the configuration space manifold equipped with a Riemannian metric.

The relationship between dynamics and Riemannian geometry is established through variational formulation of both geodesics and Newtonian dynamics. The geodesics of a Riemannian manifold are defined as the extrema of the arc length functional:

$$\delta \int_A^B \mathrm{d}s = 0 \,. \tag{1}$$

On the other hand, the trajectories, in configuration space of a mechanical system according to the Maupertuis least action principle are given by the extrema of the action integral [6]:

$$\delta \int_{\gamma(t)} 2T \, \mathrm{d}t = 0 \,, \tag{2}$$

where T is the kinetic energy and $\gamma(t)$ are all the isoenergetic curves joining two points q_0 and q_1.

Comparing (1) and (2) and setting $\mathrm{d}s = 2T\,\mathrm{d}t$, it is seen that the trajectories of a mechanical system can be viewed as geodesics of the configuration space manifold M equipped with a suitable metric which can be obtained through:

$$g_{ij} = 2a_{ij}(E - V(\boldsymbol{x})) \,. \tag{3}$$

In local coordinates, the geodesic equations of a Riemannian manifold are given by [7]:

$$\frac{\mathrm{d}^2 x^i}{\mathrm{d}s^2} + \Gamma^i_{jk} \frac{\mathrm{d}x^j}{\mathrm{d}s} \frac{\mathrm{d}x^k}{\mathrm{d}s} = 0 \,, \tag{4}$$

where s is the proper time and Γ^i_{jk} are the Christoffel coefficients and x_i's are the configuration space coordinates.

Equation (4) are also the equations of motion that can be obtained from the Lagrange equations. In local coordinates the components of the Riemann tensor are given by:

$$R^l_{kij} = \partial_i \Gamma^l_{kj} - \partial_j \Gamma^l_{ki} + \Gamma^m_{kj} \Gamma^l_{mi} - \Gamma^m_{ki} \Gamma^l_{mj} \,. \tag{5}$$

The trace $R_{kl} = R^i_{kil}$ is the Ricci tensor and the scalar $R = g^{kl} R_{kl}$ is the scalar curvature of M, which is called Ricci scalar.

It has been shown that from the curvature properties of a Riemannian manifold some relevant consequences about the stability properties of its geodesics can be derived [1,2,7,8]. This is the central point that this work is focussed on.

Now consider some region of the manifold through whose points geodesics can be drawn in various directions. Concentrate on one geodesic and its neighboring geodesics. Let s designate the parameter that locates points on the geodesic and τ is a continuous parameter that labels different geodesics. In order to measure the degree of sensitivity to initial conditions, we consider a congruence of geodesics $\gamma(\tau, s)$ and a separation vector $\boldsymbol{J}$, which is Lie dragged by the congruence (as is shown in Fig. 1) [1]. Now choose any given geodesic $\gamma(\tau_0, s)$ as the "reference geodesic" and define the parameterization

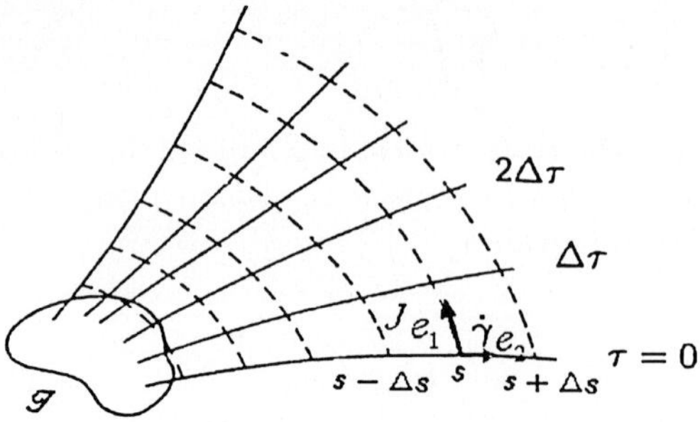

Fig. 1. bundle of geodesics and Jacobi vector, base vectors also are represented

so that $\gamma(\tau_0, s) = \gamma(\tau = 0, s)$. then the magnitude of the separation vector $\boldsymbol{J}(s)$ is defined by:

$$J(s) = \left[\frac{\partial\gamma(\tau, s)}{\partial\tau}\right]_{\tau=0} . \tag{6}$$

Using finite difference in τ, s, as is shown in Fig. 2, we can see that $J(s)$ can be interpreted as the distance between two nearby geodesics. Two geodesic coordinates $\boldsymbol{e_1}, \boldsymbol{e_2}$ which we will use as the base vectors of the decomposition of $\boldsymbol{J}$ are shown in Fig. 1.

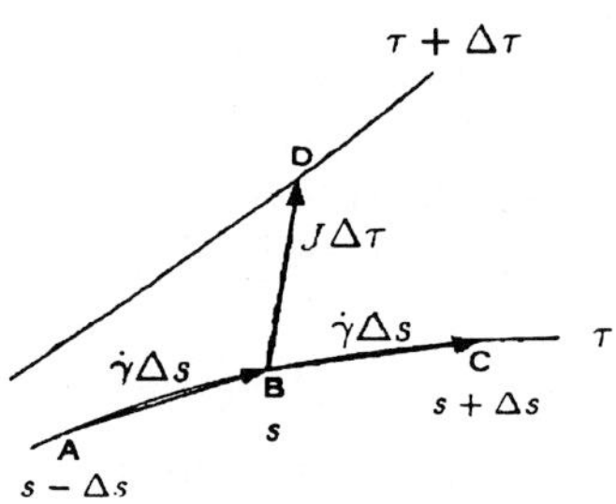

Fig. 2. The Jacobi can be used to measure the separation between nearby geodesics

By taking the second variation of $\boldsymbol{J}$ we can obtain the Jacobi equation [1,7]:

$$\frac{\nabla}{\mathrm{d}s}\frac{\nabla}{\mathrm{d}s}\boldsymbol{J}(s) + R(\boldsymbol{J}, \boldsymbol{v})\boldsymbol{v} = 0 , \tag{7}$$

where $\boldsymbol{v} = \dot{\gamma}$ and $\nabla/\mathrm{d}s = \nabla_{\dot{\gamma}}$ is the covariant derivative along the geodesic.

The Jacobi equation (7) gives the dynamics of the geodesics with respect to the "reference geodesic". Divergence of J explain the instability of the geodesic and so the relevant Hamiltonian system [1,8]. The instability of the system can be studied by means of this equation. It shows the relationship between geodesic instability and chaos.

For a general two degree of freedom system, we can assume the following decomposition for $\boldsymbol{J}$:

$$\boldsymbol{J}(s) = \sum_{i=1}^{2} \xi_i(s)\boldsymbol{e}_{(i)}(s) \ , \tag{8}$$

where $\{\boldsymbol{e}_{(1)}, \boldsymbol{e}_{(2)}\}$ is an orthonormal system of parallely transported vectors. In this reference we have [1,8]:

$$\nabla_s \nabla_s \boldsymbol{J} = \sum_{i=1}^{2} \frac{\mathrm{d}^2 \xi_i}{\mathrm{d}s^2} \boldsymbol{e}_{(i)}(s) \ . \tag{9}$$

Moreover,

$$\begin{aligned} R(\boldsymbol{v},\boldsymbol{J})\boldsymbol{v} &= \sum_j \langle R(\boldsymbol{v},\boldsymbol{J})\boldsymbol{v}, \boldsymbol{e}_{(i)}\rangle \boldsymbol{e}_{(i)} \\ &= \sum_j \langle R\boldsymbol{v}, \boldsymbol{e}_{(i)})\boldsymbol{v}, \boldsymbol{e}_{(i)}\rangle \xi_i \boldsymbol{e}_{(i)} \ . \end{aligned} \tag{10}$$

So (7) can finally be written as:

$$\frac{\mathrm{d}^2 \xi_i}{\mathrm{d}s^2} + \sum_i Q_{ij}\xi_j = 0 \qquad\qquad i,j = 1,2 \tag{11}$$

with $Q_{ij} = \langle R(\boldsymbol{v}, \boldsymbol{e}_{(i)})\boldsymbol{v}, \boldsymbol{e}_{(i)}\rangle$.

We can choose our basis $\{\boldsymbol{e}_{(1)}, \boldsymbol{e}_{(2)}\}$ such that $\boldsymbol{e}_{(1)} \perp \boldsymbol{v}, \boldsymbol{e}_{(2)} \parallel \boldsymbol{v}$ so that $g_{ij} e^i_{(\alpha)} e^j_{(\beta)} = \delta_{\alpha\beta}$ [8]; then since $\boldsymbol{v}$ is parallely transported by definition of the geodesic, $\boldsymbol{e}_{(2)}$ is parallely transported too, and if $\boldsymbol{e}_{(1)} \perp \boldsymbol{e}_{(1)}$ then $\boldsymbol{e}_{(2)}$ is also parallely transported. With respect to a local(static) reference frame they can be written as:

$$\boldsymbol{e}_{(1)} = \left\{ -\frac{\mathrm{d}x^2}{\mathrm{d}s}, \frac{\mathrm{d}x^1}{\mathrm{d}s} \right\} \ , \tag{12a}$$

$$\boldsymbol{e}_{(2)} = \left\{ \frac{\mathrm{d}x^1}{\mathrm{d}s}, \frac{\mathrm{d}x^2}{\mathrm{d}s} \right\} \ . \tag{12b}$$

They are the same as the base vectors of Fig. 1.
Now (11) becomes:

$$\frac{\mathrm{d}^2 \xi^1}{\mathrm{d}s^2} + Q_{11}\xi^1 + Q_{12}\xi^2 = 0 \ , \tag{13a}$$

$$\frac{\mathrm{d}^2 \xi^2}{\mathrm{d}s^2} + Q_{22}\xi^2 + Q_{21}\xi^1 = 0 \ . \tag{13b}$$

The components of the Riemann tensor are derived from (5). In two dimensions it can be shown that the only nonvanishing component of R is R_{1212} and for the Jacobi metric it is equal to [1,8]:

$$R_{1212} = \frac{\Delta V(\boldsymbol{x})}{2} + \frac{(\nabla V(\boldsymbol{x}))^2}{2W} , \tag{14}$$

where $W = E - V(\boldsymbol{x})$ and Δ, ∇ are the Euclidean Laplacian and gradient, respectively. We used MAPLE to calculate the Riemann tensor components analytically.

Combining (14) and (12) we obtain $Q_{12} = Q_{21} = Q_{22} = 0$ and

$$Q_{11} = \frac{1}{W^2}\left[\frac{\Delta V(\boldsymbol{x})}{2} + \frac{(\nabla V(\boldsymbol{x}))^2}{2W}\right] . \tag{15}$$

So (13a) and (13b) become:

$$\frac{\mathrm{d}^2\xi^1}{\mathrm{d}s^2} + \frac{1}{2}R\xi^1 = 0 , \tag{16a}$$

$$\frac{\mathrm{d}^2\xi^2}{\mathrm{d}s^2} = 0 , \tag{16b}$$

where R denotes the scalar curvature of configuration space manifold.

It is obvious that a negative curvature yields an unstable solution of (16a), (16b). This is the case of abstract geodesic flows of ergodic theory, which are defined on hyperbolic manifolds (more precisely, they are defined on the unitary tangent bundle of manifolds of everywhere negative curvature). However, for many systems like the restricted three-body problem, the curvature is mainly positive and the way to make the solution unstable is the rapid change of the curvature along the geodesic, which brings about parametric resonance and chaos [9]. Hyperbolicity is not the only way to make unstable solutions of (16a). If R is not constant, the loss of stability of the geodesics can also be introduced by parametric instability. Let us briefly recall what parametric instability is. If the parameters of a dynamical system vary periodically in time, then a stable solution can be made unstable even if it is stable for each value of the parameters; this is the case of a harmonic oscillator whose frequency is periodically modulated in time with a suitable period [9, 10]. It has been shown in several papers that this is the dominant mechanism of instability in several physical geodesic flows on high dimensional manifolds. In Sect. 3, we show that this is also the case for a typical two degrees of freedom system.

Passing from the proper time s to the physical time t in (16a), (16b) we find:

$$\frac{\mathrm{d}^2\xi^1}{\mathrm{d}t^2} - \frac{1}{W}\frac{\mathrm{d}W}{\mathrm{d}t}\frac{\mathrm{d}\xi^1}{\mathrm{d}t} + 2R_{1212}\xi^1 = 0 , \tag{17a}$$

$$\frac{\mathrm{d}^2\xi^2}{\mathrm{d}t^2} = 0 . \tag{17b}$$

From (17a), (17b) we see that the only component of the geodesic separation vector which conveys information about the stability behaviour of nearby geodesics is the transverse component ξ^1 .

Equations (17a), (17b) describe without approximation the stability properties of the dynamics of two-degree of freedom Hamiltonian systems.

To solve the Jacobi equation, we can use a new function:

$$Y(t) = \xi^1(t) \exp\left[-\frac{1}{2}\int \mathrm{d}t \frac{\dot{W}}{W}\right] \equiv \frac{\xi^1(t)}{\sqrt{W}} . \tag{18}$$

Inserting (18) into (17a) we obtain the Hill equation:

$$\frac{\mathrm{d}^2 Y}{\mathrm{d}t^2} + Q(t)Y = 0 , \tag{19}$$

where,

$$\begin{aligned} Q(t) &= \left[\Delta V + \frac{(\nabla V)^2}{W}\right] - \frac{1}{4}\left[\frac{\dot{W}}{W}\right]^2 + \frac{1}{2}\frac{\mathrm{d}}{\mathrm{d}t}\left[\frac{\dot{W}}{W}\right] \\ &= 2R_{1212} - \frac{1}{4}\left[\frac{\dot{W}}{W}\right]^2 + \frac{1}{2}\frac{\mathrm{d}}{\mathrm{d}t}\left[\frac{\dot{W}}{W}\right] . \end{aligned} \tag{20}$$

From (18) we have $\xi^1(t) = \sqrt{W}Y(t)$, the prefactor $\sqrt{W}$ has only bounded oscillations and ξ^1 has the same stable or unstable behaviour as $Y(t)$.

The stability (19) which is derived from the Jacobi equation is our main tool to observe the chaos and instability in a two-dimensional dynamical system (restricted three-body problem).

Whenever $Q(t)$ in (19) has a nonvanishing stochastic component the solution $Y(t)$ has an exponentially growing envelope whose growth rate provides a measure of the degree of chaoticity. This rate of growth is similar to Lyapunov exponent and can be used as an indicator of chaos that is derived from geometry. It is defined as [13]:

$$\Lambda = \lim_{t\to\infty} \ln \frac{Y^2(t) + \dot{Y}^2(t)}{Y^2(0) + \dot{Y}^2(0)} . \tag{21}$$

This quantity is a measure of the divergence of the nearby geodesics. It is similar to the largest Lyapunov exponent, but to produce this quantity, it is not necessary to linearize the equations of motion and it can be derived from an exact (19). We computed this quantity by solving (19) and in the next section we report it for each case and compare it with the classical Lyapunov exponent.

At this point one can ask what is the difference with respect to the common definition of chaos, because here apparently the same definition is given. First, it is necessary to distinguish between the usual explanation of the origin of chaos and the operational method to detect it numerically (Lyapunov exponents). Dating back to Poincaré, Melnikov, and others, the origin of chaos is attributed to homoclinic intersections of perturbed separatrices near hyperbolic points. This is a perturbative picture. It requires the use of action–angle variables and applies to quasi-integrable Hamiltonians. Moreover, for typical Hamiltonians of physical interest, the explicit change of coordinates to action–angle ones, in general, is a lengthy work that quickly becomes intractable with the increase of the dimension of the system. In practice and particularly at large dimensions, homoclinic intersections mainly gives a qualitative picture of the source of chaos. The main idea of this work is to show the tight relationship between local instability of the trajectories and local geometry (curvature) of the underlying manifold. The link between local instability and chaos is made by the compactness of the manifold where the trajectories live. The geometrical approach makes use of the natural coordinates (positions and velocities) and applies at any energy and at any strength of the nonintegrable part of the Hamiltonian, and allows a unified treatment of both the explanation of the origin of chaos and the method to measure its intensity.

3 Numerical Computation for Restricted Three-Body Problem

In order to understand the capabilities of the Riemannian approach to retrieve the information about order and chaos, we apply the previous analysis to a two-degree of freedom system. Therefore, by studying the stability of the solutions of (19) along the trajectories originating from regular and chaotic regions, it is possible to obtain qualitative information about order and chaos. In addition, we evaluate the both largest Lyapunov exponent and geometric indicator of chaos (21) approximately and make quantitative comparison between geometric analysis and the traditional one.

One of the simplest systems that can exhibit a chaotic behaviour is the restricted three-body problem, a schematic of which is presented in Fig. 3.

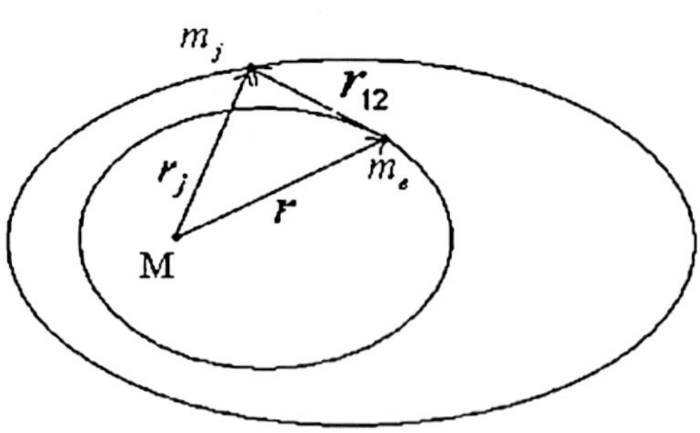

Fig. 3. Restricted three-body system

In this system we have a large body with mass M, a small body with mass m_j and a very small body with mass m_e which is much smaller than M and m_j. We assume that m_e is so small that it cannot influence m_j and so m_j have a regular planar motion around M. The motion of the smallest body get influence from both M and m_j and can have a regular or chaotic motion.

From now on we choose physical units in which we have $G = 4\pi^2$ and $M = 1$ for convenience. In these units, all the masses are actually expressed as fractions of M.

The Hamiltonian of the system in these units can be written in the form:

$$\begin{aligned} H &= \frac{1}{2}m_e\dot{x}^2 + \frac{1}{2}m_e\dot{y}^2 + \frac{1}{2}m_j(r_j^2\omega_j^2) - \frac{4\pi^2 m_e}{r} - \frac{4\pi^2 m_j}{r_j} - \frac{4\pi^2 m_e m_j}{r_{12}} , \\ T &= \frac{1}{2}m_e\dot{x}^2 + \frac{1}{2}m_e\dot{y}^2 , \\ V &= \frac{1}{2}m_j r_j^2\omega_j^2 - \frac{4\pi^2 m_e}{r} - \frac{4\pi^2 m_j}{r_j} - \frac{4\pi^2 m_e m_j}{r_{12}} . \end{aligned} \tag{22}$$

Equations of motion for m_e can be derived with the Lagrange-Euler method:

$$\ddot{\boldsymbol{r}} = -4\pi^2\left(\frac{\boldsymbol{r}}{r^3} + m_j\frac{\boldsymbol{r}-\boldsymbol{r}_j}{|\boldsymbol{r}-\boldsymbol{r}_j|^3}\right) . \tag{23}$$

We consider a circular motion around M for m_j, so the position of m_j can be written in cartesian coordinates as:

$$\begin{cases} x_j = r_j\cos(\omega_j t) , \\ y_j = r_j\sin(\omega_j t) , \end{cases} \tag{24}$$

where $\boldsymbol{r}(x, y)$ is the position vector of m_e and $\boldsymbol{r}_j(x_j, y_j)$ is the position vector of m_j which have a circular motion around M with angular speed ω_j. The biggest mass M is always in the origin. In (23) we do not see M, because we fix it as a constant. Now we can write the equations of motion for m_e, explicitly:

$$\begin{cases} \ddot{x} = -4\pi^2\left(\frac{x}{(x^2+y^2)^{\frac{3}{2}}} + m_j\frac{x-r_j\cos(\omega_j t)}{[(x-r_j\cos(\omega_j t))^2+(y-r_j\sin(\omega_j t))^2]^{\frac{3}{2}}}\right) \\ \ddot{y} = -4\pi^2\left(\frac{y}{(x^2+y^2)^{\frac{3}{2}}} + m_j\frac{y-r_j\sin(\omega_j t)}{[(x-r_j\cos(\omega_j t))^2+(y-r_j\sin(\omega_j t))^2]^{\frac{3}{2}}}\right) \end{cases} \tag{25}$$

From (3), we can obtain the metric of the ambient manifold:

$$\begin{aligned} g_{11} &= 2m_e(E - V(\boldsymbol{r}, \boldsymbol{r}_j)) , \\ g_{12} &= g_{21} = 0 , \\ g_{22} &= 4m_e(E - V(\boldsymbol{r}, \boldsymbol{r}_j)) . \end{aligned} \tag{26}$$

Or explicitly,

$$\begin{cases} g_{11} = 2m_e \left(E - \frac{1}{2} m_j r_j^2 \omega_j^2 + \frac{4\pi^2 m_e}{\sqrt{(x^2+y^2)}} + \frac{4\pi^2 m_j}{r_j} + \frac{4\pi^2 m_j m_e}{\sqrt{(x - r_j \cos(\omega_j t))^2 + (y - r_j \sin(\omega_j t))^2}} \right), \\ g_{12} = g_{21} = 0, \\ g_{22} = 4m_e \left(E - \frac{1}{2} m_j r_j^2 \omega_j^2 + \frac{4\pi^2 m_e}{\sqrt{(x^2+y^2)}} + \frac{4\pi^2 m_j}{r_j} + \frac{4\pi^2 m_j m_e}{\sqrt{(x - r_j \cos(\omega_j t))^2 + (y - r_j \sin(\omega_j t))^2}} \right). \end{cases} \tag{27}$$

The equations of motion have been integrated numerically by means of a Hamming modified predictor-corrector algorithm of fourth order, which is ideal for dynamical models. In order to perform an accurate numerical integration of the equations of motion, the time step is typically varied and adjusted in the interval $10^{-3} - 10^{-4}$. With this time step range, the error $\Delta E/E$ did not exceed $10^{-5} - 10^{-6}$. Together with the equations of motion and by means of the same algorithm, we have also integrated the Jacobi equation in its equivalent form (19) and after computing, we have computed the geometric indicator of chaos (Λ) with the use of (21). Using the same algorithm, we obtained the approximate largest Lyapunov exponent (λ) with the method of [12]. Besides its simplicity, this algorithm has the advantage of being symplectic, which means that its effect upon the equations of motion is equivalent to a canonical transformation of variables. This ensures a faithful numerical representation of a Hamiltonian flow [1].

We fix the following parameters through the computations:

$$\begin{cases} r_j = 5.203\,\text{AU}, \\ \omega_j = \frac{2\pi}{11.86}\,\text{year}^{-1}, \\ m_e = 2.96 \times 10^{-6}, \\ m_{jo} = 317.9\, m_e. \end{cases} \tag{28}$$

These are exactly the parameters of the (Sun, Earth, Jupiter) system in the units where $M_{\text{sun}} = 1$. Here, the unit of distance is the Astronomical Unit and the unit of time is the year, also m_{jo} is a constant which m_j is considered as some multiple of it.

The only variable is m_j. We considered three different values for m_j and solved the equations of motion in each case. we considered $m_j = 60\, m_{jo}, m_j = 300\, m_{jo}, m_j = 450\, m_{jo}$, with m_{jo} defined in (28). For each m_j we chose the same initial conditions $x_0 = 1, y_0 = 0, \dot{x}_0 = 0, \dot{y}_0 = 1$ for m_e. We have kept $Y(0) = 10^{-9}$ and $\dot{Y}(0) = 0$ as initial conditions of the Jacobi equation in all cases.

$Y(t)$ is proportional to the separation of the two nearby geodesics (trajectories). The divergence of $Y(t)$, in spite of $Y(0)$ being small, reveals sensitive dependence on initial conditions which is a signature of chaotic motion.

The solution of (19) for $m_j = 60\, m_{jo}$ is plotted in logarithmic scale in Fig. 4(a). As it is seen, the envelope of $Y(t)$ appears to be bounded or linearly growing for this initial condition. After solving $Y(t)$ we can compute the geometric indicator of chaos Λ. We see that it is a negative number so the geodesic flow is stable. In Fig. 4(b) the Lyapunov exponent is obtained by using a simple method which is explained in [12]. We can see that its

Lyapunov exponent is negative and it agrees with which we obtained from the geometric method (Fig. 4(a)) and the geometric indicator of chaos. In Fig. 4(c) the nonzero component of the Riemann curvature tensor of the manifold is plotted and we see that it is always positive and fluctuating. the orbit of the third mass is plotted In Fig. 4(d). We can see that because of the influence of m_j, its orbit deviates from a regular elliptic orbit.

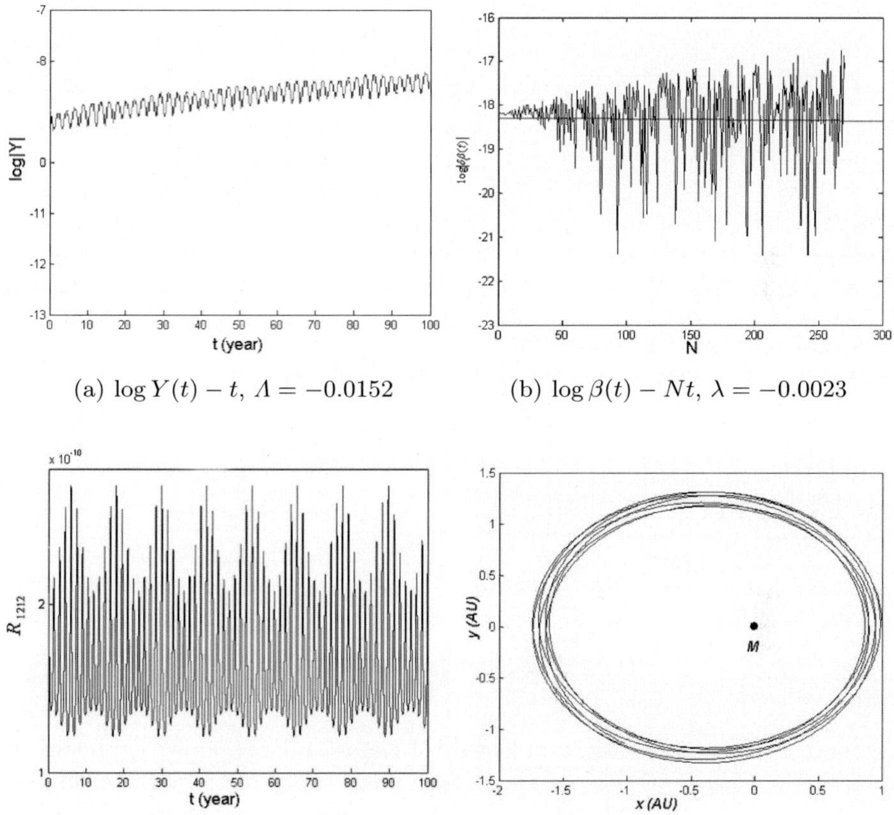

(a) $\log Y(t) - t$, $\Lambda = -0.0152$ (b) $\log \beta(t) - Nt$, $\lambda = -0.0023$

(c) $R_{1212} - t$ (d) Orbit of m_e

Fig. 4. $m_j = 60\, m_{j_0}$

The solution of (19) for $m_j = 300\, m_{j_0}$ is plotted in logarithmic scale in Fig. 5(a). As it is seen, the envelope of $Y(t)$ appears to be bounded or linearly growing for this case. The geometric indicator of chaos is computed and it is also a negative number which shows the validity of this argument. In Fig. 5(b) the Lyapunov exponent is computed. We can see that its Lyapunov exponent is negative and it agrees with which we obtained from the geometric

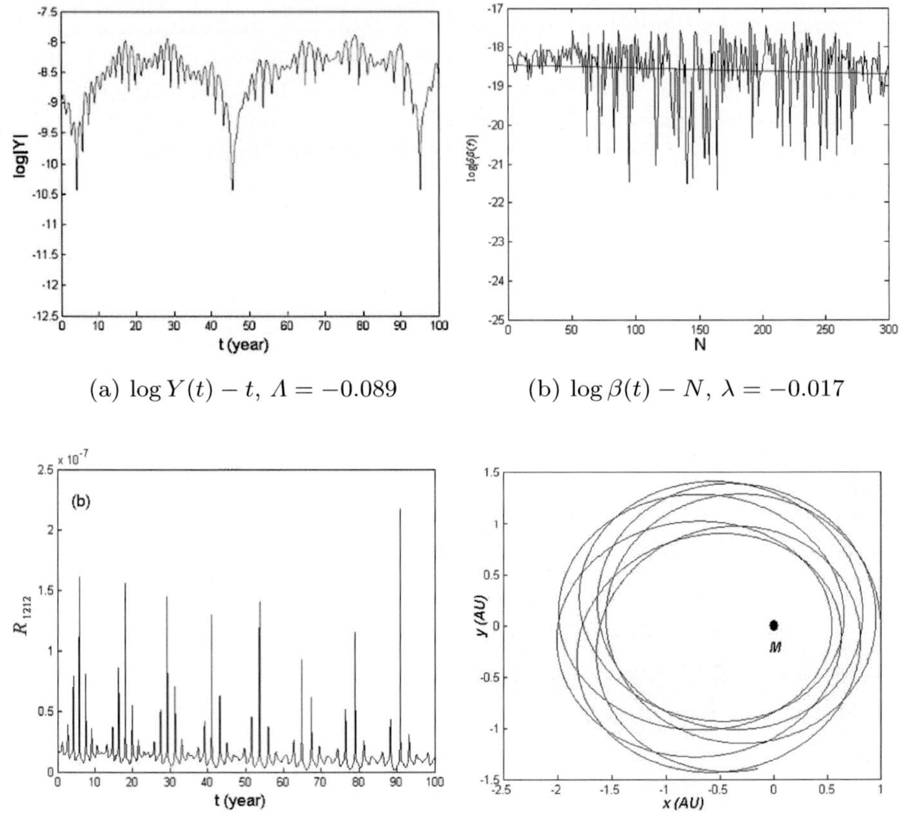

(a) $\log Y(t) - t$, $\Lambda = -0.089$

(b) $\log \beta(t) - N$, $\lambda = -0.017$

(c) $R_{1212} - t$

(d) Orbit of m_e

Fig. 5. $m_j = 300\, m_{j_0}$

method. Both the geometric indicator of chaos and Lyapunov exponent are greater than those of the previous case and it shows that by increasing m_j we can make this system chaotic. In Fig. 5(c) the nonzero component of the Riemannian curvature tensor of the manifold is plotted and it is obvious that it is always positive. The orbit of the m_e is plotted In Fig. 5(d). From this figure, one can find that this orbit is more irregular than the orbit of the previous case and the eccentricity of the orbit changes more than that of the previous case.

As the third example, we did the same steps for the third case with $m_j = 450\, m_{j_0}$ which we shall see that it is a chaotic case. We solved (19) and plotted $Y(t)$ in logarithmic scale in Fig. 6(a). In this case, we see that this function is increasing rapidly and the geometric indicator of chaos is a positive number. The magnitude of $Y(t)$ after a short time increases 10^6 times and so

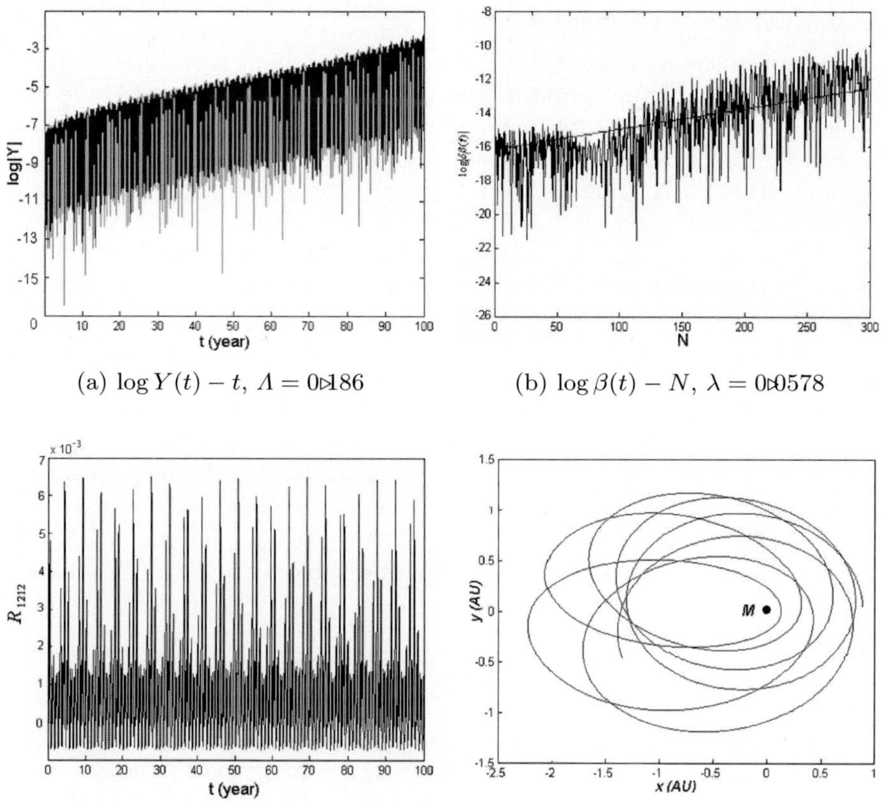

(a) $\log Y(t) - t$, $\Lambda = 0.186$

(b) $\log \beta(t) - N$, $\lambda = 0.0578$

(c) $R_{1212} - t$

(d) Orbit of m_e

Fig. 6. $m_j = 450\, m_{j0}$

the geodesics which originate from a compact region will diverge and go far away. Also we computed the Lyapunov exponent in Fig. 6(b). The Lyapunov exponent also is a positive number and it shows that we have chaotic motion. This is in agreement with which we obtained by the geometric method and by solving the Jacobi equation. The nonzero component of the Riemann curvature tensor is plotted in Fig. 6(c) and we see that it is always positive and fluctuating. So chaos is generated with the mechanism of parametric resonance. In Fig. 6(d) the orbit of m_e is plotted and we can see that it is more irregular than in the previous cases.

We have just seen that the comparison between the detailed information given by the geometry (from $Y(t)$ and Λ) reveals a complete qualitative and quantitative agreement with results obtained from the classical method (Lyapunov exponents). Here we can speak about the origin of chaos.

Evaluation of R_{1212} by using (14), and the scalar curvature $R = 2R_{1212}/W^2$, shows that both are always positive for the restricted three-body problem, no matter whether it is computed along a regular or chaotic orbit. Therefore, in the geodesic flow defined by the restricted three-body problem, chaos is due to the parametric resonance of the nearby geodesics induced by fluctuations of the positive curvature. By comparing Figs. 4(c), 5(c) and 5(c) it is obvious that in all three cases the curvature is always positive and rapidly fluctuating, but the difference between these three cases is that by increasing m_j the amplitude of the fluctuations increases. For example the magnitude of the fluctuations in the regular case $m_j = 60\, m_{j_0}$ is of the order 10^{-10} but this magnitude for the chaotic case $m_j = 450\, m_{j_0}$ is of the order 10^{-3}. It increases by the order of 10^7. So we can conclude that for occurring parametric resonance and so chaos in a system it is necessary that not only the curvature fluctuates rapidly but also its amplitude be large enough. A deep understanding of the parametric resonance can help us to understand more about the occurrence of chaos in Hamiltonian systems. It may be possible to study the chaotic criteria of a system with only studying the behaviour of the curvature of its equivalent Riemannian manifold.

Finally we can summarize the results in a table:

Table 1. Λ and λ in three cases

	$m_j = 60\, m_{j_0}$	$m_j = 300\, m_{j_0}$	$m_j = 450\, m_{j_0}$
geometric indicator of chaos (Λ):	−0.0152	−0.089	0.1860
largest Lyapunov exponent (λ):	−0.0023	−0.017	0.0578

4 Conclusion

In this chapter, we have shown that the geometric description of Hamiltonian chaos based on Riemannian geometry is in agreement with the standard approaches (Lyapunov exponents). This has been shown for a two degree of freedom system (restricted three-body problem) and the results of the stability (19), which has been derived from the Jacobi equation are computed and the behaviour of the system (chaotic or regular) specified and compared with the phenomenological standard methods. Also we used a geometric indicator of chaos (Λ) and it was shown that it can be used for predicting the behaviour of system like the Lyapunov exponents but it is derived from an exact equation (not a linearized one as in deriving Lyapunov exponents). A complete agreement is found. This method can be applied for studying chaos in any Hamiltonian system in any energy and in any range of the nonlinearity parameters and so it is very useful to apply this method to real systems like engineering problems.

The results reported in this paper are not simply alternatives to the conventional tools to describe chaos, but have a deeper meaning [13]. They provide an explanation of the origin of Hamiltonian chaos that is an alternative to the conventional interpretation, which is based on homoclinic intersections.

References

1. M. Pettini: Phys. Rev. E **47**, 828 (1993)
2. L. Casetti, M. Pettini: Phys. Rev. E **48**, 4320 (1993)
3. M. Cerruti-sola, M. Pettini: Phys. Rev. E **51**, 53 (1994)
4. R. Abraham, J. E. Marsden: *Foundations of Mechanics* (Addison-Wesley, Redwood city CA 1998)
5. J. José, E. Saletan: *Classical Dynamics: A contemporary approach* (Cambridge university press, USA 1998)
6. M. Nakahara: *Geometry, Topology and Physics* (Adam Hilger, Paris 1990)
7. M. Pettini, R. Valdettaro: Chaos **5**, 646 (1995)
8. M. Cerruti-sola, M. Pettini: Phys. Rev. E **53**, 179 (1996)
9. R. Lima, M. Pettini: Phys. Rev. A **41**, 726 (1990)
10. A.H. Nayfeh, D.T. Mook: *Nonlinear oscillations* (Wiley, New York 1979)
11. H.T. Davis: *Introduction to Nonlinear Differential and Integral equations* (Dover Publications Inc., New York 1962) pp 50–60
12. L. Hand, J. Finch: *Analytical Mechanics* (Cambridge university press, USA 1998)
13. L. Casetti, C. Clementi, M. Pettini: Phys. Rev. E **54**, 5969 (1996)

Order and Chaos in Some Hamiltonian Systems of Interest in Plasma Physics

D. Constantinescu[1] and B. Weyssow[2]

[1] Department of Applied Mathematics, Association EURATOM-MEC, University of Craiova, 13 A.I. Cuza Street, Craiova 200585, Romania
dconsta@central.ucv.ro

[2] Department of Statistical Physics and Plasmas, Association EURATOM-Belgian State, ULB CP 231, Bvd du triomphe, 1050 Bruxelles, Belgium
bweyssow@ulb.ac.be

1 Introduction

The name Tokamak, an acronym for the Russian expression TOrichnaia KAmera MAgnitnaia Katushka [1], is given to toroidal chambers in which plasma particles are confined by a magnetic field which has two basic orthogonal components, one acting in the direction of the major curvature of the torus, the other in the direction of the minor curvature of the torus.

The magnetic field line equations define a Hamiltonian system [3, 4]. In the ideal i.e. *integrable* case the helical magnetic field lines lay on constant pressure surfaces, called magnetic surfaces, having the topology of nested tori surrounding the magnetic axis. Deviations to the ideal case are always present in experiments either due to internal factors such as instabilities and fluctuations or due to external causes such as magnetic imperfections arising from the poloidal distribution of the toroidal coils. The perturbed systems, generically non integrable [3] can be studied using the Poincaré map associated to a given poloidal cross-section. This map is an area-preserving map which has to be compatible with the toroidal geometry. A typical phase portrait of the Poincaré map exhibits a complex structure of regular and chaotic zones which form a fat fractal [2]. The field line transport properties are determined by the competition between order and chaos which is observable by varying the parameters of the system.

The localization of the regions of phase-space covered by regular orbits and the description of the mechanisms for their destruction is important from both theoretical and practical points of view. From a practical point of view, the study of magnetic *internal transport barriers* relates to the more general study of particle *internal transport barriers* (ITB) which is of great importance for nuclear fusion because the particle radial transport is strongly reduced in their presence and hence the particle confinement time is increased. For magnetic ITB a reduction of particle transport is expected due to a reduction of the magnetic field line radial wandering (charge particle guiding centres follow magnetic field lines). From a mathematical point of view, a magnetic internal transport barrier is an invariant set, mainly formed by regular orbits, separating a core chaotic zone situated near the magnetic axis of the tokamak, from a peripheral chaotic zone situated near the tokamak wall. This magnetic ITB cannot be traversed by magnetic field lines, i.e. the magnetic transport through the ITB is suppressed and all magnetic field lines passing through the central stochastic zone are confined.

The aim of the chapter is to present recent results on the configuration of the transport barriers in some realistic models of magnetic field line dynamics in a tokamak geometry. This study leads to a new understanding of the dynamic of this kind of magnetic systems based on very ingenious correlations with the theory of numbers and on hidden geometrical aspects. The chapter is organized as follows. Section 2 is devoted to the presentation of the mathematical models. Some definitions and basic results from the theory of dynamical systems are presented in Sect. 3. The transport barriers as well as their localization in phase-space are described in Sect. 4. Section 5 contains a description of some scenarios of magnetic field line reconnection and of magnetic island merging. Conclusions are drawn in Sect. 6.

2 The Mathematical Models

A set of toroidal coordinates (r, θ, ζ), where ζ is the toroidal angle around the symmetry axis of the torus and (r, θ) are the polar coordinates in a (circular) poloidal cross-section at distance R_0 to the symmetry axis and having maximal radius a, is usually used for the description of the magnetic line configuration. However, since canonicity of the coordinates is needed for the derivation of the discrete dynamical systems describing the behaviour of the magnetic field lines one uses the toroidal flux $\psi = r^2/2$ instead of the radial coordinate r [3]. A Poincaré section is defined from the intersection points of a magnetic field line starting at position (θ_0, ψ_0) with the poloidal section $\mathcal{S} : \zeta = cst$. The intersection point after n toroidal turns is denoted by (θ_n, ψ_n). The third coordinate ζ does not appear here as it is a constant parameter depending only on the position of the poloidal section. The application

$$T_K : S^1 \times \mathbf{R}_+ \to S^1 \times \mathbf{R}_+ \, , \; T_K(\theta_n, \psi_n) = (\theta_{n+1}, \psi_{n+1}) \tag{1}$$

has to be an area-preserving map compatible with the toroidal geometry (if $\psi_0 = 0$ then $\psi_n = 0$ for all $n \in \mathbf{N}$ and if $\psi_0 > 0$ then $\psi_n > 0$ for all $n \in \mathbf{N}$). A map satisfying the imposed constraints is derived from the mixed generating function $F_K : S^1 \times \mathbf{R}_+ \to S^1 \times \mathbf{R}_+$ (see [3], [5]):

$$F_K(\theta_n, \psi_{n+1}) = \psi_{n+1}\, \theta_n + \alpha_0(\psi_{n+1}) + K\, P(\theta_n, \psi_{n+1}) \, . \tag{2}$$

The corresponding discrete system is

$$\psi_{n+1} = \psi_n - K \frac{\partial P(\theta_n, \psi_{n+1})}{\partial \theta_n}$$

$$\theta_{n+1} = \theta_n + W(\psi_{n+1}) + K \frac{\partial P(\theta_n, \psi_{n+1})}{\partial \psi_{n+1}} \pmod 1 \tag{3}$$

where $W(\psi) = \mathrm{d}\alpha_0(\psi)/\mathrm{d}\psi$. The analytical expression of T_K is obtained by determining θ_{n+1} and ψ_{n+1} from (3) as a function of θ_n and ψ_n. The application $W(\psi)$ is often called *the winding function*, $q \equiv 1/W$ *the safety factor* or more simply *the q-profile* and $s = \mathrm{d}\ln q/\mathrm{d}\ln\psi$ *the shear profile.* The latter quantity is positive when q is an increasing function and it is negative when q is a decreasing function of the toroidal flux (or radius). By choosing adequately the forms of $\alpha_0(\psi)$ and $P(\theta, \psi)$ in (2) some well-known models used in plasma physics are recovered. For instance, with $\alpha_0(\psi) = \psi^2/2$ and $P(\theta, \psi) = -(2\pi)^{-1}\cos(2\pi\theta)$ one obtains the Chirikov–Taylor system generated by the celebrated "standard map" [6]. With $\alpha_0(\psi) = \psi^2/2$ and $P(\theta, \psi) = -(2\pi)^{-1}\,\psi\cos(2\pi\theta)$ the "Wobig map" is obtained [7] while with $\alpha_0(\psi) = a\,(\psi - \psi^3/3)$ and $P(\theta, \psi) = -(2\pi)^{-1}\cos(2\pi\theta)$ one obtains the non-twist standard map [23,24]. None of these maps is applicable to toroidal fusion devices because either the positivity of the toroidal flux is not guaranteed or the profile for the winding function is unphysical. They should therefore be considered as academic models.

More realistic models of magnetic field line dynamics also satisfying (2) are known as the Tokamap [3] and the Rev-Tokamap [10]. They involve a monotonous and a non-monotonous winding function, respectively. The perturbation $P = P(\theta, \psi)$ is chosen in order to be consistent with the toroidal geometry. The *tokamap* is obtained by choosing in (3)

$$P(\theta, \psi) = -\frac{1}{(2\pi)^2}\frac{\psi}{1+\psi}\cos(2\pi\theta)$$

$$W_T(\psi) = \frac{1}{4}(2-\psi)(2-2\psi+\psi^2)\,, \tag{4}$$

where an index T stands for Tokamap. A magneto-hydrodynamic derivation of the realistic (decreasing) winding function W (or safety factor $q = 1/W$) is given in [8]. The tokamap is deduced also as a particular case of a map for guiding centre particles [9]. The study of reversed shear configurations can be performed using a different winding function [10]

$$W_{RT}(\psi) = w\left[1 - A\,(C\,\psi - 1)^2\right] \tag{5}$$

where the lowerscript RT stands for reversed tokamap (or rev-tokamap). The parameters are

$$A = \frac{w - w_0}{w}\,,\ C = 1 + \sqrt{\frac{w - w_1}{w - w_0}}\,, \tag{6}$$

where $w_0 = W(0)$ and $w_1 = W(1)$.

This map is $RT_K : S^1 \times \mathbf{R}_+ \to S^1 \times \mathbf{R}_+$, $RT_K(\theta, \psi) = (\overline{\theta}, \overline{\psi})$ with

$$\overline{\psi} = \frac{1}{2}\left(\psi_{\mathrm{RT}} + \sqrt{\psi_{\mathrm{RT}}^2 + 4\psi}\right)$$
$$\overline{\theta} = \theta + w\,[1 - A\,(C\,\overline{\psi} - 1)^2] - \frac{K}{4\pi}\,\frac{1}{(1+\overline{\psi})^2}\cos(2\pi\theta) \quad (\mathrm{mod}\,1) \qquad (7)$$

where $\psi_{\mathrm{RT}} = \psi - 1 - \frac{K}{2\pi}\sin(2\pi\theta)$. The winding function W has a local maximum when $\psi = 1/C$. In order to further exemplify some results to be shown in this chapter two other maps are proposed. One is the *bounded tokamap* obtained from (11) by choosing $W_{BT}(\psi) = W_T(\psi)$, as in the usual tokamap, and

$$P(\theta,\psi) = -\frac{1}{(2\pi)^2}\,\frac{\psi\,(1-\psi)}{1+\psi}\cos(2\pi\theta)$$

The analytical expression of the bounded tokamap is BT_K:

$$\overline{\psi} = \frac{\psi_{\mathrm{BT}} + \sqrt{\psi_{\mathrm{BT}}^2 + 4\psi\,(1 - \frac{K}{2\pi}\sin(2\pi\theta))}}{2\,(1 - \frac{K}{2\pi}\sin(2\pi\theta))}$$
$$\overline{\theta} = \theta + W_{BT}(\overline{\psi}) - \frac{K}{4\pi^2}\cos(2\pi\theta)\,\frac{\overline{\psi}^2 + 2\,\overline{\psi} - 1}{(\overline{\psi}+1)^2} \quad (\mathrm{mod}\,1) \qquad (8)$$

where $\psi_{\mathrm{BT}} = \psi - 1 - \frac{K}{2\pi}\sin(2\pi\theta)$. The main difference between the tokamap and the bounded tokamap is that BT_K has two invariant geometrical circles; one $\psi = 0$ near the magnetic axis of the tokamak and the other $\psi = 1$ that defines the tokamak wall. The latter transport barrier is induced by the perturbation, not by the winding function W.

The second proposed map, the *degenerate tokamap*, is obtained from (11) by choosing

$$P(\theta,\psi) = -\frac{1}{(2\pi)^2}\,\frac{\psi}{1+\psi}\cos(2\pi\theta)$$
$$W_{DT}(\psi) = \frac{1}{4}\,(2-\psi)\,(2 - 2\psi + \psi^2) + \psi^2\left(2 - \frac{7}{4}\,\psi\right). \qquad (9)$$

The analytic expression of this map is $DT_K : S^1 \times \mathbf{R} \to S^1 \times \mathbf{R}$:

$$\overline{\psi} = \frac{1}{2}\left(\psi_{\mathrm{DT}} + \sqrt{\psi_{\mathrm{DT}}^2 + 4\psi}\right)$$
$$\overline{\theta} = \theta + W_{DT}(\overline{\psi}) - \frac{K}{4\pi}\,\frac{1}{(1+\overline{\psi})^2}\cos(2\pi\theta) \quad (\mathrm{mod}\,1) \qquad (10)$$

with $\psi_{\mathrm{BT}} = \psi - 1 - \frac{K}{2\pi}\sin(2\pi\theta)$. The perturbation is the same as for the tokamap but the winding function is modified. It is still a monotone decreasing function but such that $W'_{DT}(1/2) = W''_{DT}(1/2) = 0$ (a prime denotes a

$\partial/\partial\psi$ derivative) in contrast to the tokamap which has $W_T'(\psi) < 0$ for all $(\theta, \psi) \in S^1 \times [0, 1]$. The conditions $W_{DT}(0) = W_T(0)$ and $W_{DT}'(0) = W_T'(0)$ ensure that the tokamap and the degenerate tokamap have similar behaviour near the polar axis $\psi = 0$. All four maps described in this chapter are submitted to the condition $K < 2\pi$ which must hold since the analytical properties of the maps have to be at least of class C^1.

3 Definitions and Basic Results

The prototype of area-preserving maps which will be studied in the sequel is $f : S^1 \times \mathbf{R} \to S^1 \times \mathbf{R}$, $f(\theta, \psi) \overset{\text{not}}{=} (\overline{\theta}, \overline{\psi})$ defined by the system of implicit equations (thus the notation $\overset{\text{not}}{=}$):

$$\begin{aligned} \overline{\psi} &= \psi - K\, g'(\theta)\, h(\overline{\psi}) \\ \overline{\theta} &= \theta + W(\overline{\psi}) + K\, g(\theta)\, h'(\overline{\psi}) \qquad (\mathrm{mod}\, 1) \end{aligned} \tag{11}$$

where $g : S^1 \to \mathbf{R}$, $h : \mathbf{R} \to \mathbf{R}$ are bounded functions of class C^2 (two times derivable) and g is a 1-periodic function in θ.

- An **orbit** of (θ_0, ψ_0) is the set $\{(\theta_n, \psi_n), n \in \mathbf{N}\}$ such that $(\theta_{n+1}, \psi_{n+1}) = f(\theta_n, \psi_n)$.
- A **lift** of f, i.e. $F : \mathbf{R} \times \mathbf{R} \to \mathbf{R} \times \mathbf{R}$, $F(\theta, \psi) \overset{\text{not}}{=} (\overline{x}, \overline{\psi})$, can be obtained from (11) by removing the mod 1 in the second equation.
- The **winding (or rotation) number** of an orbit is defined as

$$\omega = \lim_{i \to \infty} \frac{\overline{x}_i}{i} \tag{12}$$

  if the limit exists.
- The orbit of (θ_0, ψ_0) is periodic of period n if $f^n(\theta_0, \psi_0) = (\theta_0, \psi_0)$. The rotation number of a periodic orbit is a rational. The stability of the periodic point (θ_0, ψ_0) is characterized by the eigenvalues of the Jacobian matrix $J_{(\theta_0, \psi_0)} f^n$ which are called the multipliers of (θ_0, ψ_0) and will be denoted by λ_1 and λ_2. Because f is an area-preserving map ($\det(J_{(\theta,\psi)} f) = 1$ for all $(\theta, \psi) \in S^1 \times \mathbf{R}$) their product $\lambda_1 \lambda_2$ is 1. **Elliptic** periodic points have complex multipliers (and $\lambda_2 = \overline{\lambda_1}$), while **hyperbolic** periodic points have real multipliers $\lambda_1 = 1/\lambda_2 \neq \pm 1$ and **parabolic** periodic points have $\lambda_1 = \lambda_2 = \pm 1$.
- A set homeomorphic to a circle that is mapped onto itself by f is called a rotational **invariant circle**. All orbits contained in an invariant circle densely fills it and they have the same winding number, which is generically irrational. An orbit for which there is no winding number is chaotic.
- A **transport barrier** is a f-invariant subset of $S^1 \times \mathbf{R}$ separating two other f-invariant subsets of $S^1 \times \mathbf{R}$. The transport barriers cannot be crossed by the orbits starting from its neighbourhood.

– The map f is a **positive twist map** (respectively negative) if

$$\frac{\partial \overline{\theta}}{\partial \psi}(\theta, \psi) > 0 \quad \text{(respectively } \frac{\partial \overline{\theta}}{\partial \psi}(\theta, \psi) < 0\text{)}$$

for all $(\theta, \psi) \in S^1 \times \mathbf{R}$.

– A map for which $(\partial\overline{\theta}/\partial\psi)(\theta, \psi) = 0$ for some $(\theta, \psi) \in S^1 \times \mathbf{R}$ is called a **non-twist map**.

The study of twist area preserving maps initiates with the works of H. Poincaré (see [17] pp 38). Their dynamics is described by the Poincaré–Birkhoff theorem and the KAM (Kolmogorov–Arnold–Moser) theory.

The unperturbed map corresponds to $K = 0$. It is an integrable map in the sense that the orbit of (θ_0, ψ_0) is included in the (geometrical) circle $\psi = \psi_0$ and it is periodic if $W(\psi_0) \in \mathbf{Q}$, or quasiperiodic (dense in the circle) if $W(\psi_0) \notin \mathbf{Q}$. Every circle $\psi = \psi_0$ with $\psi_0 \notin \mathbf{Q}$ is a transport barrier.

In the case of the perturbed map ($K > 0$) some orbits are dense in rotational invariant circles which are transport barriers. The elliptic periodic points are surrounded by stability islands. The hyperbolic periodic points having same rotation number are connected by their stable and unstable manifolds. If these manifolds do not intersect transversely the dynamics near them is regular (the orbits of the neighboured points are dense in rotational circles or in curves surrounding elliptic periodic points). If the manifolds intersect transversely, a chaotic layer is formed in a zone containing their (topological) closure. The phase portrait of the tokamap (which is a negative twist map [8]) is presented in Fig. 1 where the coordinates θ and ψ are represented on the horizontal and vertical axis , respectively. Near the vertical line $\theta = 0.5$ the elliptic periodic points with the rotation number 1/1, 2/3, 1/2, 1/3, 1/4, and 1/5 can be identified. Some chaotic layers surrounding the hyperbolic periodic points are observed in the upper part of the figure. These chaotic layers are separated by invariant circles. A large chaotic zone is observed in the central part of the phase space. In this region some manifolds of hyperbolic points having the rotation number 1/3, 1/2 respectively 2/3 overlap. Since the tokamap is almost integrable near the polar axis $\psi = 0$ a regular zone is also observed in the lower part of the phase space.

For $K \ll 1$ the maps given by (11) are close (at least in a C^0 topology) to the integrable map corresponding to $K = 0$, hence they are almost integrable. In this case the main part of the orbits is still regular and the eventually formed chaotic layers are very thin (being separated by very frequent rotational circles).

The Poincaré–Birkhoff theorem [11] ensures that to each rational number n/m in an appropriate interval corresponds a pair of periodic orbits of rotation number n/m in the perturbed map: an elliptic one and an hyperbolic one. The points of these orbits interlace and they are included in the so-called n/m-type Poincaré–Birkhoff island chains.

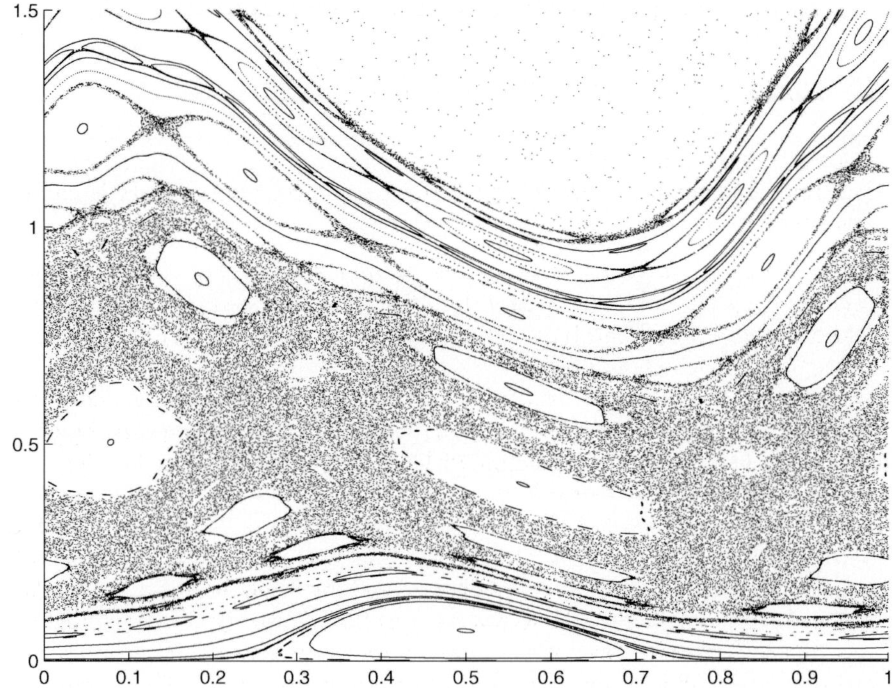

Fig. 1. Phase portrait of tokamap (a twist map) corresponding to $K = 4.5$ and $w = 1$

The KAM theory provides the conditions for the existence of invariant circles in a slightly perturbed integrable twist map: the invariant circles having diophantine rotation number are more robust than the others [12].

For larger values of K this theorem may not be applicable, even if numerical evidences indicate that some resistant barriers have diophantine rotation number [8,13]. Some conditions for the existence of invariant circles were proposed in the eighties by J.M. Greene [14], S. Aubry [15] and J.N. Mather [16]. In this case the enlargement of the chaotic zone can be explained by the converse KAM theory (cone crossing criterion, folding properties, overlapping criterion etc.) [17].

The non-twist maps started to be explored only in the last decade, although the richness of new phenomena exhibited by such maps (the existence of twin Poincaré–Birkhoff chains with the same rotation number, reconnection, meandering tori) were pointed out early in the eighties [18]. The phase portrait of rev-tokamap (a non-twist map) is presented in Fig. 2. The twin Poincaré–Birkhoff island chains having the rotation number 1/2 can be observed. The lower one is contained in a regular region, but the upper one is surrounded by a chaotic layer. The hyperbolic points of twin Poincaré

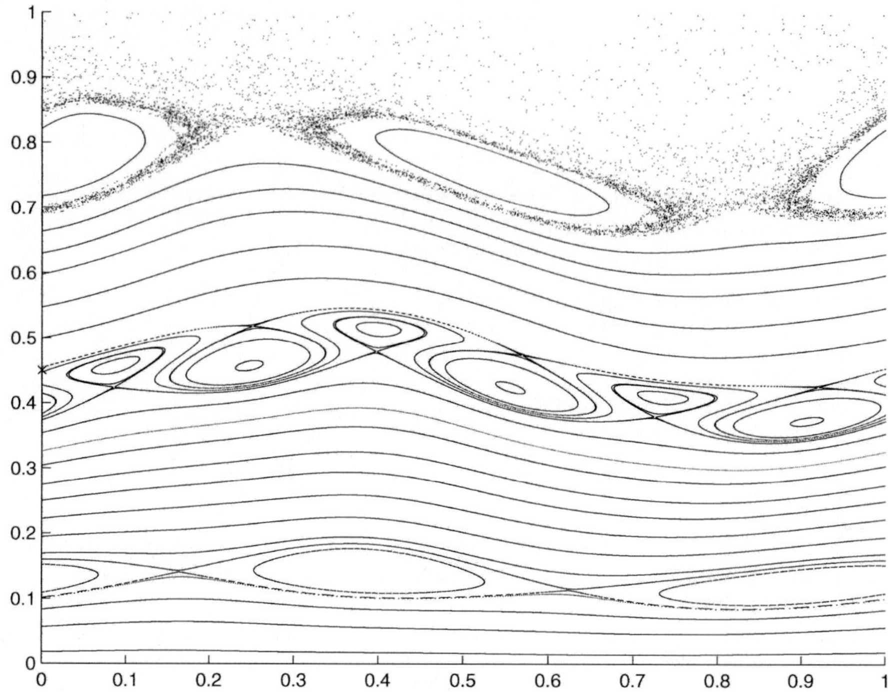

Fig. 2. The phase portrait of the rev-tokamap (a non-twist map) with parameters $K = 1.95$, $w = 0.67$, $w_0 = 0.3333$, $w_1 = 0.1667$

chains with the rotation number 2/3 are connected both by homoclinic and heteroclinic connection. Between them a meandering circle is observed.

The non-twist maps are used to describe the behaviour of magnetic field lines in reversed tokamaks [3], chaotic advection in fluids [19,20], chaotic ionization in atomic physics [21], oscillations in astrophysics [22]. It is also to be noted that non-twist maps appear naturally in iterations of twist maps because the composition of twist maps is generally not a twist map. There exist only few theoretical results concerning the dynamical behaviour of these maps. However, even if the Aubry–Mather variational theory for twist maps does not generalize to non-twist maps, an analogue of the KAM theory has been developed [23–25]. The regions of instability for non-twist area-preserving maps are studied in [26]. The non-twist maps with reversing symmetry group studied in [27] and an analytical method for determining the reconnection threshold was proposed in [28].

The tokamap and the bounded tokamap and negative twist maps. They deviate the lines $\theta =$ cst to the left.

The rev-tokamap and the degenerate tokamap are non-twist maps because the equation $(\partial\overline{\theta}/\partial\psi)(\theta,\psi) = 0$ has solutions in $S^1 \times \mathbf{R}_+$. Their properties

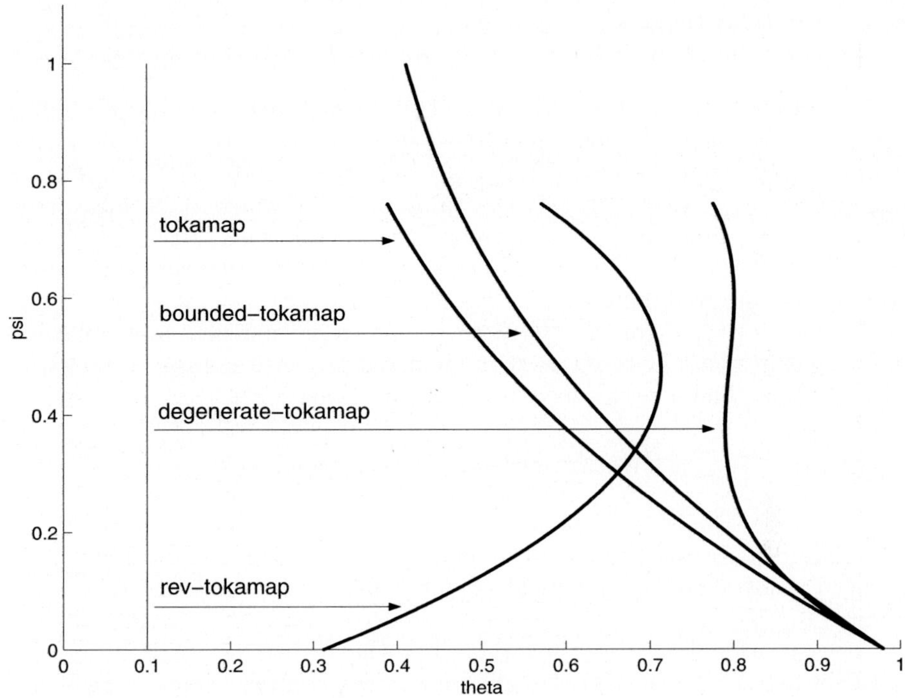

Fig. 3. The effect of tokamap, bounded tokamap, degenerate tokamap (all corresponding to $K = 5.9$, $w = 1$) and rev-tokamap ($K = 5.9$, $w = 0.67$, $w_0 = 0.3333$, $w_1 = 0.1667$) on the vertical line $\theta = 0.1$

are different because the winding function of the rev-tokamap is a quadratic function having a maximum, while the winding function of the degenerate tokamap is a decreasing function (having an inflexion point $\psi = 1/2$).

Figure 3 shows the line $\theta = 0.1$ and its images produced by the tokamap, the bounded tokamap, the degenerate tokamap and the rev-tokamap. The image is tilted to the left by the tokamap and the bounded tokamap showing they are negative twist maps. For the two remaining maps, the rev-tokamap and the degenerate tokamap, the image is no more a graph of a function depending on ψ. This results from non-twist property of the maps.

In order to describe the transport barriers in non-twist maps we introduce the following definitions.

Definition 1. *Let* $f : S^1 \times \mathbf{R} \to S^1 \times \mathbf{R}$, $f(\theta, \psi) \overset{\text{not}}{=} \left(\overline{\theta}, \overline{\psi}\right)$ *be a non-twist map defined by (11).*

- *The* ***regular set*** *of* f *is* $C_{\text{reg}} : W'\left(\overline{\psi}\right) = 0$.
- *The* ***shearless orbit*** *has a rotation number which is a local extremum of all rotation numbers of orbits in the map. The closure of the shearless orbit, denoted by* C_{sh}*, is called the* ***shearless curve.***

- *The* ***critical twist set*** *of f is the curve C_{nt} : $(\partial\overline{\theta}\partial\psi)(\theta,\psi) = 0$.*
- *The* ***non-twist annulus*** *is the closure of all orbits starting from C_{nt}.*

The unperturbed system ($K = 0$ in (11)) defined with a quadratic winding function satisfies the obvious equalities $(\partial\overline{\theta}/\partial\psi)(\theta,\psi) = W'(\psi) = W'\left(\overline{\psi}\right) = 0$ which imply that the curves $C_{\mathrm{nt}}, C_{\mathrm{reg}}$ and C_{sh} coincide i.e. $C_{\mathrm{nt}} = C_{\mathrm{reg}} = C_{\mathrm{sh}} : \psi = \psi_{\max}$ where $\psi_{\max}$ is the value of ψ for which W is maximum. If the winding function has no extremum (as in the case of the degenerate tokamap) the shearless curve does not exist but the regular and the critical twist curves are well defined.

In the perturbed case ($K \neq 0$) these curves are different and a natural question arises on how to relate the dynamical properties of the system with each of them. These curves have the following equations:

$$C_{\mathrm{nt}} : W'\left(\bar{\psi}\right) + kg\left(\theta\right) h''\left(\bar{\psi}\right) = 0\ , \tag{13}$$

$$C_{\mathrm{reg}} : W'\left(\overline{\psi}\right) = 0\ , \tag{14}$$

The analytical form of the shearless curve (if it exists) cannot usually be given because it involves the numerical computation of the (perturbed) winding number.

The shearless curve is f-invariant but the regular set and the critical twist set are not f-invariant. For this reason the non-twist annulus (which is f-invariant) has to be considered.

If the equation $(\partial\overline{\theta}/\partial\psi)(\theta,\psi) = 0$ has a unique solution for every $\theta \in S^1$, which is a typical situation, the critical twist set of f is a rotational circle. The non-twist annulus collects all the points having a non-twist dynamic. The phase space splits in three invariant zones on which the map f acts separately.

In Fig. 4 the non-twist annulus and the chaotic invariant zones of the rev-tokamap (a non-twist map [10]) are represented. The map BT_K acting below or above the non-twist annulus has opposite twist properties due to the opposite monotonicity of W. The map DT_K of the degenerate tokamap case has negative twist properties both below and above the non-twist annulus because W_{DT} is a monotonous function.

The existence of the shearless curve is a non-twist essential phenomenon hence the shearless curve is contained in the non-twist annulus.

4 Transport Barriers

In the twist systems the transport barriers which are usually analysed are invariant circles. However, large transport barriers can be obtained by using appropriate perturbations which create some zones where the twist map is almost integrable. This is what is happening with the bounded tokamap (8).

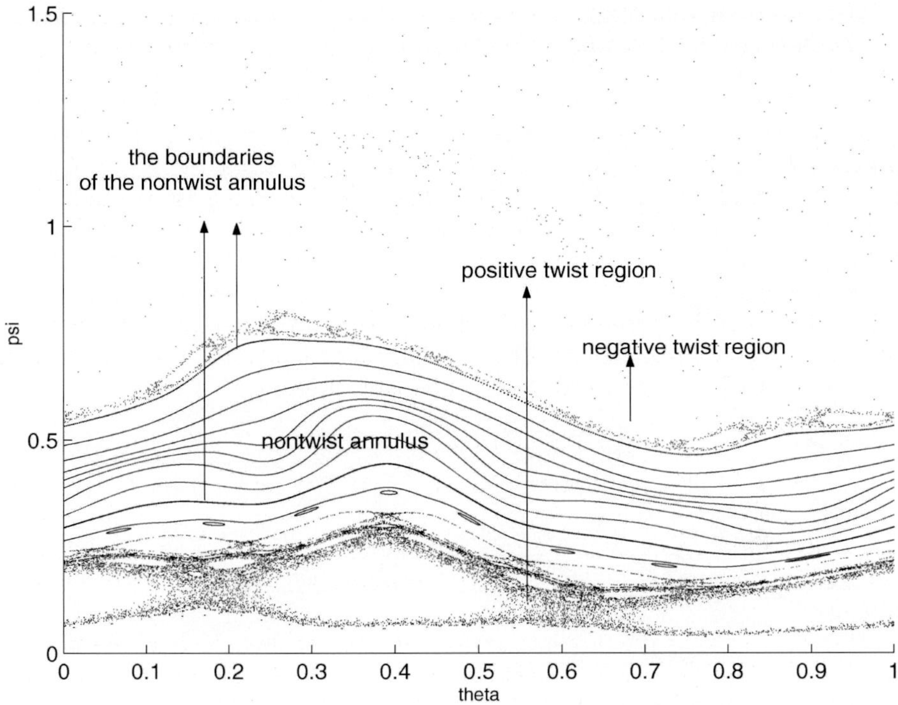

Fig. 4. The non-twist annulus and the invariant chaotic regions in the rev-tokamap model ($k = 2.5$, $w = 0.67$, $w_0 = 0.3333$, $w_1 = 0.1667$)

Proposition 1. *In an annulus surrounding the circle $\psi = 1$ the map (8) is almost integrable.*

Proof : In an annulus surrounding the invariant circle $\psi = 1$ the map (8) is close to the map $I_K : S^1 \times \mathbf{R} \to S^1 \times \mathbf{R}$, $I_K(\theta, \psi) \overset{\text{not}}{=} (\theta', \psi')$ defined by

$$\begin{aligned} \psi' &= \psi \\ \theta' &= \left(\theta + W(1) + \frac{K}{8\pi^2} \cos(2\pi\theta) \right) \qquad (\text{mod } 1) \end{aligned} \tag{15}$$

because

$$\begin{aligned} d^2((\theta', \psi'), (\overline{\theta}, \overline{\psi})) = & \left(\frac{K}{2\pi} \frac{\overline{\psi}\,(1 - \overline{\psi})}{1 + \overline{\psi}} \sin(2\pi\theta) \right)^2 \\ & + \left(W(\overline{\psi}) - W(1) + \frac{K}{16\pi^2} \frac{\overline{\psi}^2 + 2\overline{\psi} - 3}{(\overline{\psi} + 1)^2} \right)^2 \end{aligned}$$

is a small quantity for $\overline{\psi} \approx 1$ (which means $\psi = 1$ because $\overline{\psi} = 1$ if and only if $\psi \approx 1$). (Q.E.D.)

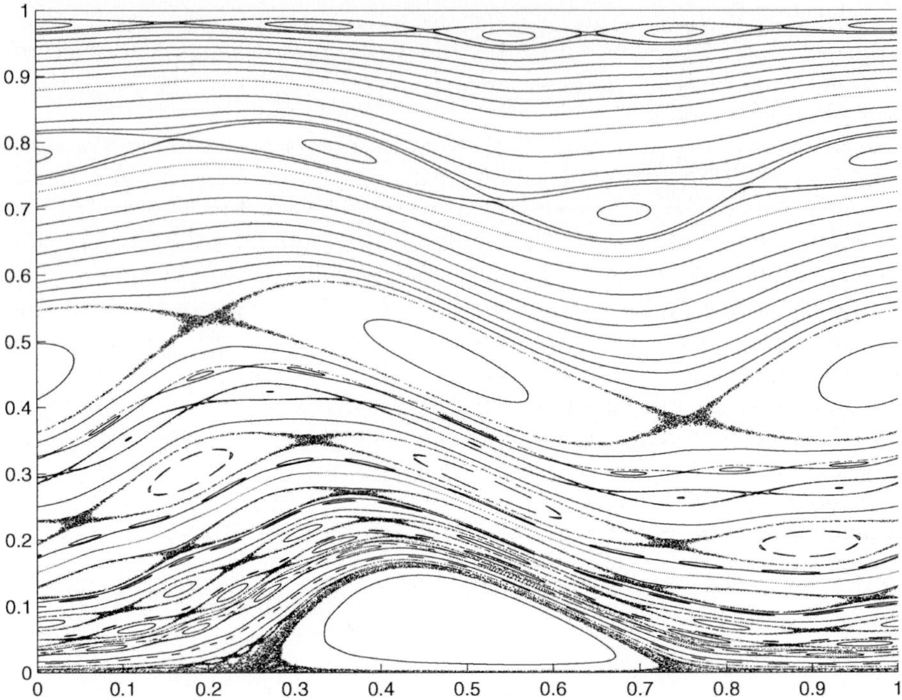

Fig. 5. The phase portrait of the bounded tokamap ($K = 5.9$, $w = 1$)

It results from this that a transport barrier having positive area appears in a region surrounding the circle $\psi = 1$. In Fig. 5 some regular orbits can be observed near the circle $\psi = 1$. The chaotic layers formed around the Poincaré–Birkhoff chains having the rotation numbers 1/2, 3/7, 2/5, 1/3, 2/7, 1/4, 1/5 etc. are getting closer and closer to $\psi = 0$ but are separated by invariant circles. We note that for $K = 5.9$ (the value considered in Fig. 5) there are no transport barriers in the tokamap model.

Proposition 2. *The region where the non-twist map f given by (11) is closest (in C^0 topology) to a rigid rotation is an annulus intersecting the curve C_{reg}.*

Proof: Consider $(\theta_0, \psi_0) \in S^1 \times \mathbf{R}_+$ and the map f_c defined by the implicit equations

$$\begin{aligned} \psi' &= \psi \\ \theta' &= \theta + W\left(\overline{\psi}_0\right) . \end{aligned}$$

It results that

$$\begin{aligned} d^2((\overline{\theta}, \overline{\psi}), (\theta', \psi')) = {} & \left(K\, g'(\theta)\, h(\overline{\psi})\right)^2 \\ & + \left[W(\overline{\psi}) - W(\overline{\psi}_0) + K\, g(\theta)\, h'\left(\overline{\psi}\right)\right]^2 . \end{aligned}$$

Now, because $W(\overline{\psi}) - W(\overline{\psi}_0) = W'(\overline{\psi}_0)(\overline{\psi} - \overline{\psi}_0) + \frac{1}{2}W''(\overline{\psi}_0)(\overline{\psi} - \overline{\psi}_0)^2 + \cdots$ is smaller near C_{reg} (here $W'(\overline{\psi}_0) = 0$) it results that f and f_c restricted to the disk surrounding (θ_0, ψ_0) are closer in C^0 topology if $(\theta_0, \psi_0) \in C_{\text{reg}}$ than in all other cases. (Q.E.D)

It is important to mention that the winding function W and the functions g and h which define the perturbation are independent objects. It results from the integrability of f_c that the region where f has a chance to be almost integrable (hence to have regular dynamics) is an annulus intersecting C_{reg}. Such an annulus does not exist for large perturbations (i.e. $\left|Kg'(\theta_0)h(\overline{\psi}_0)\right|$ or $\left|Kg'(\theta_0)\,h\left(\overline{\psi}_0\right)\right|$ are large for all $(\theta_0, \psi_0) \in C_{\text{reg}}$) because f and f_c are not close enough to ensure the regularity of f. It can be very large for small perturbations, usually when $K \ll 1$.

Figure 6 shows the phase portrait of the degenerate tokamap (a non-twist map) for $K = 5.9$. The central region of the phase space presents a clear regular zone. It is the annulus where the map is almost a pure rotation.

In order to relate the non-twist properties of a map to the existence of an annulus where the map is almost integrable we consider the rev-tokamap system.

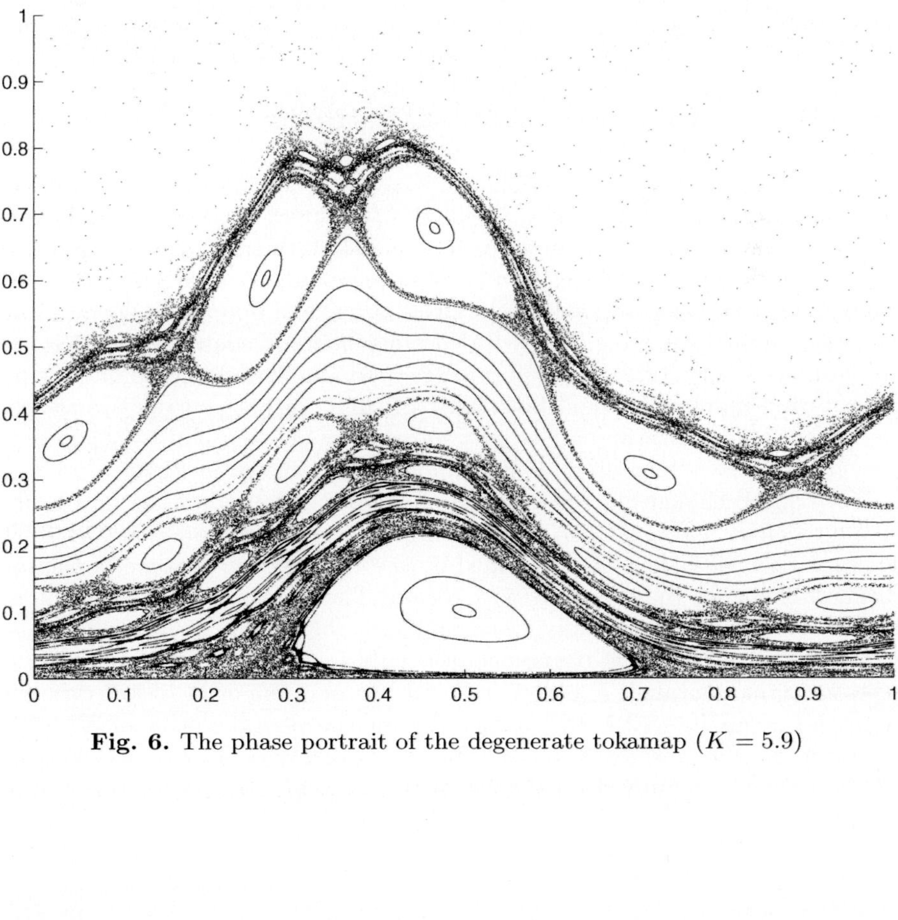

Fig. 6. The phase portrait of the degenerate tokamap ($K = 5.9$)

Using the explicit form of W_{RT} one obtains $C_{\mathrm{reg}} : \overline{\psi} = 1/C$, i.e.

$$C_{\mathrm{reg}} : \psi = \frac{1}{C} + \frac{K}{2\pi\,(1+C)} \sin(2\pi\theta)$$

The implicit (13) cannot be solved analytically for the rev-tokamap hence C_{nt} is computed numerically (using the Broyden method with the error 10^{-15}). Computer experiments point out that C_{reg} and C_{nt} are very close even for large values of $K \in [0, 2\pi)$. This effect is due to the fact that $\left|K g\,(\theta)\, h''\,(\bar{\psi})\right| < \left[2\pi\,(1+1/C)^3\right]^{-1}$ is usually small. We can conclude that in the non-twist annulus the rev-tokamap is almost integrable.

The analytical result concerning the regular annulus is in agreement with the experimental observations in tokamaks which point out the existence of a regular zone surrounding the so-called "zero shear region" i.e. the shearless curve.

In order to give a different perspective on the objects C_{reg} and C_{nt}, we consider their intersections with the line $\theta = 0.5$. These points are represented in Fig. 7 for a fixed winding function characterized by the parameters $w = 0.67$, $w_0 = 0.3333$ and $w_1 = 0.1667$ and for various values of K in the range $[0.6, 4.5]$. For small values of K the non-twist annulus is thin (it reduces to the curve $\psi = 1/C$ for $K = 0$). Its size increases as K increases hence the non-twist dynamical property becomes more important. The shearless curve and the regular curve are located between the non-twist annulus. The steep jump observed in the position of the shearless curve for $K \in [1.8, 1.9]$ is explained by the collision-annihilation of some periodic points having a rotation number slightly less than 0.67.

The non-twist annulus is, as shown above, a useful analytic tool, but the computer experiments have also revealed a transport barrier containing the non-twist annulus for small values of K. For $K = 0$ the map is integrable and the non-twist annulus reduces to a curve. For small perturbations ($K < 2.5$) the regular zone extends down to the circle $\psi = 0$ and no chaotic zone can be easily identified in this region. By further increasing K some invariant circles are destroyed, a chaotic zone is formed around the hyperbolic points having the rotation number 1/2 and both sides of the transport barrier are clearly identified. The transport barrier gets narrower while the non-twist annulus gets larger as K is increased. This phenomenon is shown in Fig. 7.

Very sharp changes in the position of the upper bound of the ITB is observed when $K \in (1.2, 1.3)$, $K \in (1.7, 1.8)$, $K \in (2.7, 2.8)$ and $K \in (3.4, 3.5)$. These abrupt variations are explained by the chaotisation of some Poincaré–Birkhoff chains in the negative twist region. The Poincaré–Birkhoff chain of 1/3-type enters in the globally stochastic zone for a value of K in the range $K \in (1.2, 1.3)$. The same phenomenon occurs for the periodic orbit of 1/2-type for K in the range $K \in (1.7, 1.8)$, for the periodic orbit of 4/7-type for K in the range $K \in (2.7, 2.8)$ and finally for the periodic orbit of 3/5-type for K in the range $K \in (3.3, 3.4)$. The width of the islands in a Poincaré–Birkhoff chain generally decreases when the period increases; for this reason

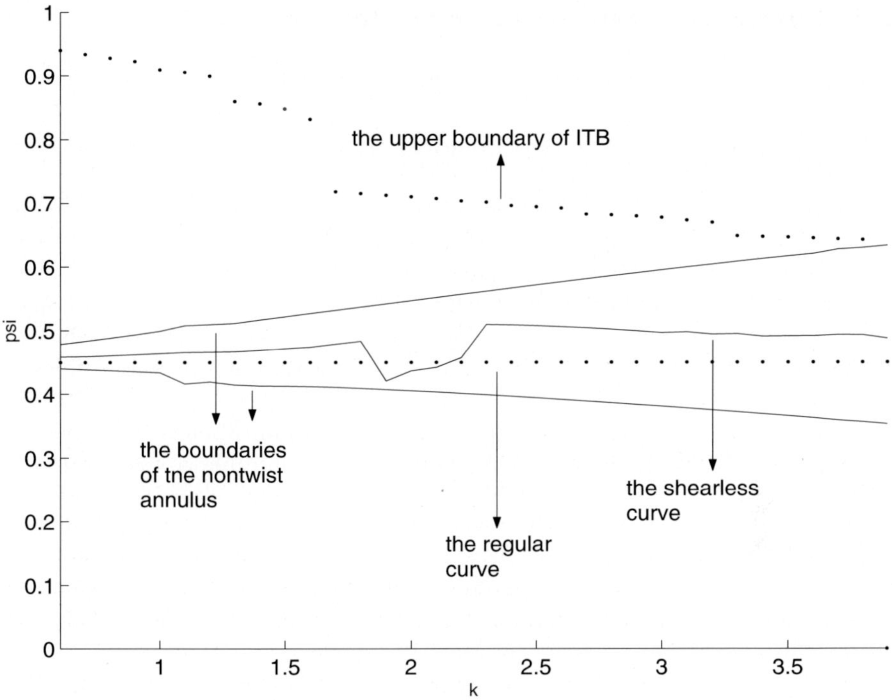

Fig. 7. The position of ITB, of the boundaries of non-twist annulus, of the shearless curve and of the regular curve on the line $\theta = 0.5$ for the rev-tokamap with the same winding function ($w = 0.67$, $w_0 = 0.3333$, $w_1 = 0.1667$) and various values of $K \in [0.6, 3.9]$

the effect of the chaotisation of 1/2-type periodic orbits ($K \in (1.7, 1.8)$) on the position of the ITB upper bound is more important than for the other cases.

Many other periodic orbits also enter into the globally stochastic zone, but these cannot be observed in Fig. 7 because their periods are large and the corresponding island chains are very thin.

The rev-tokamap has positive twist property below the non-twist annulus and negative twist property above it. As long as an invariant circle is contained in these regions their destruction can be explained by arguments used in the theory of twist maps, the KAM theory or the cone crossing criterion for example. More interesting is the case of invariant circles contained in the non-twist annulus.

5 Reconnection and Transport Barriers

Reconnection is a global bifurcation in the dynamics of a non-twist area preserving map. It is a change in the topology of the invariant manifolds of the hyperbolic points of two Poincaré–Birkhoff chains having the same rotation number m/n.

We shall study the reconnection process in the rev-tokamap model. In the degenerate tokamap model this phenomenon does not occur because there are not twin Poincaré–Birkhoff chains (the winding function is monotonous). Reconnection appears in systems with fixed winding function W whose maximum value decreases to a value slightly larger than m/n as the stochasticity parameter K is increased or for fixed K as the maximum value of W is decreased.

The general scenario for reconnection is as follows: before reconnection the Poincaré–Birkhoff chains having same rotation number are separated by invariant sets (usually invariant circles), at the reconnection threshold the hyperbolic periodic points of the two chains are connected by their manifolds and after reconnection two dimerized island chains are formed. Each of them contains a set of hyperbolic periodic points connected either by heteroclinic connections or by homoclinic connections surrounding elliptic periodic points. These dimerized islands are separated by meanders.

In the generic position before reconnection the periodic points situated approximately at the same θ have opposite stability (one is elliptic, the other is hyperbolic). In this case, the twin Poincaré–Birkhoff chains approach each other until the hyperbolic points get connected by a heteroclinic connection. This is the typical behaviour in the rev-tokamap model.

Starting from a nongeneric position before reconnection, the periodic points situated approximately at the same θ have the same stability. Before reconnection a saddle-centre bifurcation occurs in a Poincaré–Birkhoff chain i.e. a new pair of periodic points having the same rotation number as the preexistent ones is created. By modifying the parameter the new points annihilate the old elliptic orbit. Finally only one elliptic orbit remains in a generic position with the other Poincaré–Birkhoff chain. From this moment the generic route to reconnection is followed.
We shall describe the reconnection of the 2/3 Poincaré–Birkhoff chains for $K = 2$ and $K = 4$ for decreasing values of w:

- The value $K = 2$ is small enough to ensure that the reconnection occurs in the regular zone:
  - For $w > w_r \approx 0.6702$ the island chains are in the regular zone in a generic position and they are separated by invariant circles;
  - For $w = w_r \approx 0.6702$ the reconnection occurs;
  - For $w < w_r \approx 0.6702$ the dimerized islands are separated by meanders.

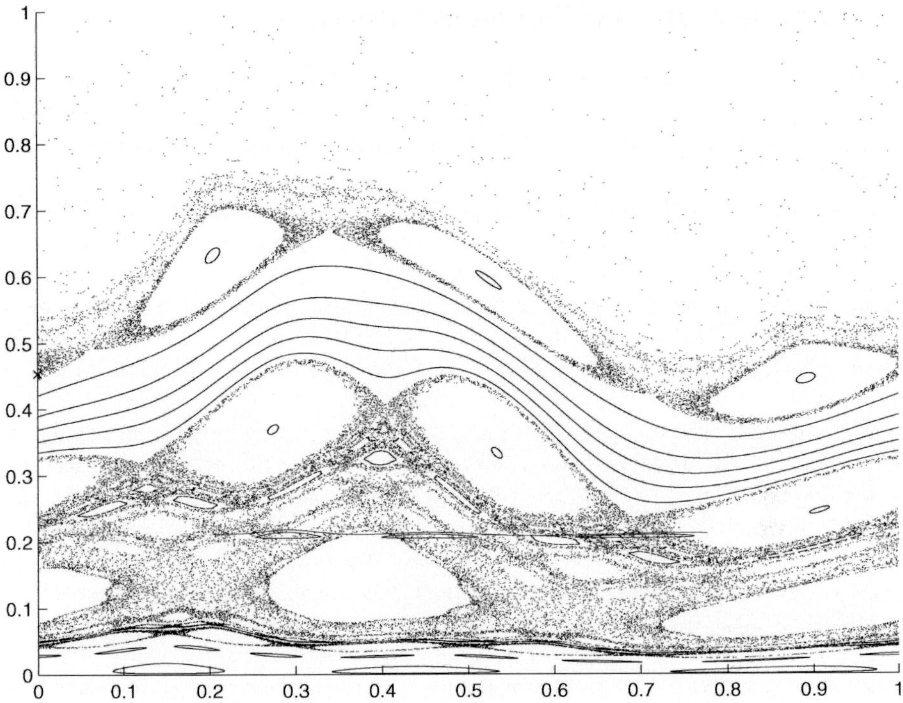

Fig. 8. Twin Poincaré–Birkhoff chains from the rev-tokamap model ($K = 4$, $w = 0.71$, $w_0 = 0.3333$, $w_1 = 0.1667$)

- For $K = 4$ the perturbation is stronger. The reconnection occurs in a chaotic layer produced by the manifolds of the hyperbolic points (these manifolds intersect transversely):
  - For $w = 0.7$ the island chains are outside the regular annulus in a nongeneric position (see Fig. 8).
  - For $w = 0.685$ a saddle-centre bifurcation occurred in the upper island chain. Because at the bifurcation threshold the periodic points which bifurcate is parabolic (its multipliers are $\lambda_1 = \lambda_2 = 1$) the map is almost integrable around it, hence the upper chain enters the regular zone (Fig. 9)
  - For $w = 0.68213$ the reconnection occurs (Fig. 10) in a chaotic layer.

By decreasing w the periodic points of the upper island chain situated approximately on the same vertical approach each other, collide and annihilate. The other island chain enters the regular zone and remains there until the collision–annihilation of its points occurs.

The main conclusion of this section is that the reconnection takes place inside the non-twist annulus (because it is specific to the non-twist dynamics)

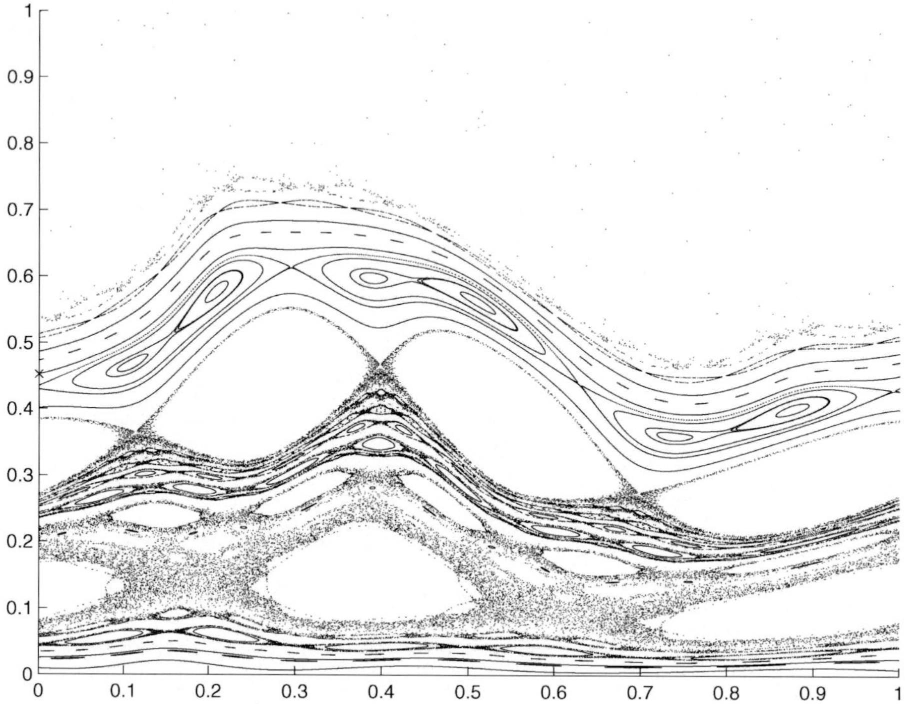

Fig. 9. The saddle-centre bifurcation occurred in the upper Poincaré–Birkhoff chain (rev-tokamap $K = 4$, $w = 0.685$, $w_0 = 0.3333$, $w_1 = 0.1667$)

in a zone which can be regular or chaotic, depending on the stochasticity parameter.

In our last example, the shearless curve disappears at the end of the reconnection process (because it would be located between the dimerized island chains in a chaotic region) but a regular zone i.e. a transport barrier can however be identified (Fig. 9). Therefore the shearless curve of the rev-tokamap is not the most resistant invariant circle. This result, in contradiction with what is observed for the non-twist standard map [19], is explained by the fact that the latter map has time reversal and symmetry properties which are not possessed by the former. For this reason the rev-tokamap exhibits more general dynamical properties.

6 Conclusions

Four Poincaré maps giving the intersection points of a magnetic field line with a given poloidal cross section are presented: the tokamap representing usual tokamak conditions, the rev-tokamap for a tokamak running with reversed

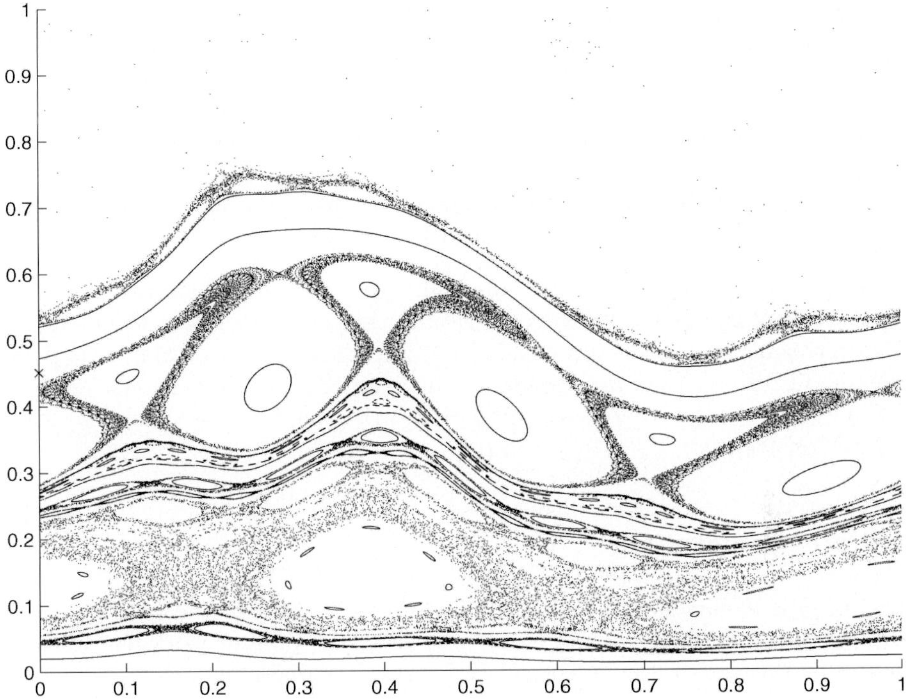

Fig. 10. The rev-tokamap phase portrait at the reconnection threshold ($K = 4$, $w = 0.68213$, $w_0 = 0.3333$, $w_1 = 0.1667$)

shear conditions, the bounded map with the outmost tokamak magnetic surface forced to be a smooth surface and the degenerate tokamap. Each of the four maps develops its own magnetic structure which are explained using some of the new results on non-twist maps reported here.

To be specific, the maps may be characterized in terms of the ITB width, the formation of the ITB and the process of magnetic reconnection:

(a) Previous studies of the ITB width considered as important three particular curves: the regular curve, the non-twist curve and the shearless curve which coincide in the unperturbed case for non-twist maps. The detailed study of the four maps for different parameter values has definitively showed that the regular curve indeed relates to the regular dynamics but that the shearless curve does not.
(b) The mechanisms leading to the formation of ITB's with positive area has been clarified in two situations which had not been considered previously: twist systems, like for instance the bounded tokamap, support the formation of large ITB because of their proximity to an integrable system and non-twist systems with monotonous unperturbed winding function, like the degenerate tokamap (see Sect. 4).

(c) The principal difference between twist and non-twist maps lies in the possibility in the former to proceed with a reconnection mechanism. This process of reconnection is described in details in two limits, for small perturbations (in a regular zone) and for large perturbations (in the chaotic zone). The main consequence of a reconnection in a chaotic layer is that in the rev-tokamap system the shearless curve is not always the most resistant invariant circle (this result differs from the usual behaviour of models that have a symmetry group).

Maps for magnetic field line behaviour for tokamak geometries have already given a number of unexpected results that could facilitate the analysis of experimental results, even considering particles instead of magnetic field lines in ITB's. Further progresses are however still awaited in the area of guiding centre maps where some difficulties arise due to more stringent topological constraints.

Acknowledgement

The authors would like to thanks the members of the International working group "Fusion BFR", for their great interest in the present work and in particular R. Balescu, J. Misguich and E. Petrisor for many detailed discussions on non-twist maps.

References

1. J. Wesson: *Tokamaks* (Cambridge University Press, Cambridge 1987) pp 21–30
2. R. Eykholt, D.K. Umberger: Physical Review Letters **57**, 2333 (1986)
3. R. Balescu: Phys. Rev. E **58**, 371 (1998)
4. P.J. Morrison: Physics of Plasmas **7**, 2279 (2000)
5. S.S. Abdulaev: Nucl. Fusion **44**, S12 (2004)
6. B. Chirikov: Phys. Rep. **52**, 265 (1979)
7. H.Z. Wobig: Naturforsch, Teil A **42**, 1054 (1987)
8. J.H. Misguich, J.D. Reuss, D. Constantinescu, G. Steinbrecher, M. Vlad, F. Spineanu, B. Weyssow, R. Balescu: Annales de Physique **28** , 1 (2003)
9. B. Weyssow, J.H. Misguich: Europhys. Conf. Abstracts J 23 795 (1999)
10. R. Balescu: Phys. Rev. E **58**, 3781 (1998)
11. G.D. Birkhoff: *Dynamical Systems* (AMS Colloq. Publ., vol 9, revised 1966)
12. J.K. Moser: Nachr. Akad. Wiss. Gottingen, Math-Phys. KI. II. **1**, 1 (1962)
13. R.S. MacKay: *Renormalisation in Area-Preserving Maps* (World Scientific Publishing Co. Pte. Ltd 1993) pp 78–84
14. J. Greene, J. Math. Phys. **20**, 1183 (1979)
15. S. Aubry: The twist map, the extended Frenkel-Kontorova model and the devil's staircase. In *Proceedings of the conference on Order and Chaos, Los Alamos (May 1982), Physica* **7D** (1983), pp 240–258
16. J.N. Mather, Topology **21**, 457 (1982)

17. R.S. MacKay, I.C. Percival: Comm. Math. Phys. **98**, 469 (1985)
18. J.E. Howard, S.M. Hohs: Phys. Rev. A **29**, 418 (1984)
19. D. del-Castillo-Negrete, P.J. Morrison: Phys. Fluids A **5**, 948 (1993)
20. E. Shuchburgh, P. Haynes, Phys. Fluids **15**, 3342 (2003)
21. C. Chandre, D. Farelly, T. Uzel: Phys. Rev. A **65**, 1 (2002)
22. A. Munteanu, E. Petrisor, E. Garcia-Berro, J. Jose: Comun. Nonlinear Sci. and Numerical Simulation **8**, 355 (2003)
23. D. del-Castillo-Negrete, J.M. Greene, P.J. Morrison: Physica D **91**, 1 (1996)
24. D. del-Castillo-Negrete, J.M. Greene, P.J. Morrison: Physica D **100**, 311 (1997)
25. A. Delhams, R. de la Llave: Siam J. Math. Anal. **31**, 1235 (2000)
26. C. Simo: Regular and Chaotic Dynamics, **3**, 180 (1998)
27. E. Petrisor: Int. J. of Bifurcation and Chaos **11**, 497 (2001)
28. E. Petrisor: Chaos, Solitons and Fractals **14**, 117 (2002)
29. E. Petrisor, J.H. Misguich, D. Constantinescu: Chaos, Solitons and Fractals **18**, 1085 (2003).

Index

Understanding Complex Systems

Edited by J.A. Scott Kelso

McDaniel, R.R.Jr.; Driebe, D.J. (Eds.)
Uncertainty and Surprise in Complex Systems:
Questions on Working with the Unexpected
X, 200 p. 2005 [3-540-23773-9]

Kerner, B.S.
The Physics of Traffic:
Empirical Freeway Pattern Features,
Engineering Applications, and Theory
XXIII, 682 p. 2004 [3-540-20716-3]

Kocarev, L.; Vattay, G. (Eds.)
Complex Dynamics in Communication Networks
X, 361 p. 2005 [3-540-24305-4]

Jirsa, V.K., Kelso, J.A.S. (Eds.)
Coordination Dynamics: Issues and Trends
XIV, 272 p. 2004 [3-540-20323-0]

Kleidon, A.; Lorenz, R.D. (Eds.),
Non-equilibrium Thermodynamics
and the Production of Entropy
XIX, 260 p. 2005 [3-540-22495-5]

Ausloos, M.; Dirickx, M. (Eds.),
The Logistic Map and the Route to Chaos:
From The Beginnings to Modern Applications
XVI, 411 p. 2006 [3-540-28366-8]

Printing: Krips bv, Meppel
Binding: Stürtz, Würzburg